A BIOGEOGRAPHICAL ANALYSIS OF THE CHIHUAHUAN DESERT THROUGH ITS HERPETOFAUNA

BIOGEOGRAPHICA

Editor-in-Chief

J. SCHMITHÜSEN

VOLUME IX

DR. W. JUNK B.V., PUBLISHERS THE HAGUE 1977

A BIOGEOGHAPHICAL ANALYSIS OF THE CHIHUAHUAN DESERT THROUGH ITS HERPETOFAUNA

by

David J. Morafka

DR. W. JUNK B.V., PUBLISHERS, THE HAGUE 1977

ISBN 90 6193 210 6

© 1977 by Dr. W. Junk bv, Publishers, The Hague

Cover Design Max Velthuijs, The Hague

Subadult Bolson Tortoise, *Gopherus Flavomarginatus.*

This specimen was a female, straight plastron length 190 mm.
It was collected by author at Rancho Las Lilas, Coahuila (Lat: 26°40′N Long: 103°25′W)
September 9, 1973.

To: DARIUS COUCH, AARON and DOROTHY MORAFKA, JAY SAVAGE and J. TED VIDMAR.
Who, by their actions, acknowledged beauty and demonstrated courage beyond the reach of the written word.

Contents

ACKNOWLEDGEMENTS

In the pursuit of my dissertation research I have enjoyed the benefits of compassion, generosity, and the professional insights of colleagues, students, and the citizens of the states of Arizona, New Mexico, Texas, and the United States of Mexico. The efforts of so many people on my behalf is very deeply acknowledged. Those individuals cited here were particularly crucial to the success of this dissertation and personally sustaining to me as devoted friends and colleagues. Many others, from children to old men, some nameless to me, also have my gratitude.

My deepest gratitude is extended to Professor JAY M. SAVAGE, chairman of my dissertation committee at the University of Southern California. His guidance, and even more his example of personal and professional courage, integrity, and clarity of purpose will be a personal inspiration throughout my life. I am also very thankful to Professors BASIL G. NAFPAKTITIS, EDWIN M. PERKINS, and FINLEY L. RUSSELL for their incisive advice, support and well placed criticisms.

Several individuals and institutions made my field efforts not only possible, but profoundly fulfilling personal and professional experiences. Foremost, I wish to thank Dr. TIRSO EDWARDO CANO GOMEZ and his wife DIANE of the city of Chihuahua for their friendship, hospitality, and support through my field work. I also wish to acknowledge the considerable assistance extended to me by the faculty and students of the Escuela de Ganaderia of the Universidad de Chihuahua. Their assistance included field guides and summer housing in Chihuahua. Collections were made with the permission of the Mexican Secretaria de Industria y Commercio and the Secretaria de Fauna y Silvestre. Advice from Professors GOMEZ-POMPA and VILLA of the Universidad Autonoma Nacional de Mexico was also of great value. Numerous citizens and officials of the Mexican Republic assured both my personal safety in the field and the success of my efforts. I especially thank the dozens of Mexican children who enthusiastically contributed rare specimens of exceptional value to my collection.

My several field companions and assistants are worthy of special praise. WILLIAM T. ROGERS, field assistant for the summer of 1970, undertook the collection and preservation of hundreds of specimens, transcribed my verbal remarks and scrawl into a magnificent work of literature, appropriately titled 'field notes.' My museum and field assistant of 1971, JEROMY JACOBS, at the age of 16 undertook a continuous 12 to 16 hour work day in my behalf for four long and difficult months, without pay. JEROMY's performance was far beyond expectation.[1]

Dr. JAMES HENDRICKSON of California State University, Los Angeles, joined me in the field during the summer of 1971. Through his friendship, botanical knowledge, and the use of his vehicle, insights were developed and investigations undertaken that would otherwise have been impossible for me. I thank him for thousand different acts of kindness and intellectual sharing. I also extend thanks to Dr. MARSHALL JOHNSTON for his extensive botanical advice.

A large number of museum curators, their staffs and one major private collector

---

1. I extend my gratitude also to WILLIAM KENT for his assistance in the field during the winter of 1970.

(ERNIE A. LINER) made crucial information available to me while also extending very gracious welcome. The collections visited are listed under sources, methods, and analytical procedures. Some of the many individuals who should be acknowledged include the following: JAMES ANDERSON, ROBERT BEZY, C. BRYCE BROWN, JAY COLE, WILLIAM DEGENHARDT, JAMES DIXON, HARRY GREENE, GEORGE JACOBS, JOE LA POINTE, ALAN LEVITON, ERNIE LINER, ROBERT MARTIN, C. JACK MCCOY, ROBERT MURPHY, the late JAMES PETERS, F. HARVEY POUGH, DOUGLAS ROSSMAN, WAYNE SEIFERT, PHILLIP SMITH, ROBERT STEBBINS, DAVID WAKE, J. MARTIN WALKER, RICHARD WEBB, LARRY DAVID WILSON, RICHARD WORTHINGTON, RICHARD ZWEIFEL, and GEORGE ZUG. RALPH AXTELL is particularly acknowledged for distributional information from his personal collections, for the corrections of several factual errors and for numerous valuable criticisms of text style.[2]

I would also like to thank my fellow graduate students of the Biology Department of the University of Southern California for the advice, support, assistence, and a general spirit of brotherhood. In particular RONALD HARRIS is to be acknowledged for his superb and extensive assistance with map and illustration construction as are KATHERINE LINK, GREG BERNIE, SHIRLEY TORSTENSON, and MIKE MIYAMOTO. ROBERT SMITH provided the program and supervision necessary to carry out the computer dendrogram analysis. CARL LIEB' supplied invaluable advice in regard to Texas and Mexican skink distributions. He also served as accurate and valuable critic in the development of the biogeographical discussion. Drs. NORMAN SCOTT, JOHN MEYERS, and MICHAEL MARTIN are also thanked for their assistance. The Chairman of the Department of Biological Sciences, University of Southern California, Dr. BERNARD ABBOTT is gratefully acknowledged for his many kindnesses and financial support.

Financial support from the University of Southern California may only be described as massive. It included a three year (1968-1971) National Defense Education Act Fellowship, a National Health Institute Biomedical Research Grant, and museum tour travel grant awarded jointly by the Graduate School and the Department of Biological Sciences. In addition, the latter department supported my graduate studies through three terms by awarding me Teaching Assistantships (Fall 1967, Spring 1968, and Fall 1971). My graduate education would not have been possible without this support, and I thank all parties responsible.

Travel costs for field work during the summers of 1970 and 1971 were met in great part by two separate grants for pre-doctoral research (from the Society of Sigma Xi). Their support is most appreciated.

Photographic and typing preparation of the manuscript, figures, and maps were funded by Department of Biological Sciences, California State College, Dominguez Hills. I particularly thank CLIFF BROWN for his aid.

I wish to thank my typists, SHEILA DOUCETTE, SUE VAN ROSSEN, JOANNE FITCH, and MARILYN CHEUNG for service far and above the call of duty. Lastly, EVE WYNANT is thanked for a superb job of editing and proofing final drafts.

---

2. In addition, CRAIG MEYER critically corrected chapters on climate and physiography.

I. INTRODUCTION

The Mexican Plateau, in its magnificent dimensions and material wealth, stood among the first and perhaps most alluring discoveries of European explorers. Buried deeper in the verbal histories of a now vanquished people, the American Indians, must be the primordial human awareness of the inverted complex triangle that dominates the Mexican topography, climate and biota. It always has been viewed by man as a source of wealth and a center of authority. The plateau is the pillar upon which all Mexican conquerors have erected their capitols, tilled their crops and mined for their treasure, and from which they dispersed the forces of their authority. Ironically, the same size and diversity that give the plateau its value, also make it an immense barrier. Its broad desert and three to five thousand meter high crests constitute severe obstacles in the path of North American man.

What has just been said of mankind in general, can be applied to the biologist in particular. He too has termed the goliath southern plateau as the crucible of the arid biotas of the continent (i.e., 'Madro-Tertiary'). The biologist found the plateau to be a region of tremendous richness and diversity. But he also has been inhibited both physically and intellectually by its high mountain and vast desert barriers. Even as biological awareness of the importance of the plateau grows, that awareness remains confined to fragmentary and circumstantial evidence until the barriers are crossed and the actual contents of the region defined as to their nature and distributions.

This dissertation is directed at one of the most important and least understood components of the plateau, its Chihuahuan Desert. The investigation has assumed four specific tasks: (1) to define the Chihuahuan Desert in biologically and physiographically realistic terms; (2) to translate that definition into an explicit geographical base map and quantitatively test the predictive value of that map against a specific set of biological systems, the amphibians and reptiles; (3) to assess the relative differentiation and affinities of the Chihuahuan herpetofauna; and (4) to reconstruct this history of the Chihuahuan Desert and its herpetofauna through geologic time to its evolution to a contemporary state of being.

The investigation then is a biological profile of the Chihuahuan Desert, a defining profile in terms of content, characteristics, and spatial limits, but also defining as an ongoing process through time.

The analytical effort required for this undertaking may also be broken down into four explicit endeavors: (1) the definition of components and conditions of the desert, climate, physiography, and vegetation, and their spatial distributions; (2) the determination of the herpetofauna of the Chihuahuan Desert and their total spatial distributions; (3) quantification and evaluation of the degrees of correlation between information from sources (1) and (2) above; and (4) the determination of the most probable historical events which account for the patterns and correlations observable in the present.

Past studies are few in number both generally and for the herpetofauna in particular (the reader is reminded that the largest reptile in temperate western North America, the Chihuahuan Bolson Tortoise — *Gopherus flavomarginatus* — was described in 1959). The desert has been analyzed and defined ecologically, at least in English literature, largely from its unstable and atypical grassland ecotones lying within United States boundaries. The resulting distortion in the literature is much like an analysis of the contents of an egg extrapolated from a sample of its shell. In addition, even investigators who have gone below the United States border have usually confined themselves to the arbitrary borders of local Mexican political states. The resulting accounts, though often well documented and detailed, are accurate only locally and without perspective as to the variation or real limits of the systems they investigate. Finally, much exclusive terminology has set up false alternatives for the biologist describing or defining desert conditions. In particular the concepts of desert (sclerophyll) and grassland have been used as almost antithetical. It is clear now that many grasslands are both physiologically and geographically qualified as true and typical participants in a climatic desert biota. Finally, many deserts have been defined in totally subjective and nonbiological terms, forcing too many researchers to employ artificial and inadequate geographical units.

A synthetic evaluation, unbiased geographically or rhetorically has been attempted in this investigation. It is sincere, conscientious, and imperfect. I am confident that it will receive both the concern and the criticism that will nourish and mature the beginnings founded here.

David Morafka, Ph.D.
Assistant Professor of Biology
California State College
Dominguez Hills

II. SOURCES, METHODS, AND ANALYTICAL PROCEDURES

This presentation is a resume of data sources, techniques of base map construction and data compilation, and the philosophical bases and statistical procedures with which the final analysis was undertaken.

Data Sources

Information was drawn from three primary sources: field work, a literature search for information on Chihuahuan physiography, ecology, and herpetofauna, and a museum tour of major systematic collections in the United States. The author undertook field work with two major objectives in mind. Firstly, field work was directed toward establishing a general geographical and ecological orientation to the Chihuahuan Desert. This orientation was documented by a continuous field journal and approximately 1000 Kodachrome slides. The journal recorded weather, soil, elevation, vegetation, slope, and hepetofauna for all localities visited. Abrupt shifts in the dominant flora were fixed geographically and compared to desert limits mapped in the available literature. Gradual but sustained transitions in dominant plant cover were discerned and evaluated with the aid of simple sampling techniques. Sets of ten random-toss meter squares were compared between localities for shifts in the percentage area occupied by plant species. Edaphic and seasonal (pre and post summer precipitation) variations in vegetation were also noted.

The second purpose of the field efforts was to establish a preserved collection of the indigenous herpetofauna in order to determine which species were ecological as well as geographical inhabitants of the desert and to close distributional gaps left by museum and literature searches. A species account was maintained in which each specimen was assigned a field number and locality. Elevation and ambient air temperature were also recorded.

Field work took place June 21 through September 15, 1970 and July 1 through September 30, 1971. During the cumulative six months over 25,000 kilometers of road was traveled (sometimes repeatedly). Within the United States, south-eastern Arizona, most of New Mexico, and Texas west of the Pecos River were visited. Travel in Mexico included the eastern half of Chihuahua, Coahuila, northeast Durango, Zacatecas, San Luis Potosi, southwest Nuevo Leon, Aguascalientes, Guanajuato, and Queretero (See Map 1). During December 15, 1970 through January 1, 1971, Chihuahua, Durango, and Coahuila were sampled briefly. These efforts resulted in a collection of over 1200 specimens representing one hundred species of amphibians and reptiles. The collection will be deposited as a unit in the Carnegie Museum, Pittsburgh, Pennsylvania (field numbers are designated DJM). *Cnemidophorus inornatus* have been deposited in the Los Angeles County Museum of Natural History, and *Crotaphytus collaris* in the Museum of Natural History of the University of Kansas.

The second source of information was a literature search for information relative to the definition and delimitation of the desert in its physiography, geological history, climate, and ecology. The search was then extended to a more detailed analysis of its vegetation associations, and ultimately to its herpetofauna, extant and fossil, and their ecological and geographical distributions. Texts, field notes, articles, and unpublished manuscripts were all utilized. The range maps and locality listings from the following texts were of particular importance: CONANT (1975), RAUN & GEHLBACH (1972), SMITH & TAYLOR (1966) and STEBBINS (1966). Available technical literature in both English and Spanish was checked.

A base map, Map 2, of the Chihuahuan Desert was constructed from a set of defining climatological, physiographical, and vegetative parameters. Criteria were drawn from pertinent literature and direct field observation. The details of base map construction will be discussed in a separate chapter.

A museum tour provided the final and most important source of distributional information. Using the base map as a delimiting reference, I attempted to record all records of all species of amphibians and reptiles occuring within one hundred kilometers of the mapped limits of the desert and its filter barriers. This hundred kilometer corona around the desert provided crucial data on the composition of the peripheral fauna, and thus a basis for objective discrimination of desert from non-desert herpetofaunas. Redundant records were not recorded (nor those within 15 kilometers of a previous record unless they extended known peripheral range). All museum records that extended known distributions were confirmed by personal examination of the specimens on which they were based. The same procedure was followed in the treatment of complex or recently revised systematic groups (i.e., *Cnemidophorus, Eumeces, Tantilla*, and *Thamnophis*).

The following systematic collections were examined (in descending order by the approximate size of their Chihuahuan collections): American Museum of Natural History, Museum of Zoology of the University of Michigan, United States National Museum (Smithsonian Institution), Museum of Natural History of the University of Kansas, Field Museum of Natural History (Chicago), Texas Natural History Collection of the Texas Memorial Museum at Austin, Museum of Natural History of the University of Illinois, Carnegie Museum at Pittsburgh, Museum of Vertebrate Natural History at the University of New Mexico, Los Angeles County Museum of Natural History, Strecker Museum at Baylor University (Waco), Texas Cooperative Wildlife Collection of Texas A & M, Museum of Vertebrate Zoology at the University of California (Berkeley), California Academy of Sciences, ERNIE A. LINER Collection (private) (records sent), Museum of Natural History at the Louisiana State University (Baton Rouge), Fort Worth Natural History Museum, Cornell University, New Mexico State University at Las Cruces, University of Texas at El Paso, University of Southwestern Louisiana (Lafayette), University of Texas at Arlington, Dallas Museum of Natural History, Academy of Natural Sciences of Philadelphia.

The accrued distributional data resulting from the museum search was so extensive as to make an explicit listing of localities impractical. The compiled information is represented on the dot maps for each species. Locality listings by species will be prepared separately at a later date. Significant range extensions may be documented and summarized in a separate publication.

Methods of Data Compilation and Presentation

Data from field, literature, and museums were combined and presented in a sequential format as follows: (1) a descriptive definition of the Chihuahuan Desert, its major subdivisions, and periphery was developed in terms of climate, physiography, and vegetation; (2) this definition was translated onto a base map of the desert; (3) documented localities, represented by dots, were plotted onto the base map for each species occurring within the defined desert or its 100 kilometers corona; and (4) based on the dots, the general habitat association of the species, and the continuity of the regional topography, a hypothetical range was constructed and delimited by heavy borderline.

This format for the presentation of distributions was chosen to achieve several ends. The maps simultaneously display the relationships between documented localities (dots) and hypothetical range (borderline) and between range and the limits of the desert. Thus, by simple inspection, the reader is in a position to evaluate the degree of speculation involved in a range construction, and the goodness of fit between species distributions and the limits of the desert. Hypothetical ranges were constructed using a combination of "least polygons" and physiographical barriers.

Analytical Procedures

The objectives of the analysis were four-fold: (1) to test the biological validity of a synthetic spatial definition (map) of the Chihuahuan Desert; (2) to determine the internal homogeneity of the desert (i.e., establish subdivisions); (3) to establish and quantify affinities with geographically or paleoecologically related biotic units (i.e., provinces); and (4) to construct a paleoecological history for the Chihuahuan Desert, a temporal process, evaluating and contributing evolutionary models for present internal and external biotic relationships. The steps of this procedure follow a logical sequence, but only step one is critical to the pursuit of the remaining three approaches for investigation.

HAGMEIER & STULTS (1964) cited a procedure in their numerical analysis of North American mammal distributions which is worthy of explanation here. Their modified format is as follows: (1) selection of taxa; (2) definition of a single initial area; (3) determination of the distribution of the selected taxa within the initial area; (4) determination of primary areas by disruptions of maximal IFC (Index of Faunal Change) between squares in a grid across the initial area, using the distributions of the selected fauna; and (5) dendrogram analysis of affinities between the faunal assemblages of the primary areas, using the Coefficient of Community as a measure of faunal resemblance. This approach provides a strong quantification and standardization in the construction of primary areas. I have adopted the HAGMEIER & STULTS procedure in this study with the addition of one further step. This sixth step is a historical analysis of the faunal relationships between primary areas depicted by the dendrograms and of the elements producing the constituent fauna within these provinces.

PETERS (1971) employed a modified form of this procedure — noting three ubiquitous preliminaries for quantified biogeographical analysis. They are: (1) selection of taxa of a given rank from a particular group of organisms; (2) definition of primary areas; and (3) the determination of distributions of the selected taxa (1) in the primary areas (2).

I shall now undertake a brief review of each of these steps to explain crucial operational decisions.

The choice of taxa was based on several considerations. First, my previous professional orientation has been strongly herpetological and these are the groups which are closely bound by both physiology and motility to the landscape, closely reflecting both ecological and historical (paleoecological and paleogeographical) influences. Furthermore, these particular stocks have well defined species and in fact osteological and karyotypic phylogenies are available for several major genera (*Phrynosoma, Cnemidophorus* and *Sceloporus* in part). The chosen herpetofauna, over 170 species in the involved portion of the Mexican Plateau, is sufficient for quantitative analysis, but it is not so diverse as to overburden a researcher limited to three years of active investigation. Finally, amphibians and reptiles from the study region were available in museum collections and accessible in the field in adequate numbers for the construction of functional distribution maps.

The definition of the initial base region of study was a particularly difficult task, one in which lies tremendous potential for arbitrary and subjective distortion. I asked the question, "How is the Chihuahuan Desert to be defined in biologically meaningful terms that may be translated into an operational base map?" I have attempted to enhance the value of the defined area by considering a broad range of apparently congruent parameters: climate, physiography (soil, topography, and drainage), and climax vegetation. I have purposely omitted any consideration of the herpetofauna in the construction of the base map in order to avoid circular definitions. The herpetofauna was set aside as a test of the effectiveness of the other factors in delimiting a distinguishable faunal assemblage. The base map construction is described in explicit detail in the subsequent section, Definitions, Terminology, and the Base Map of the Chihuahuan Desert. This defined area thus serves both to summarize descriptive information about the desert and to supply a basis for subsequent analysis and evaluation.

The third stage, construction of specific distributional maps, has just been discribed under Methods of Data Compilation and Presentation. Range maps employed were similar to those used by HAGMEIER & STULTS (1964) in that they depict general distribution. However, the maps used here provide more accurate or at least uniform information in that all ranges were constructed by the author himself, all from the same general sources of information, and all were made without regard to artificial political boundaries (in contrast to the U.S.-Mexican border which limited the study of HAGMEIER & STULTS). Furthermore, ranges were constructed without conscious consideration of previous and subjectively defined biotic provinces. Hypothetical range limits were drawn conservatively, based on known localities and sharply defined associations with known vegetation, or on the presence of specific and physiologically imposing barriers (i.e., mountain ranges).

The use of IFC values in defining primary areas-provinces is well defended by HAGMEIER & STULTS (1964). This quantification of fauna change, especially when coupled with a grid analysis using standard units of area, provides a basis for both relative and absolute standard definitions of provinces. This strategy of definition goes a long way toward overcoming the objections of PETERS (1955) which make the analogy between faunal provinces and arbitrary subspecies.

PETERS' specific illustration was a particularly poor test in that he was examining a subjectively defined province in a complex tropical continuum where many temperate units of ecological measure, including the community concept, were simply not applicable on a practical level (HOLDRIDGE, 1967). Ultimately the grid served two related functions: to test the herpetofauna against the base map and secondarily to define primary areas. The particular application of IFC values used here will be described in Ch. IX, A Spatial Analysis of the Chihuahuan Herpetofauna.

The evaluation of faunal resemblance is a particularly complex and crucial step of this investigation. Several procedural explanations are especially important here. Firstly, all primary areas, having been defined on the range of the total terrestrial herpetofauna, were now redefined in terms of their typical or "participant" herpetofaunas. In practice, this means that all relict assemblages, occurring in exclusive association with relict vegetation were eliminated from comparison. In the case of the Chihuahuan Desert primary area all strictly montane and riparian faunas were removed from consideration. Similarly, isolated desert faunas were deleted from comparisons involving more mesic provinces. Secondly, not all primary areas under comparison were defined quantitatively (using IFC values or even other techniques). HAGMEIER & STULTS (1964) indicate that the herpetofaunal provinces of SAVAGE (1960) correspond closely to their mammalian superprovinces. I have assumed a general applicability of their provinces to the province under examination. This assumption is further supported by the virtually perfect correlation between the Chihuahuan Desert province here defined by herpetofauna and the mammalian Mapimian province defined by HAGMEIER (1966). Subjectively constructed Mexican provinces are more difficult to study directly or by comparison. I have generally used the provinces defined by STUART (1964) as the most compatible with the available evidence from physiography, vegetation and faunal distributions. Nonetheless, these Mexican provinces are poorly defined and comparisons made with them are somewhat expedient and of limited significance. Ideally, all North America should be subjected to grid analysis of its herpetofauna before provincial areas are designated. Unfortunately, present knowledge of Mexican faunal distribution is inadequate to complete the task at this time. Thus the compromise is made between objective and subjective units undertaken here.

Another major consideration in determining faunal resemblence between primary areas is the choice of coefficients. SIMPSON (1960) reviewed most of the simple fractional measures previously employed. For simple presence or absence comparisons, he favors his own Similarity Coefficient:

$$\text{S.C.} = (C/N_1) \times 100$$

C = number of shared species

N_1 = number of species in the smaller of two samples

over the classical Coefficient of Community of JACCARD (1902):

N_2 = the larger of two samples.

$$\text{C.C.} = \frac{C}{N_1 + N_2 - C} \times 100$$

SIMPSON favored the former value, especially in comparisons where the discrepancy between two sample sizes is great. He suggests that unless the discrepancy is intrinsically important, the C.C. will exaggerate differences between faunas of common derivation. For example, a depauperate relict fauna might have a C.C. value of only 20% when compared with the fauna of its province of origin. In contrast, the S.C. for the same comparison could be as high as 100%. Both coefficients tend to mask information. The C.C. gives the best ecological information based on presence of absence of the existing differences between samples; in other words where discrepancies in faunal diversity (species density) are important. The S.C., in contrast, would only obscure important ecological differences. However, the S.C. value does, as SIMPSON stated, effectively demonstrate derivation between faunas of common derivation, and is therefore a valuable indicator of historical biogeographical relationships. Since this investigation of the Chihuahuan Desert involves both spatial (ecological) and historical (evolutionary-paleoecological) relationships both coefficients have been used. Since each coefficient tends to counter the distortion produced by the other, relationships that are confirmed by both values are considered here to be particularly significant. The Coefficient of Community has been particularly emphasized in the spatial analysis because it supplies an objective basis for evaluating and comparing primary faunal areas as defined by HAGMEIER & STULTS (1964).

A final consideration regarding faunal comparisons between primary areas is the method of generating dendrograms. Dendrograms used here were generated from the program titled CLASS written in the language PL1. The program was authored by ROBERT SMITH, graduate student in biology at the University of Southern California. The University of Southern California Computer Center processed all the data used in this study. The program itself constructed dendrograms using the techniques for clustering by average linkage described in SOKAL & SNEATH (1963). These authors, in reviewing methods of generating dendograms recommend group average clustering as the least biased approach. The group average sorting process intrinsic to CLASS groups coefficients in progression starting with the two which have the highest resemblance. It continues down the scale of resemblance (100% to 0) averaging and reevaluating each new group it has combined with those remaining until all values are accounted for. C.C. and S.C. values are always sorted independently from one another, and separate dendrograms are constructed for each set of coefficients from the same samples.

The final stage in this analysis is synthetic reconstruction of the biogeographical history of the primary areas considered and the herpetofaunal elements. The procedures of deriving the evolutionary and distributional assemblages using a synthesis of direct fossil evidence, paleogeographical evidence, paleoclimatic indicators, geofloral assemblages with contemporary associations and distribution have been well explained and applied by SAVAGE (1960), specifically to American desert herpetofaunas. The principle of biological uniformitarianism is as relevant to biogeography as it is to geology. UVARDY (1969) further reviews this specific concept in dynamic biogeography. ESTES (1970) offers additional affirmation that valid correlation may be extrapolated from geofloral assemblages to coexisting fossil herpetofaunas. The process of extrapolation is most effective if three conditions are fulfilled. These are the presence of some direct indication as to the actual herpetofauna (preferably from the region under consideration, or at the same geochronological level in an adjacent locality) and secondly, the presence of several independent lines of circumstantial evidence. By this latter point I mean, for example, independent climatic indicators from the fossil matrix, the geofloral assemblage, and from marine sediments at parallel latitudes. The more independent

10

lines of confirmation, the more confidence should be placed in the extrapolation. A third provision is that the reconstruction utilizes large assemblages of organisms, such as faunal elements, rather than be directed toward a particular single species lacking a fossil record. This last concept, essentially that there is safety in numbers, is based on the consideration that the system is speculative, and the reconstruction of evolutionary patterns and dispersal may be valid as general trends for a compound assemblage, but pretentious in their accuracy if confined to a single species. As I am fully aware of the speculative nature of these extrapolated historical assemblages, I have confined myself to the largest and most generalized units, the Historical Elements, in the historical analysis presented here.

The processes of historical biogeography demand some broad and subjective decisions. I have attempted to quantify some of my evaluations by the use of percentages. I have also corrected the relationships between primary areas by removing sibling species or specific historical components to determine how these contemporary resemblances and distinctions are derived. Ultimately, the criteria for judging the models and hypothetical history presented here, will be their probability in relation to the available evidence and in comparison with alternatives. Whenever clear choice was possible, I have chosen the simplest and least speculative mode of explanation.

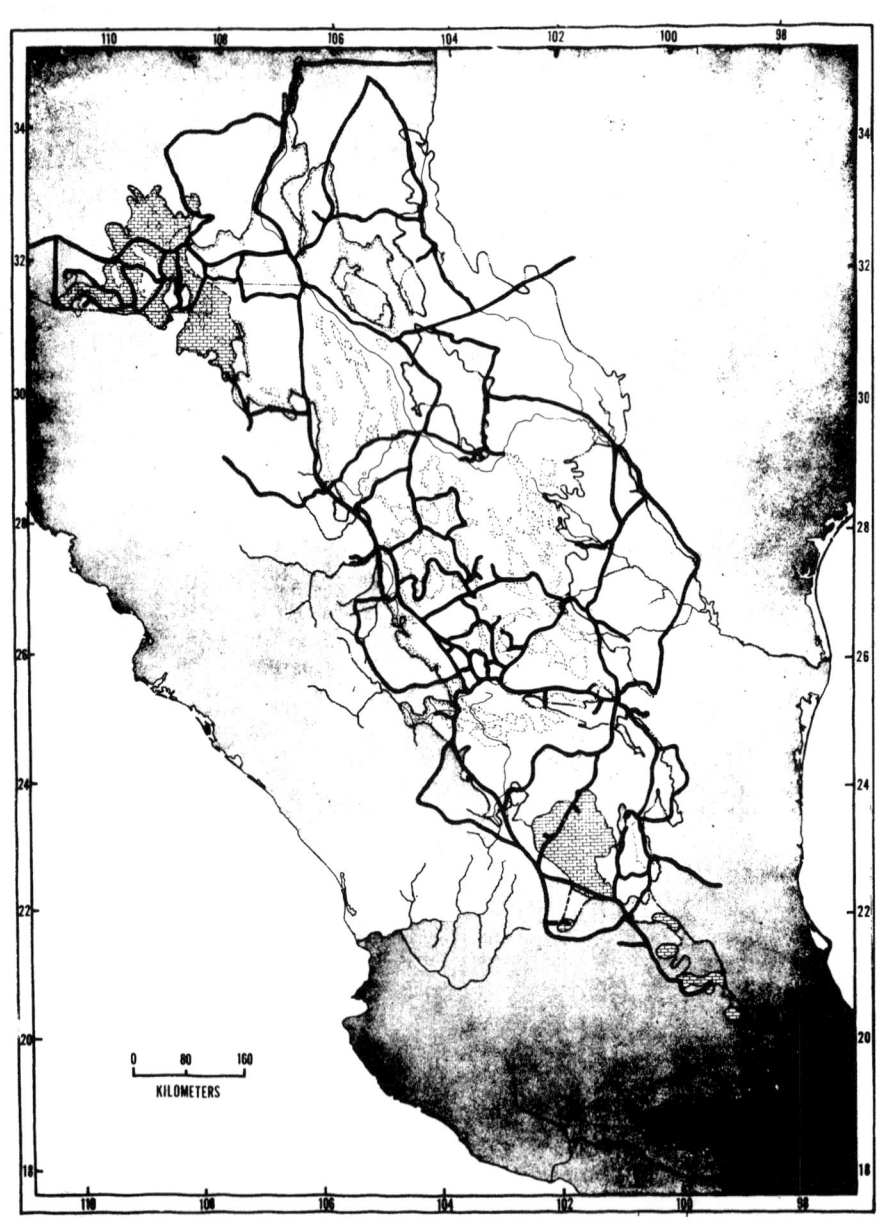

Map 1. Routes traveled by the author (1970-1971).

III. CLIMATE

Ultimately, the major deserts of the earth's surface are the results of climate. While edaphic phenomena generate locally xeric conditions or extend the limits of deserts, only major climatic trends can regulate the soil, topography, vegetation, and faunas worthy of geographical definition.

Existing Conditions

Deserts are specifically determined by the quantity and endurance of free water on their surfaces. This water determines topography, soil structure, microhabitats, and the kinds of vegetation and fauna that can survive in a particular region. Geography defines a desert as a region where evaporation (including transpiration) exceeds precipitation.[1] The most conspicuous immediate consequences of this condition are the usual absence of permanent bodies of standing water. In addition, stream flow is so ephemeral and reduced in discharge that closed basin drainage results. Rapid evaporation loss also produces an atmosphere of reduced relative humidity and restricted cloud cover. The result is a harsh and extreme temperature regime with intense solar radiation penetration. This effect provides a system of positive feedback, more intense solar radiation, to produce drier deserts, which are in turn even more exposed by their greater aridity which in turn engenders still greater penetration by solar radiation. High temperature is a factor of secondary importance, but arid polar regions should be excluded as true deserts. They are very different environments in all factors but free water.

Climatic definitions of a desert can be made regionally specific. The Chihuahuan Desert is the expression of a set of climatic phenomena which both in degree and temporal rhythm define an entity distinguishable from other components of the Great North American Desert. By the criteria of the HOLDRIDGE (1967) (climatic-geographical) life zone classification system, the Chihuahuan Desert is a Temperate to Tropical-Transition Warm Desert Scrubland. According to the Koeppen system of climate classification (from TAMAYO, 1962) this region is a hot desert with a winter dry season. In this same system the adjacent Tamaulipan scrubland and southern portion of the Sonoran Desert would both be considered hot steppe or semiarid climates. THORNTHWAITE's (1931) classification of climates categorizes this desert as temperate, very dry with mild winters (the northwest edge being dry with extreme winters).

More specific dimensions to the Chihuahuan climate can be provided. Drawing from the data of ESCOTO (1964), TAMAYO (1962), U.S. Weather Bureau Annual State Summaries (1971) for Texas, New Mexico and Arizona, and GARCIA, SOTO & MIRANDA (1961), 30 stations were chosen in order to construct a general model of Chihuahuan Desert climate. Ten stations were taken from the Trans-Pecos region at 30-32 degrees north latitude, ten from the Mapimian region at

---

1. Often an arbitrary geographical definition of less than 250 mm (10 inches) is set to distinguish deserts (JAEGER, 1957).

26-28 degrees north latitude, and ten from the Saladan region at 22-24 degrees north latitude. Based on this compilation, the Chihuahuan Desert was determined to have an Average Annual Temperature of $19.2°C$, with a range of 16-22°C, an Average Annual Precipitation of 320 mm with a range of 75-400 mm (between 300-400 mm the manifestation of a desert scrub or mesquite grassland is determined by local P/T ratios, soil structure, and slope). The biological impact of the climate is determined in part by the relationship between precipitation and the water loss to evaporation and transpiration. This can be evaluated crudely and indirectly by determining the ratio of Average Annual Precipitation to Average Annual Temperature (P/T ratio) and also by the percentage of precipitation falling during the coolest (winter) portion of the annual temperature cycle.

Chihuahuan stations report P/T ratios (mm/°C) averaging 14.3 with a range of 7-21. Winter precipitation averages 10% of the total with the range between 3-20%. While the P/T ratios are relatively high for a North American Desert (indicating a relatively mesic climate), the very low percentage of winter rainfall indicates that the desert is watered at the very time when evaporative loss is greatest. Summer precipitation does, in fact, account for 60-80% of the total.

TAMAYO (1962) indicated in his climatic maps that the portion of the Mexican Plateau exposed to an Average Annual Relative Humidity of less than 50% corresponds to the Chihuahuan Desert as does his isopleth delimiting those areas averaging less than sixty overcast days annually. Eighty percent of the precipitation is between May and October. Variability of rainfall is greater than 40%. This high variability contributes further to physical harshness of the system.

Despite the huge geographical expanse of the desert, it shows surprising internal uniformity in the climates of its physiographical subdivisions. A brief comparison based on the three sets of ten stations is offered below as follows:

A Comparison of Climatic Characteristics between Physiographical Subdivisions of the Desert.

Sub-division	Latitude	Elevation	Average Annual Temp.	Seasonal Variation[2]	Average Annual Prec.
Trans-Pecos	30-32°	1000-1500 m	16.2°C	20.8°C	295 mm
Mapimian	26-28°	700-1400 m	20.9°C	13.7°C	292 mm
Saladan	22-24°	1500-1700 m	17.7°C	7.8°C	371 mm[3]

Average Annual Temperature for each of these subdivisions lies within a 15% of the general average of 19.2°C; regional Average Annual Precipitation also falls within 15% deviation of that for the desert as a whole. In contrast, the seasonal variation in monthly temperature averages differs radically between northern, central and southern portions of the desert. This seasonal variation, 20.8°C at 30-32° North Latitude has deteriorated at the rate of 1.3°C per degree latitude toward the Equator. At the 22-24° North Latitude, monthly seasonal variance averages only 7.8°C and locally may be as low as 6°C (see Table 1 and Figure 1).

At this point in the investigation of the intrinsic nature of the Chihuahuan

2. Based on difference between warmest and coldest average monthly temperatures.
3. Based on stations associated with human populace, usually in exceptionally mesic or peripheral localities, therefore probably 20 to 50 mm higher than true Saladan Desert.

Desert climate, one might well consider whether the region is temperate or tropical in its climate. By simple geography, the desert is essentially temperate, except at its southern extreme, that 10 to 20% lying south of the Tropic of Cancer (22° 37' North Latitude). HOLDRIDGE (1967) defined the tropics biologically as those climates sustaining an average annual temperature at sea level of 24°C or higher. He would correct localities above sea level in elevation at the rate of −6°C/1000 m. By a narrow margin, one to three degrees after elevational correction, the vast majority of Chihuahuan stations could be thus considered tropical. I find one severe flaw in the HOLDRIDGE definition which is of particular pertinence to the problem being considered. Namely, the definition totally ignores the importance of seasonality in determining a tropical temperature system. A temperate climate has a clear seasonal temperature cycle with a dramatic difference in temperature averages between hottest and coldest months.

A comparison between two stations at northern and southern edges of the desert will serve well to illustrate both the difficulty in applying HOLDRIDGE'S definition and the tremendous southward reduction in seasonality.

A Comparison in Temperature Regimes at Northern and Southern Edges of the Chihuahuan Desert.

Station	Latitude	Elevation	Average Annual Temp.	HOLDRIDGE Bio. Temp.	Annual Variation[4]
Carlsbad Caverns Nat'l Park, Eddy Co., New Mexico	32°11'	1364 m	16.1°C	24.3°C	18.5 (7-25.5)
Venado, San Luis Potosi, Mexico	22°30'	1750 m	17.7°C	28.5°C	9.9 (11-20.9)

By the HOLDRIDGE definition both stations have corrected biotemperatures of 24°C or higher, qualifying them as "tropical". Yet there is virtually twice the seasonal variation between hottest and coldest months at the more northern locality. Biologically, the differences are very real between these two "tropical" settings. Carlsbad has genuine winter with repeated frosts and extended biological inactivity. The southern (El Saladao) desert and the Mapimian portion as well have sufficiently mild winters to sustain continuous activities for many species. However, winter conditions make these activities more sporadic and tend to favor the juvenile or small heliothermic lizards. Vegetative growth might be expected to follow similar patterns. At neither locality do summer temperatures reach a monthly average of 30°C, indicating that there is no summer inhibition to the growing season.

In conclusion, the Chihuahuan Desert is in essentially the same position climatically that it is geographically — borderline between tropical and temperate. Biotemperatures, corrected for elevation, would qualify most of the desert as tropical according to the HOLDRIDGE definition, but often with only one degree centigrade or less to spare. A comparison of seasonal temperature variation indicates

4. From hottest to coldest monthly average.

major latitudinal differences within the desert. If an arbitrary line were to be drawn between tropical and temperate conditions bissecting the desert, it would most logically fall along the westward Anticlinorium of Arteaga of the Sierra Madre Oriental which separates the Mapimian from the Saladan region at about the 25th parallel north. Due to its high elevation, the Saladan region presents stark cool landscape, but it has the highest biotemperatures (corrected for elevation) of the whole desert and a most marked reduction in seasonal temperature variation, virtually always less than $10°C$. Thus, it is best viewed as a tropical highland plateau.

Causal Factors

This discussion will be a review of the climatic forces molding the Chihuahuan climate in its qualitative, regional and temporal aspects. It will conclude with a comparative analysis of the Chihuahuan climate in relation to other portions of the Great North American Desert.

Three climatic factors have largely determined the nature and position of the Chihuahuan Desert. They are as follows: the dry hot air subsidence on the Mexican Plateau as a result of subtropical convergence, the deep continental position of the plateau, and the high Sierra Madre astride the eastern and western edges of the plateau, inducing rainshadows across the interior basins and ranges. The desert is not simply the result of mountain induced rainshadow alone; this is affirmed by the presence of adjacent arid lowlands on the Pacific and Gulf Coast Plains. Wind patterns, elevation, latitude, surface configurations and incident solar radiation further determine local climates and season patterns within the desert.

Subsidence of sub-tropical calms is the most fundamental factor involved in the generation of the Chihuahuan Desert. The pattern is determined not by local features alone, but by fundamental temperature and air current systems intrinsic to the earth's atmosphere. Hot air masses from the earth's thermal equator rise, cool, adiabatically dry, and gradually subside to the north at about 23 to 25 degrees North Latitude (Hadley's Cell). The cooling dry convection masses heat adiabatically, and then as they descend parch much of the land between these parallels. They contribute further to aridity by blocking moisture-laden coastal weak trade winds coming from the coasts (in this case the Pacific Ocean and Gulf of Mexico). The resulting dry season extends from late October to early June. The cooling subsiding air masses settling on the central plateau produce a high pressure system that operates as a foil, insulating against the warmer mesic currents of low pressure impinging from both coasts.

Late June through early October is characterized by a reversal in relative pressures and their interactions. The air masses derived from the subtropical convergence become low pressure in summer relative to the high pressure storm fronts and cyclones moving along both coasts. Due to this difference in summer pressures, the subtropical calms actually serve to draw moister cooler trades over the region of the plateau, deflecting them inland from their clockwise movements around the Gulf of Mexico. As these moister masses are drawn over the low pressure region of the plateau, they are forced to rise and cool by the local heating of convection currents thrust upward from the surface of the landscape. The result of this elevation and cooling is the fragmentation of the moist masses and their subsequent cooling and precipitation in form of local thunderstorms. Most air masses are derived from trade winds, but others represent the peripheral influences of Caribbean cyclones (basically rapid whirling hot wet air). The cyclones whirl up to 180 kms/hr along a northwest axis center between 6 and 20 degrees north parallel. The summer low pressure calms also deflect Pacific storms from a northeastern route to a more southeastern passage, thus bringing them over the desert, but they make only a feeble contribution to the summer precipitation. The Sierra Madre Occidental provides a more uniform barrier to Pacific storms and high pressure systems than does the discontinuous Oriental to the Gulf Coast storms. The activities of local hot convection currents also contribute to the summer wet season on the Mexican Plateau. A circadian rhythm of local currents in hot basins is characterized by gusts of hot air upslope in the mornings, followed by an afternoon descension of cooled air often continuing into evening and associated with thundershowers, or even hailstorms. These local phenomena are major aspects of summer climate in the Chihuahuan Desert and may have very impressive manifestations, ranging from flash floods to allegedly 120 kms/hr winds.

During the eight to nine month dry season, the desert does not stand completely inviolate to precipitation and storms. These rare winter events may be induced by powerful cyclones which succesfully deflect part of their force northwest from the Gulf lowlands and northeast from the Pacific. Incursions of cold North Polar air move south from the middle latitude West Wind Belt in the form of winter frontal storms. The latitudinal position of the insulating subtropical calms simultaneously moves south. The result is that northwestern Chihuahuan Desert, the Arizona filter barrier and the western half of the Trans-Pecos region are subjected to extremely cold winters with winds and storms from the north and northwest. These northern storms induce sudden temperature drops, dust storms and 150 kms/hr winds, and even snow storms. Occasional northern storms penetrate as far south as Zacatecas and San Luis Potosi. In the winter these northern storms so dominate northern Chihuahua that sand crests and their blowouts shift orientation, temporarily burying vegetation (LE SUEUR 1945). The remainder of the year brings winds from the east and southeast, reversing the orientation of the dunes and excavating the vegetation. With the exception of this northwestern edge of the desert, the entire Mexican Plateau is dominated by winds from the east and southeast.

What has been discussed so far is the interaction of landmasses, wind currents, and oceanic bodies. What remains unexamined are the thermal conditions which generate local temperatures through an interaction of solar radiation (determined by latitude and local atmosphere), and elevation of the land surface. Like most deserts, relative humidity is low (under 50%) and overcast days rare (under 60), so solar radiation is high. The Chihuahuan Desert is very high in elevation (1000-1700 m). The impact of this high topography is so great as to reverse the expected influence of latitude across the 12 degrees parallel expanse of the desert. Table 1 and Figure 1 illustrate the nature of this interaction. From about 34 degrees North to 26 degrees North, average annual temperatures increase generally toward the equator. The very warmest aspects of the Chihuahuan Desert are in the low valleys of the Rio Grande and Cuatro Cienegas Basin between 27-29 degrees North Latitude. At the very northern and especially northwestern edge of the desert, high elevation and high latitude combine to reduce temperature dramatically. Thus a 6.6°C increment in average annual temperature between Truth or Consequences, New Mexico and Ojinaga, Chihuahua is the result of a four degree

Table 1. Correlations Between Latitude, Elevation, and Temperatures of Selected Chihuahuan Desert Climatic Stations.

Station	Latitude	Elevation	Extremes[1] Min.	Max.	Ann. Temp.	Corrected[2] Temp.
1. Truth or Consequences, Sierra Co., N.M.	33 14	1427 m	−12.2	39 °C	15 °C	24 °C
2. Lordsburg, Hidalgo Co., N. M.	32 18	1308 m	−11.7	43 °C	15 °C	23 °C
3. Carlsbad Caverns, Natl. Park, Eddy Co., N. M.	32 11	1365 m	−12.8	38.5°C	18.5°C	24.3°C
4. El Paso, El Paso Co., Texas	31 48	1206 m	−10.0	40.6°C	18.4°C	24.3°C
5. Ojinaga, Chihuahua, Mexico	29 33	841 m	−10.0	50 °C	21.6°C	26.8°C
6. Sierra Mojada, Coahuila, Mexico	27 11	1250 m	−10.0	43 °C	21.2°C	28.4°C
7. Cuatro Cienegas, Coahuila, Mexico	26 58	742 m	− 2.0	44 °C	23 °C	27.5°C
8. Carillo, Chihuahua, Mexico	26 58	742 m	− 2	44 °C	23 °C	27.5°C
9. Lerdo, Durango, Mexico	25 32	1140 m	− 4.2	39 °C	21.3°C	28.2°C
10. Flor de Jimulco Coahuila, Mexico	25 07	1220 m	− 9	45 °C	21.3°C	28.4°C
11. Matahuala, San Luis Potosi, Mexico	23 38	1615 m	− 4.4	39.8°C	20.3°C	30.1°C
12. San Luis Potosi, San Luis Potosi, Mexico	22 09	1877 m	− 2.7	37.3°C	17.6°C	28.7°C

1. Some U. S. values based on one year only (1971).
2. Calculated at −6 C per 100 m elevation, biotemperature.

decline in latitude multiplied by a 600 m drop in elevation. When temperatures are corrected for elevation, the steady increase toward the south is still apparent, but the much reduced rate of increase indicates how the added influence of elevation has acted upon the actual temperatures.

South of the 25th parallel, elevation so outstrips latitude as an influence that real mean temperatures actually decline. The 26th parallel runs along the Parras Bolson and Anticlinorium of Arteaga which mark an abrupt uplift of the plateau floor. The southern desert ascends about 700 m in elevation over a distance of four parallels to the south. If the rate of temperature decline is estimated at $6°C/1000$ m rise in elevation (taken from HOLDRIDGE, 1964), and the increase of temperature per descending northern parallel at about $0.5°C/$parallel degree, it can be predicted the mean annual temperatures would decline about $0.7(6) + 4(0.5) = 2.2°C$ per degree south decrease in temperature. The general formula for this estimate being:

Change in temperature of station II over station I= $0.5(A) + (−6)(B)$
A: number of degrees latitude between stations I and II (II-I)

Table 2. A Comparison of the Climatic Characteristics of the North American Deserts.

Desert	General Chihuahuan	Trans-Pecos	Mapimian	Saladan	Magdalena
Characteristic					
1. Average Annual Temperature (C)	19(16-22)	16.5	20.9	17.7	22(20-25)
2. Annual Daily Maximum and Minimum Temperature	−10 to 50	−10 to 50	−5 to 45	−5 to 40	0 to 50
3. Degrees (C) Between Hottest and Coolest Monthly Averages	14.3(7-20)	20.8(15-23)	13.7(10-15)	7.8(6-10)	12(8-15)
4. Average Annual Precipitation (mm)	75-300 (300-400 peripheral)	295(X)	242(X)	371(X)	25-250
5. Variability of Rainfall (%dev. from Annual X)	40 to 60	40 to 60	40 to 60	40 to 60	40 to 60
6. Percentage of Precipitation Falling in Winter	3 to 25	12.3(8-20)	7.6(5-12)	6(4-9)	19.7(2-36)
7. P/T ratio (mm/C) Based on Annual Averages	14.3(7-21)	13.7(X)	12	17.5	5.6(1-10)
8. % Rel. Humidity	50(40-60)	50(40-60)	50(40-60)	50(40-60)	60(50-70)
9. Wind Dir. Summer	SE, E	SE, E	SE, E	SE, E	S
Winter	N, SW, SE	NE, SW	SW, E	SW, SE	N
10. Solar Radiation (Ann. Langleys in gm/cal/cm^2)	500+	500+	500+	500+	500

B: number of kilometers elevation difference between stations (II-I)
(plus for decrease, minus for increase for both A and B).

Local climatic conditions, especially relative humidity, days overcast (number of), precipitation, and the interactions between local topography and convective wind currents still can alter specific local conditions by several degrees centigrade. Nonetheless, the simplistic formula offered above can be used as a low resolution indication of the conditions in which elevations do outstrip latitude in determining the geographical trends of thermal conditions.

In summary, the Chihuahuan Desert is a thermal cradle of cool northern and southern arches suspending a hot level central floor as illustrated by Figure 2. The northern Arch, the Trans-Pecos Desert is marked by a steep increment in average annual temperatures toward the south, a condition accelerated by the effects of declining latitude and elevation operating in tandem. Seasonal cycles are severely pronounced here. The floor of the cradle is the Mapimi Bolson. It is relatively low and level with a gradual increase in average annual temperatures toward the south being produced by declining latitude alone. Local low pockets within this central floor provide the desert's highest temperatures, with average annual temperatures reaching 23°C and summer extreme maxima exceeding 50°C. Seasonality it is moderate. The southern Arch, the Saladan Desert, climbs so abruptly that elevation

Table 2a. A Comparison of the Climatic Characteristics of the North American Deserts (continued)

Desert	Sonoran	Coloradan	Mojave	Great Basin	Tamaulipan	Sp. Chihuahuan Relict (Toliman, Que., Mexico)
Characteristic						
1. Average Annual Temperature (C)	23(20-25)	22(20-25)	19.2 (15-22)	12(8.3-16.7)	22.2(20-25)	18.8
2. Annual Daily Maximum and Minimum Temperature	0 to 50	0 to 50	−5 to 45	−15 to 40	0 to 50	−5 to 40
3. Degrees (C) Between Hottest and Coolest Monthly Averages	20.5 (19-24)	19.6 (18-24.5)	23.7 (22-28)	27.5 (25-30)	11 (9-15)	8
4. Average Annual Precipitation (mm)	250-500	50-250	50-150 (rarely to 300)	150-300	560(400-800)	350
5. Variability of Rainfall (% dev. from Annual X)	30-40	40-60	35-40	15-30	20-40	−
6. Percentage of Precipitation Falling in Winter	19(2-41)	40(30-55)	45.5	50	10(5-20)	7
7. P/T ratio (mm/C) Based on Annual Averages	9.6(3-15)	3.65(2-6)	5.8(2-11)	15 est.	20(10-25)	18.7
8. % Rel. Humidity	55(40-70)	35(30-40)	25(20-30)	50(40-60)	65(60-70)	−
9. Wind Dir. Summer	S, SE	SW	NW, NE	NW, S	N, S, E	NE, SE
Winter	N	N, E	NW	S, W	E	N
10. Solar Radiation (Ann. Langleys in gm/cal/cm^2)	500	500-550	500-550	400-500	400-500	−

outstrips the influence of declining latitude, in fact reversing its effects for several hundred kilometers down the southern Mexican Plateau. Seasonal oscillations are dampened to virtually tropical conditions in this section of the Chihuahuan Desert.

Comparative Climates of North American Deserts

This discussion is a comparative climatic characterization of the Chihuahuan Desert. This can be done in both an absolute statistical sense and by relative comparisons with other North American deserts. There are three points which I hope to stress in these definitions and comparisons: (1) the Chihuahuan Desert is a distinctive and definable climatic entity; (2) the Chihuahuan Desert has climatic subdivisions which are just as distinctive as those of the North American Desert west of the Continental Divide; and (3) there are symmetrical climatic parallels between desert subdivisions east and west of the Continental Divide (see JAEGER, 1957 for good, general comparisons).

Table 2a provides concise statistical definition of the Chihuahuan Desert climate and supplies a quantitative basis for comparisons.

The Chihuahuan Desert can be characterized as a moderately warm and relatively mesic desert region. The impact of a relatively high precipitation is reduced by its seasonality, 60% to 80% falling during the heat of the summer months. Still, the precipitation/temperature ratios are so high that much of the desert, especially on its northern and southern peripheries, undergoes repeated and reversible shifts in vegetation between desert scrub and semi-arid grassland. The desert is in delicate balance, not only between hot and cold desert and between scrub and grassland

Table 3. A Quantitative Evaluation of Climatic Differences Within and Between Western (Sonoran) and Eastern (Chihuahuan) North American Deserts.
Explanation: Numbers on the left hand margins represent the same climatic characteristics as listed in Table 1.
In the first table the same system is used with the Sonoran Desert proper providing base values.
In the second table, below the Mapimi region is used as the source of base values; deviations for other deserts of less than 15% are indicated by an 'X'; deviations from the base value of more than 15% are indicated by a "—".

Table 3a. The Western (Sonoran) Deserts.

Characteristics/Deserts	Sonoran	Colorado	Magdalena	Mojave	Great Basin
1.	X	X	X	−/X	−
2.	X	X	X	X	−
3.	X	X	−	X	−
4.	X	−	−	−	−/X
5.	X	−/X	−/X	X	−
6.	X	−	−	−	−
7.	X	−	−	−	−
8.	X	−	X	−	X
9.	X	X	X	−/X	−
10.	X	X	X	X	X/−
Points of Comparison:	10	10	10	10	
#of Deviant Values:	5	5	6	8	
Deviant Percentage:	50%	50%	60%	80%	

21

Table 3b. The Eastern (Chihuahuan) Deserts (using the Mapimi as a base)

Characteristics/Deserts	Mapimi	Saladan	Trans-Pecos	Tamaulipan
1.	X	–	–	X
2.	X	X	X	X
3.	X	–	–	X
4.	X	–	X/–	–
5.	X	X	X	–
6.	X	–/X	–	X
7.	X	–	X	–
8.	X	X	X	–
9.	X	X	X	X/–
10.	X	X	X	–
Points of Comparison:		10	10	10
# of Deviant Values:		5	3	5
Percentage:		50%	30%	50%

Table 3c. A Climatic Comparison of Mapimi, Magdalena (Peninsular Baja California), and Mojave Deserts.
Explanation: As for Table 3b, here using the Mapimi Desert values as the base. The symbol 'R' has been added here to indicate a reversal between compared conditions.

Characteristic/Deserts	Mapimi	Magdalena	Mojave
1.	X	X	X
2.	X	X	X
3.	X	X	–
4.	X	–	X
5.	X	X	–
6.	X	–	– (R)
7.	X	–	–
8.	X	–/X (18%)	–
9.	X	–/X	–
10.	X	X	X
Points of comparison:		10	10
# of Deviant Points:		5-(2)	6
Percentage Difference:		50% (30%)	60%

Table 3d. A Climatic Comparison of the Great Basin, Trans-Pecos, and Saladan Deserts.
Explanation: As for Tables a and b, but using the Great Basin Desert as the source of base values.

Characteristic/Deserts	Great Basin	Trans-Pecos	Saladan
1.	X	–	–
2.	X	X	X
3.	X	– (25%)	–
4.	X	X	– (20%)
5.	X	–	–
6.	X	– (R)	– (R)
7.	X	X	X/–
8.	X	X	X
9.	X	– (R)	– (R)
10.	X	X/–	X/–
Points of Comparison:		10	10
# of Deviant Values:		5	6
Percentage Total:		50%	60%

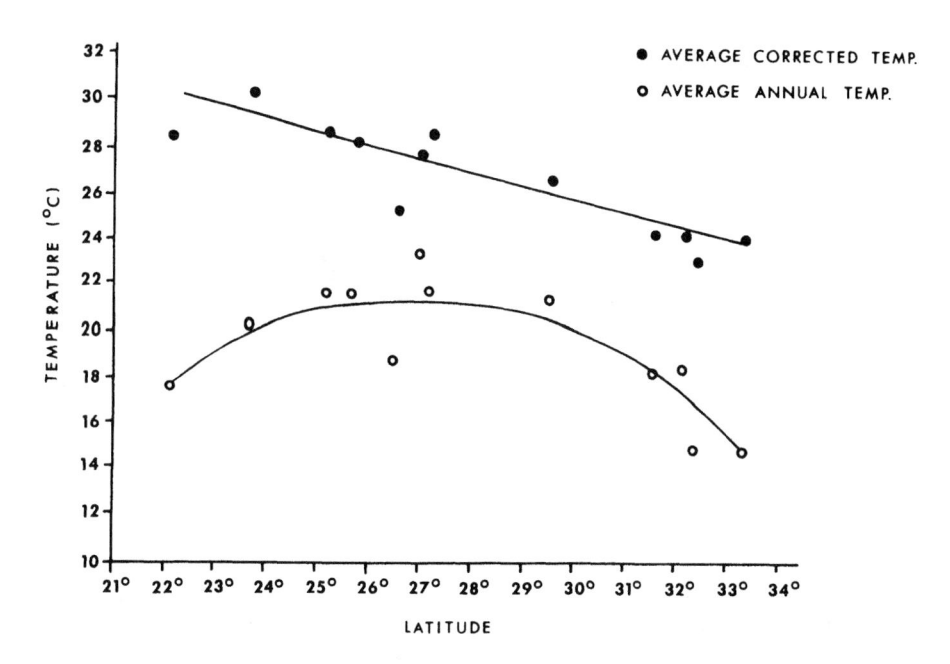

Fig. 1. Correlations Between Latitude, Elevation, Average Anual Temperature, and Corrected-Temperatures.

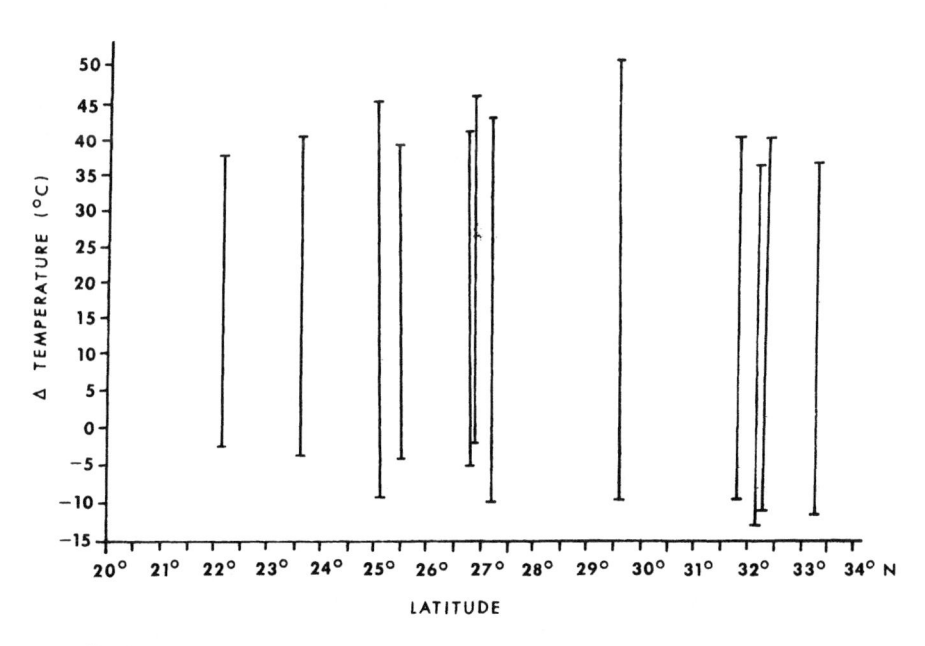

Fig. 2. Correlation Between Elevation and Range of Annual Temperature Extremes.

steppe, but between temperate and tropical as well. Biotemperatures could be used to arbitrarily define the desert as tropical highland, often with only a single centigrade degree to spare. Seasonality of temperatures is far more pronounced at the northern end of the Mexican Plateau than in the south. It is far more responsive to real conditions to describe the desert as transitional between temperate and tropical climates.

Expanding from the descriptive characterization provided above and the more explicit values assembled in Table 3, a comparative résumé can be made of the Chihuahuan Desert's internal divisions, and to other units of the North American Desert.

The North American Desert can be bisected into western and eastern components, partitioned by the Continental Divide. GEHLBACH (1967) argued for this partition based on vegetation, and it has obvious merit on both climatic and simple geographical grounds. The western division has been richly endowed with regional names familiar to most English and Spanish-speaking North Americans. The terms Mojave, Colorado, Sonora all carry geographical, and more vaguely, ecological implications to most Americans. Unfortunately, they lack uniformly defined geographical boundaries and characteristics, both in vernacular usage and often in biological and geographical literature as well. The situation is further complicated by ambiguous terms (i.e., the Colorado Desert is applied both to southeastern California and southeastern Utah) and very localized names, applied solely on the basis of locally spectacular topography (i.e., the Painted Desert, the Anza-Borrego Desert, and Death Valley). If this situation did not generate enough confusion, the desert east of the Continental Divide is virtually devoid of a comparable geographical terminology. The so-called Chihuahuan Desert is applied to this entire vast region, obscuring the fact that it has three climatic, geographical, and biologically distinct subdivisions. Each of these divisions are as distinct from one another as any of the named western divisions.

That equally distinct subdivisions exist as climatic entities east and west of the Divide is established by the comparisons provided by Table 3A and 3B. The Mapimi and Saladan Deserts of the east differ from one another in 5 of 10 climatic characters by more than 20%, which is equivalent to the differences between Sonoran, Colorado, Baja Calfornia (Magdalena) Desert as shown by Table 3A-D.

In addition, both eastern and western regions are bordered by peripheral arid regions to which the vernacular term "desert" is freely applied, but which differ strikingly in some of their climatic features. The Great Basin "Desert", geographically part of the western desert system, shows an 80% lack of congruence with the climatic character of the Sonoran Desert. It has significantly lower average temperatures, higher seasonal oscillations, and inferior impinging solar radiation. While the high aridity and P/T ratios still qualify this region as true desert in my opinion, it is definitely a cold desert and not climatically part of the Sonoran system.

On the Gulf Coast plain adjacent to the eastern deserts is another arid borderland of dubious status. This is the Tamaulipan Desert or Thornscrub. Like the Great Basin, it receives an inferior annual solar radiation relative to adjacent deserts. While its temperature regime is like the Chihuahuan Desert, this lowland plain also is endowed with far greater absolute annual precipitation than any North American desert. The Tamaulipan Thornscrub is best viewed as a subtropical mesquite grassland savanna, not as a desert.

24

Comparisons between eastern and western subdivisions of the North American Desert reveal that the Mapimi Desert of the east is slightly closer climatically to the Magdalena Desert of Baja California than it is to the geographically adjacent Saladan Desert to its immediate south; see Table 3B and 3C.

Similar parallels can be made between the Mapimi (east) and Mohave Deserts (west) and between the Trans-Pecos and Saladan Deserts (east) and the Great Basin (west) in Table 3C and 3D. This latter set expresses in climatic similarities in only three or four characters on a ten point scale. However, some of the negative comparisons are reversals of one another, especially the wind directions and the percentage of annual precipitation incurred during winter. In later ecological comparisons of dominant vegetation and faunal affinities, these climatic parallels and reversals will be instrumental in deducing the reversely symmetrical distribution of North American Desert communities.

IV. THE PHYSIOGRAPHY OF THE CHIHUAHUAN DESERT

Geography and Dimensions

The Chihuahuan Desert, set in an immense arid plateau in the core of southern North America, is the largest desert on that continent. The grandeur of its dimensions can be appreciated by comparisons with the better known western deserts of the United States. The Mojave, Great Basin and Colorado Deserts combined are still inferior to the Chihuahuan in surface area occupied. In total surface area, it is about 450,000 km^2.

Using political states as convenient units of geographical reference, the desert is situated in the Rio Grande drainage of New Mexico, and western Texas in the United States, and in the Mexican States of Chihuahuan, Coahuila, northeast Durango, southwestern Nuevo Leon, and San Luis Potosi. A filter barrier with significant Chihuahuan biotic components extends into southeastern Arizona and extreme northwestern Sonora, and relict pockets of Chihuahuan communities occur in the Mexican States of Aguascalientes, Hidalgo, and Queretero.

The desert extends from latitude 35 North, south through the Tropic of Cancer to the 22nd parallel. Desert relicts in Hidalgo project the limits south another degree. Longitudinal limits can be set at approximately the 101st meridian at the eastern edge and 108th meridian in the west. The northwestern filter barrier continues Chihuahuan components another three degrees west into south central Arizona.

The dimensions of length and width of the Chihuahuan Desert make explicit its huge size. A southeastward diagonal from the northwestern extreme of the desert at Albuquerque, New Mexico to its southeastern edge at San Luis Potosi, would run a full 1700 kilometers.

The disjunct Chihuahuan relicts in Hidalgo and Queretero would add another 200 kilometers to the diagonal. The 104th meridian which runs approximately down the center length of the desert intercepts this region for 1050 kilometers, which may indicate average straight length. The greatest width of the desert stretches at slightly southern diagonal to east-west from about the Continental Divide in Grant County, New Mexico to Devil's River in Val Verde County, Texas. This constitutes a width of about 550 kilometers, with the addition of the Arizona filter barrier contributing another 200 kilometers. Much of the desert south of the Rio Grande drainage maintains an average width of approximately 300 kilometers. By comparison the greatest length of the Great Basin Desert is about 500 kilometers (see BANTA, 1962b) and its greatest width about 800 kilometers; the Mojave Desert has a maximum length of approximately 500 kilometers and its greatest diameter is less than 450 kilometers.

Topography

In elevational aspect, the Chihuahuan Desert is situated on a dry plateau tilted upward along its southern end and extreme northern periphery and lying between the Continental Divide and the Gulf Coast Plain of North America. The Rio Grande drainage has dissected and lowered the former bolsons of the northern desert down 500 meters elevation. Most of the northern and central desert lies above 1000 meters and below 1500 meters. South of the Parras Bolson of southern Coahuila, the desert climbs into valley floors from 1500 to 1800 meters in elevation. Along the lateral walls of the plateau south of the Parras Bolson, desert gives way to chaparral and Pinyon-Juniper Woodland at about 1750 meters. For the most part the desert is encapsuled by higher slopes and ranges rarely limited by a lower minimum elevation. However, along its eastern edge the Sierra Madre Oriental, and its Coahuila Folded Belt derivative, provide gaps leading from the central plateau down to the Tamaulipan or Gulf plain. One such gap, along the Rio Grande River (vicinity of Devil's River, Texas) at about 30 degrees North Latitude, results in a transition from desert to grassland and thornscrub conditions at about 600 meters elevation. At about 23 degrees North Latitude another gap occurs in the vicinity of Cerritos, San Luis Potosi. Here a transition from desert to dry subtropical woodland and thornscrub occurs between 1400 and 1200 meters.

Ranges within the desert generally rise 500 to 1000 meters above the valley floors resulting in ridges of 2000 to 3000 meters elevation. Peaks astride the western wall of the plateau (Sierra Madre Occidental), reach above 3000 meters and those of eastern perimeter, Sierra Madre Oriental, may rise above 4000 meters elevation. Standing on a 1000 to 2000 meter high plateau with parallel serrations running nortwest-southwest, the Chihuahuan Desert presents an arid topography strikingly like that of the Basin Range Province of the western United States and the Great Basin Desert in particular. However, the landscape is essentially a karst, a now water impoverished limestone plain, largely devoid of stable stream channels, lakes, or external drainage, but endowed rather with solution caverns, springs, mountain blocks, bolsons, mesas, dunes, and ephemeral lakes. Soils are strongly mineral in nature, and include sands, alkaline and gypsum, desert pavements, caliches, gray and red desert soils (Sierozems), clay silts, and magnificent bajadas of volcanic and limestone clastics, sometimes comprised of gravels a hundred meters deep. Along the southern and western edges of the desert, the monotony of the karst is interrupted by volcanic extrusions of rhyolite, andesite, and basalt (rarely).

Major topographical features of desert include delimiting geographical borders, internal and intruding mountain ranges, dominant soil conditions, major cavern and spring systems, external drainage systems, and outstanding bolsons with their associated ephemeral lakes, and internal drainage systems.

Before proceeding with an analysis by specific topic, some background must be provided regarding the platform upon which all else is set, the central plateau of Mexico. This plateau, designated alternately as the Mesa del Norte, or the Altiplanicie Septentrional, runs for nearly 2500 kilometers from the disjunct ranges merging with the Rocky Mountains to north in Texas, Arizona, and New Mexico, southward to the edge of the Isthmus of Tehuantepec.

ATWOOD (1940) classified the Mexican Plateau as part of the Cordilleran Province among the physiographical divisions of North America. The southernmost part of a series of six plateaus terminating in the Yukon, this plateau joins with the Colorado, Great Basin, Snake River, Columbia, and British Columbia plateaus to form a ruggedly montane western North America. Ranges within this province generally run northwest-southeast. The history of the Mexican Plateau is intimately involved with orogeny of the Rocky Mountains which they resemble in structure as well as in origin. The orogenic events responsible for the contemporary topography being with the Laramide Revolution of the Upper Cretaceous Period. From this period through early Tertiary, the epicontinental seas of central North America were driven from their geosynclinal beds as these were progressively folded and obliterated. These beds were replaced by exposures of diorite and quartziferous monzonite. These tectonic events determined the general coastlines and major plateaus and mountain ranges for western and central North America. They involved massive deformations of Mesozoic marine sediments, especially limestones, gypsums, and marls which still dominate the Chihuahuan landscape today. In addition, at the northern end of the Mexican Plateau there is an involvement of Paleozoic sediments in folding and faulting responsible for the existing mountain ranges. The plateau as a whole rests on typical continental granitic basement. There may be local exposures of Pre-Cambrian and Paleozoic strata lying beneath the predominent Mesozoic components throughout the region.

Geological History

The subsequent discussion will trace very briefly the historical derivation of the contemporary features of the plateau. The following references on Chihuahuan tectonics are offered both as a guide to a more technical introduction and as an acknowledgement of sources. Organized by political division, they are as follows: for Texas: BOSE & CAVINS (1927), KING (1935), and MAXWELL, LONSDALE, HAZZARD & WILSON (1967) (for Big Bend National Park); for New Mexico: JONES, HERNON & MOORE (1967), and KOTTLEWSKI (1958); for southeastern Arizona: COOPER & SILVER (1964); for Mexico (the first four references all in Handbook of Middle American Indians): MALDONADO-KOERDELL (1964), STEVENS (1964), TAMAYO & WEST (1964), and WEST (1964), ARELLANO (1951), HUMPHREY (1956), IMLAY (1936), and WEIDIE & MURRAY (1967). ATWOOD (1940) and EARDLEY (1951) provide relevant accounts within reviews of the whole continent. The text just noted parenthetically, and in a more general sense, KING (1959) stand as the outstanding sources of information regarding the physiography of the Mexican Plateau.

The geological events of the Cenozoic are largely repetitions and rearrangements of the fundamental tectonic patterns established in the Laramide Orogeny. Laramide folding may have continued through the Paleocene and was apparently followed by a relatively quiet later Paleocene marked by erosion and aggregation of basin sediments. The Miocene Epoch was marked by renewed mountain building and extensive volcanic activity. The basin and range topography previously established was amplified in its relief by extensive faulting. The result was a series of ranges (horsts) exaggerated in their features by fault dropped basins (grabens) running parallel to them. These abrupt faulted ranges are characterized by their northwestern oriented axis, deep canyons and sharp-crested spurs. The Miocene also ushered an episode of widespread volcanic activity of the Mexican Plateau. Flows, dikes, and plugs of igneous extrusions, especially rhyodacite, rhyolite, and andesite, were predominant. Volcanic activity was most pronounced along the western edge of the desert (the eastern foothills of the Sierra Madre Occidental) and along the southern periphery. Sediments of clay and sandstone were also deposited in the middle Tertiary, but these were of local importance only.

The Pliocene Epoch was marked by erosion of the ranges and alluvial aggregation in the basins. Massive aggregations of alluvial clastics in the form of gravel and sand marked the epoch. Volcanic activity was renewed in the Upper Pliocene, enhancing the formation of the massive New-Volcanic Axis which lies essentially just south of the desert (19 to 21 degrees north latitude). Pliocene flows within the desert were much more restricted than those of the Miocene and involved regions primarily in the northwestern edge. The flows themselves consisted largely of dacite and rhyolite. The Upper Pliocene also witnessed a further development of range-basin relief with episodes of simple faulting and upthrusting, again less extensive and less deforming than those of the Miocene. The relative tectonic stability of Pliocene conditions probably established systems of aggregating clastics that have maintained continuous development into the present time. Likewise zonal development of current soils may have been initiated during this epoch. Current basic hydrological patterns could have become established, as well.

Pleistocene events brought further erosion, basaltic volcanics, cyclic lake and river formation and regression, and high altitude glaciation. Possibly continuing from Pliocene origins (STRAIN, 1966 — for the Pecos) massive alluvial fans coalesced (during xeric intervals) to form bajadas covering the pediments of ranges with as much as 100 meters of gravel and rock. The most severely eroded ranges were reduced to knolls buried almost completely with alluvium. The typically closed bolson floors, barrials, were gradually raised by the aggregation of the clastic fragments of adjacent deteriorating ranges. The result of this closed bolson erosion system is not merely a descent toward sea level but rather a gradual leveling (wearing down and filling up) of the plateau into fairly uniform peneplain.

Pleistocene volcanics brought basaltic flows to the surface. These flows, exposed to the forces of erosion for less time, and poorer in nutrient potential for vegetation than the preceding andesitic extrusions have left considerable stretches of the northwestern desert (especially the Malpais of central New Mexico) in a biologically devastated state through the present time.

Depending on absolute precipitation and temperature governed precipitation-evaporation ratios, the barrials of the desert bolsons are capable of holding a variety of contents. The fundamental flooring of the barrial is usually an impervious clay silt carried down from adjacent slopes as colloidal suspension in feeble runoffs. Over this substrate a salt precipitate may form from rapidly evaporated runoff water collecting in the floor of the bolson, or if precipitation is sufficient to overcome the evaporative capacity of the desert, an ephemeral lake may form.

Major lakes of more permanent nature were established in the floors of major bolsons as a result of blocked rivers coming from the Sierra Madre Occidental or the Rio Grande from the north. The deteriorating lake of the Parras Bolson, Laguna Mayran, evidently underwent radical fluctuations in size, expanding with pluvial conditions in Pliocene, Pleistocene, and post-glacial times. Its major source, the Rio Nazas, drains east from the Sierras and has undoubtedly experienced numerous alterations in headwaters, outlets, contributing branches, and impinging precipitation on its watershed. The retreating lakes of northwest Chihuahua, Lago Palomas being an example, are fed by Sierran streams at present; they may have received the flow of the Rio Grande or its overspill during mesic episodes of the Pleistocene.

High altitude glaciation, even during glacial maxima, was probably of restricted and indirect influence on the desert portion of the plateau. WEST (1964) suggested by extrapolation that glaciers occurred above 3800 meters along the crests of the Mexican Plateau. Ridges even approaching this height do not occur within the present confines of the desert (Cerro Potosi, Nuevo Leon is the only peripheral mountain of adequate height).

Erosional forces involved in erosion and aggregation include temperature fracturing of parent rock, gravity fall, rain and flash flood movements of alluvium and downslope creep of weathered particles. Wind seems to be important only in the central and northwestern sections of the desert and is primarily associated with gypsum and silica dune formation, usually in association with barrial lakes.

Adjacent Topographical Features

The geomorphic history of plateau applies with modification to limiting contiguous landforms of the desert as well. On the west the desert is bordered by high (above 1500 meters elevation) alluvial basins, these often draining into the barrial lakes on the desert floor. These high basins support a deeper, more organic soil and a vegetation of grassland, mesquite, and pinyon-oak woodland. West of these high non-desert valleys lies the Continental Divide and the Sierra Madre Occidental. This major mountain block, rising 2500 to 3000 meters in elevation, is built of middle Tertiary volcanics, especially rhyolite and andesite. It runs south from the United States boundary to the Rio Mesquital in southern Durango, while the ranges further south are structurally part of the basin range topography of the plateau itself. Other authors would extend it south to the Santiago River in Jalisco.

To the south of the desert portion of the plateau lies the Neo Volcanic Axis, a still active volcanic strip (originating in middle Pleistocene) across the southern plateau. It lies about one to two degrees south of the desert edge, with the intervening region marked by transitions in vegetation (desert to Mesquite-grass), topography (karst to volcanic), and soils (sierozem to lithosols and chernozem).

The Sierra Madre Oriental forms the eastern perimeter of the desert, while a westward extension nearly severs the region in half. This Sierra, unlike that of the western perimeter, is comprised of a parallel series of gigantic anticlines of stratified limestones, a more folded and thrusted continuation of karst formations of the plateau itself. Its southern portion, running northwest in parallel to the basin ranges of San Luis Potosi, undergoes a dramatic contortion to the west in southern Coahuila, forming the Anticlinorium of Arteaga. This huge curved anticline consists of the Sierras del Parras and Jimulco and is associated with a massive and ancient Geosyncline, the Parras Bolson, lying parallel on its immediate north. The Parras Geosyncline may date back to Cretaceous times with a history of repeated inundations, first by marine embayments and later by Neogene lakes.

While the Anticlinorium protrudes westward across the southern third of the desert as a massive forested peninsula, another arm of the Sierra Madre Oriental continues northwest to complete the desert's eastern wall. This northern arm, the Coahuila Marginal Folded Belt, extends from the curve of the Anticlinorium north to the Chisos Mountains of Big Bend National Park in Texas. The topography of the Belt reflects diastrophic thrust faulting and recumbent folding of extreme intensity. Its ridges often rise 2000 to 3000 meters above sea level, but near its southern base heights of 3000 to 4000 meters are reached. The belt is characterized by parallel series of steep sinuous ridges and valleys stretching in tandem with plunging anticlinal ridges. While the montane topography of this region is dramatic, it is also marked by repeated disjunctions, and Chihuahuan fauna and flora have enveloped much the belt region, especially toward the north. Thus the desert has, in effect, overcome the ridges themselves but is limited by the more humid lowland climate, soil, and biota that occupies the piedmont of the adjacent Gulf Coast Plain. Three major passes along the eastern desert perimeter should be mentioned. At the northern edge the Rio Grande has forced passage through massive limestone blocks between the Chisos and Carmen Mountains, creating canyons, flood plains and valley dropping 500 meters below the surrounding landscape. At the point of flexure of the Anticlinorium of Arteaga is a second major passage, the Paso de Aguila. Here erosion, acting of the narrow ridge at the southern base of the Coahuila Folded Belt, has brought the eastern and western pediments of the Sierra Madre in direct contact. The result is a gradual east tilting downslope descending into the Gulf Coast Plain from about the 1500 meter level on the plateau side of the pass.

As the plateau tilts upward the desert gives way to acacia scrub in the south. Beyond the southeastern edge of the desert, the Rio Verde (in the state of San Luis Potosi) drains toward the Gulf of Mexico. Its upper branches cut deep canyons in the limestone slopes of the Sierra Madre. Still further south in Queretero, the Rio Extorax extracts another deep transverse cut across the Sierra Madre. Both rivers are branches of the Rio Panuco, and both create steep slope soil and drainage conditions of their sheltered canyon walls where relictual and depauperate Chihuahuan Desert biota still survive. These relictual situations exist as nearly vertical plains inpocketed on walls of these 100 meter deep chasms.

The northern edge of the desert gives way to three major physiographic regions. On the northeast lies the Great Plains with its immense level prairies, grassland vegetation, dark deep organic soils (chernozems), and Mississippi oriented drainage systems. The Rocky Mountains neighbors the desert to its north. This mountain system extends with general continuity through central New Mexico, where it becomes fragmented and is joined to both converging Sierra Madres by series of disjunct intermediate ranges, mostly under 3000 meters at their crests. Finally, the lower desert occupied valleys of the Rio Grande are terminated on their northwestern aspects by the ranges and tablelands of the Colorado Plateau. At the extreme northwestern edge of the desert there is a complex biological filter barrier along the region of the Continental Divide, especially the 1200 to 1500 meter plateau that lies in the gap between the Colorado Plateau and the Sierra Madre Occidental. Here, too, occasional ranges, all under 3000 meters, intervene.

Two aspects of contemporary physiography are worthy of more detailed examination because of their direct relevance to the structure of the desert. These are soil and drainage.

Soils

The present structure and distribution of Chihuahuan Desert soils is the result of interactions between parent materials, topography, climate, and vegetation, modulated in quality and quantity by the factor of time. Parent materials available in this region consist predominately of limestones and marls. As these outcroppings weather, the fundamental matrix of calcium carbonate can be redistributed as solution in addition to water moved clastic fragments. Soils from this source have rich nutrient potential for prospective vegetation. Volcanic extrusives, andesite and basalt, provide a lesser, more localized, but still significant source of mineral rich substrate (in relation to vegetation). Old sandstones and rhyolite provide the small remainder of parent material; the latter sources providing a nutrient poor, acidic silaceous substrate.

Of soil molding factors of the desert, the climatic interactions of temperature and precipitation are most important. In some portions of the desert parent rock fracturing is accelerated by frosts. Precipitation is characteristically feeble and falls primarily during the hot summer months. Despite evaporative loss, its erosional effects are considerable. The result, topographically, is an abundance of eroded closed drainage basins, or bolsons, because local streams lack sufficient power to form spillways or overflows out of the bolsons. These bolsons thus aggregate huge coalescing slopes of alluvium, and their lower slopes and floors are typically encrusted with a calcium carbonate. This crust or caliche, is often supplemented with superficial deposits of salts and gypsum, also reflective of feeble runoff and rapid evaporation resulting in extensive precipitative surfaces.

The basin and range topography of the Chihuahuan Desert combines with climate to produce the results just discussed. Closed drainage bolsons form receptacles for aggregating coalluvial soils. The low dips on their floors act as collecting trays for clay silt and the calcium and salt precipitates of evaporating runoff water. The steep faulted slopes of the adjacent ranges provide ready parent material sources for clastics lost to the adjacent valleys by gravity and runoff water. Where the topography and watershed provide sufficient resources, river systems actively erode parent material and disperse alluvial gravels, sands and silts along their channels and adjacent flood plains. Topography also determines the importance of another climatic erosive force, wind. In the broad open plains of the desert northwest, the Regosol dunes running across a one hundred fifty kilometer east-west axis south of Juarez, Chihuahua are illustrative. Except for these exposed plains in the northwest, wind appears to be an erosive agent of relatively minor importance in the desert.

Vegetation, itself reflective of climate and topography, exercises a powerful influence on soil structure. In classical desert conditions, vegetation is generally sparse, small, sclerophytic, and widely spaced. Thus it fails to develop a leaf litter or humus of any appreciable depth. Organic matter, except for the roots, is lost to wind and runoff water. As a result the soil remains largely inorganic. Poor vegetation also results in rapid evaporation, superficial precipitation and the production of superficial or shallow soil horizon. Grassland conditions, even within climatic deserts, provide a radical change in the situation. Here a thick surface of dense vegetation stabilizes the accumulation of organic material, holding it against wind and runoff. In fact, it reduces the flow of runoff and rate of evaporation. The result is that both the organic component of the soil increases in depth, and that surface water is allowed to percolate downward, deepening the C soil horizon, and fomenting soil zonation (A,B,C). More water and organic matter generate more humic acids to break down the remaining organic soil elements into ions valuable to plant nutrition. This same process is sometimes enhanced along much the same lines in the coniferous woodlands and forests bordering the desert where rich, deep organic soils are associated with thick mats of leaf litter and humus.

Table 4 provides a minimal résumé of the spectrum, importance and distribution of Chihuahuan soil types. It may be concluded that the Chihuahuan Desert is dominated by calcification-process soil, a hardly surprising state of affairs for a residual karst land. The hardpan layer is superficial or very nearly so, almost always within 35 cm of the surface. Above the hardpan is a sierozem soil, usually a gray desert soil containing some organic material and clay. This hardpan consists of a layer of water impervious clay silt, topped with a mineralized encrustation of calcium carbonate and salt precipitates. Both components of the hardpan are the result of soil percolating water which has left precipitate and colloidal clay residues after reaching a maximum depth. This depth is determined by the amount of runoff water, the rate of evaporation (drawing water back to the surface) and the previous layering of a hardpan which may block further

Table 4. Chihuahuan Desert Soil Classification (based on WEST, 1964).

General Category	Great Soil Group	Distribution
A. Zonal Soils		
1. Calcification-Process Soil	Sierozem (gray, red, and brown desert soils; caliche or desert pavement, at the surface)	Dominant soil of desert barrials and lower slopes
2. Podzolization Process Soils	Chernozem; Podzolic soils (red, gray and brown)	Peripheral to desert in conifer dominated Sierra Madres and relict forest on high ranges within the desert
B. Interzonal Soils		
1. Halomorphic	Solonchak Solonetz Soloth	Most frequent in lowest aspects of basins and bolsons. Throughout the desert, but especially Bolson de Mapimi, Bolson de Parras, El Salado and N.W. Chihuahua.
C. Azonal Soils	1. Lithosols	Local igneous extrusive pockets, especially in southern Zacatecas, San Luis Potosi, New Mexico and northeast Durango.
	2. Regosols	Gypsum dunes, esp. White Sands Natl. Monument, N.M.; Salina del Rey, Coah.; and Cuatro Cienegas, Coah.; Silaceous dunes: Medanos south of Juarez, Chih.; Bolson de Parras, Coah.
	3. Alluvial Soils	Alluvial fans and bajadas throughout the desert; along river washes and flood plains; across the floors of some aggregational bolsons.

Table 5. Hydrographical Divisions of the Chihuahuan Desert.

Major Drainage System	Components	Geographical Region
A. Cochise Filter Barrier		
I. Pacific Drainage (Gulf of Calif.)	1. Gila River	Filter barrier of desert, Cochise County, Arizona (and adjacent New Mexico)
	2. Rio Yaqui	Filter barrier of N.W.
	a. Papogochic	Chihuahua and N.E. Sonora
	b. Bavispe	Mexico
	c. Moctezuma	
B. Trans-Pecos Region		
I. Atlantic Drainage	1. Rio Grande (Bravo)	Trans-Pecos, Texas, central New Mexico
	a. Pecos River	
	b. Devil's River	
	c. Cochos Rio	Northeast Chihuahua
	1) San Juan	
	2) San Pedo	
	3) Parral	
	4) Florido	
II. Interior Drainage	1. Rio Guzman	Northwest Chihuahua
	2. Rio Santa Maria	
	3. Rio Patos	
C. Bolson de Mapimi		
I. Atlantic Drainage . (only on eastern periphery of the Bolson)	1. Rio Sabinas 2. Rio Salado	Northeast Coahuila
II. Interior Drainage	1. Rio Nazas	Northeast Durango and
	a. Rio Oro	southwestern Coahuila,
	b. Rio Ramos	into Lago Mayran.
	2. Rio Aguanaval (lower part)	Extreme eastern Durango and southwest Coahuila into Laguna Viesca.
D. El Salado		
I. Interior Drainage	1. Rio Aguanaval upper part including:	Northwestern Zacatecas
	a. Rio Florido	
	b. Las Cruces	
	c. Sain Alto	
E. Southern Desert Relicts		
I. Atlantic Drainage	1. Rio Panuco	Southeastern San Luis,
	a. Moctezuma	Potosi, Guanajuato and
	1) Extorax	Queretero
	b. Tamuin	
	1) Verde	
	2) Santa Maria	

downward penetrance. Because Sierozems or desert soils provide such direct reflections of climatic and biotic environment, they are powerful tools in constructing the geographical limits of a desert.

The Sierozem of the Chihuahuan Desert surface is locally complemented with halomorphic soils, characterized as Solonchak, Solonetz and Soloth in progressive order regarding the degree of water leaching of salts. These salt or alkaline flats are generally associated with barrial valley floors and closed or bolson drainage.

The remainder of the desert proper is surfaced with azonal soils, essentially the raw clastics of parent material, moved slightly spatially but qualitatively unmodified. In this category, alluvial soils are the most important component since these aggregating gravels form the fans and bajadas (sloping from the angle of repose), most of the desert basins and bolsons. In bolson situations, coalescing bajadas are actually in the progress of filling and raising bolson floors. Regosols are represented by the wind structure sand dunes of silica and gypsum found in the northern two-thirds of the desert. Lithosols are confined to extrusive volcanic outcroppings, and steep slope exposures where gravity and climate conspire to inhibit the formation of any deep or mature soil.

Dark organic soils with horizons much deeper characterize the grassland and woodland (or coniferous forest) which replaces the desert on its periphery and in montane pockets within its borders. Slight changes in climate, conducive to a transition from desert scrub to grassland can have radical effects on the soil structure. For reasons already discussed, the dense matting of grass accelerates the aggregation of organic material and the breakdown of inorganic substrate. Vegetation, operating as function of slope, substrate, temperature and precipitation has wrought major transformations in soils of the Mexican Plateau. WEST (1964) documented one such transformation between Matehuala, San Luis Potosi to the state of Queretero. Soil, climate, and vegetation change in coordinated fashion from desert to grassland and eventually forest conditions. Thus podzolic soils, involving organic aggregation and humic acid breakdown, can serve as valuable indicators of non-desert and transitional conditions.

Hydrology

The contemporary hydrology of the Chihuahuan Desert is a continuation of fundamental patterns probably established in the Late Tertiary Period and modified in recent times by regional perturbations in the volume of stream flow. The flow in turn is contingent on the topography of the watershed, the permeability of the substrate, the impinging precipitation, and the vapor loss determined by temperature, soil, and vegetation (most of the following discussion is from TAMAYO & WEST 1964).

There are three major hydrological regions to the desert and two minor peripheral components as well. The five units and their component branches have been summarized in Table 5.

The northernmost unit, the Cochise Filter Barrier, is the only Chihuahuan component lying west of the Continental Divide, and its drainage is the Pacific Ocean, especially into the Gulf of California. Local drainage in this region also include closed basins feeding local ephemeral lakes and alkaline flats, Wilcox Playa[1] being a conspicuous illustration of the latter.

---

1. The Wilcox Playa was part of a more permanent and extensive glacial and pluvial lake.

The Trans-Pecos region contains a hydrological system centered around the Rio Grande. This dominant river runs southeast into the Gulf of Mexico, first transversing the Chihuahuan Desert and ultimately crossing the Gulf Coast plain to the sea. Its headwaters start at 4000 meters in the Rocky Mountains of Colorado. In its continuing course, the Rio Grande is sustained by a drainage basin of 472,000 square kilometers and carries an average annual discharge of 1,213,511,000 cubic meters (WEST, 1964). The river has two major peaks in discharge. The first occurs in April and May when Rocky Mountain snow melt feeds the head waters and combines with that of eastern Sierra Madre Occidental (via the Rio Conchos). The second occurs during the Mexican Plateau and Gulf Coast rainy season of July through September. The Rio Conchos is the most important tributary of the Rio Grande to protrude deep into the desert. Fed by the Sierra Madre Occidental watershed including springs and summer precipitation, it provides 18% of the total Rio Grande discharge. The northwestern or upper Rio Grande probably forged its union to the Conchos and Pecos branches of the drainage systems in the Middle Pleistocene Epoch. Previous to its union, these headwaters apparently were closed basin drainages, with their overflow spilling off into the interior drainages of northwest Chihuahua, Laguna Santa Maria and Laguna Guzman in particular. These deteriorating lakes now receive only ephemeral discharge from the Sierra Madre to their west. Thus even the presently interior drainage systems of the Trans-Pecos region are intimately related to the Rio Grande. The Pecos River is quite probably an earlier headwater for the system, easily dating back to a Pliocene establishment.

The Bolson de Mapimi is essentially a system of internal drainage at the central core of the desert. External drainage (to the Rio Grande) is confined to the extreme eastern periphery (Rio Salado and Sabinas, and now including the canal-drained Cuatro Cienegas Basin), and by definition should be considered outside the limits of the Bolson proper. The headwaters (the Rio Oro and Ramos) of the Rio Nazas drain the eastern slopes of the Sierra Madre in Durango. The current channel runs east from the Sierras, down to the lowlands of eastern Durango where its silts and waters are responsible for the rich agricultural district, La Comarca Lagunera, and finally terminates in the deteriorating alkaline Laguna de Mayran, near San Pedro, Coahuila. During the past, differing climatic conditions and watershed structure have influenced the course and discharge of the river. At one time the Nazas evidently maintained sufficient discharge to join the Rio Grande system, as still reflected in its fish fauna. At other points in its history, the river may have terminated in Laguna de Tlahualilo, Durango and Laguna de Viesca in Coahuila. Its present terminal in the Parras Bolson has clearly undergone repeated major oscillations in shoreline since Upper Tertiary times, possibly encompassing most of the Bolson at its maximum. The present average annual discharge of the Nazas, 1300 million cubic meters, is more than combined flows of all the other internal drainage systems of the Chihuahuan Desert. Laguna de Mayran was still an impressive lake into the 17th Century, A.D., but more recent irrigation and dam projects (the huge Presa de Lazaro Cardenas has a capacity of 3000 million cubic meters) have taken a brutal toll on the desert lake. The result is steady increase in alkaline precipitates and decline in water volume.

While the Rio Nazas provides fairly continuous surface flow, the only other major river of the Mapimian region, the Rio Aguanaval, is an intermittent stream at best. Its upper course is better described as ephemeral with much of the flow lost as it sinks into coarse gravels. Flow is so seasonal and limited that discharge sufficient to pursue the length of its channels, from the Sierra Madre Occidental (east of Durango and Zacatecas) to Laguna de Viesca, Coahuila, is possible only during the peak of the July to September wet season. The Aguanaval parallels the Nazas, and like the Nazas, its terminal Laguna is deteriorating into a salt and alkaline playa, as a result of irrigation diversions.

The two remaining internal drainages within the Bolson de Mapimi are the Palomas area centered around the saline playa, Laguna del Rey, in southwestern Coahuila, and Cuatro Cienegas Basin, in central Coahuila. The former is a region of devastating aridity lacking major streams. The latter probably had repeated outward drainage to the Rio Grande via the Rio Salado, and is fed by springs and underground flow.

To the south of the Mapimian region is a residual karst land referred to as El Salado. Except for the peripheral arroyos of Las Cruces and Sain Alto of the Rio Aguanaval, the region is devoid of rivers, streams, or even extensive arroyo channels. The arid climate is combined with an extreme karst topography so that water percolates into underground flow. Surface water is so feeble and moves below the surface so quickly that even arroyo channels are short and poorly scoured.

As already noted, southeastern extensions of the Chihuahuan Desert occur as relicts along the steep slopes of the upper branches of the Rio Panuco, itself a member of the Gulf Coast drainage.

In summary, it can be said that the hydrology of the Chihuahuan Desert region is dominated by the Rio Grande and its present and past tributaries, even as far south as the now interior Rio Nazas. Internal drainage systems are of major importance, especially in the Mapimian region, but large stretches of the desert are totally devoid of stable, well-developed drainage patterns. External drainage, other than the Rio Grande, seems to be involved in only a few very peripheral situations, primarily at the northwest and southeast extremities of the desert.

V. VEGETATION

Paleoecological history, edaphic phenomena, and huge geographical expanse have combined to produce in the Chihuahuan Desert a complex but predictable vegetation reflecting dramatic edaphic phenomena. Geographical disruptions and climatic changes have superimposed regional faciations. The Chihuahuan Desert can be characterized and mapped on the basis of a brief list of dominant and endemic shrubs. However, the fine discrimination of boundaries and comprehension of the internal structure of this desert demand a knowledge of its vegetation in terms of an edaphic mosaic. The Chihuahuan Desert is a region characterized by edaphic phenomena. Non-desert relict vegetation contributes even further to the complexity.

Any biologist viewing the Chihuahuan Desert as an operational construct should bear in mind the following practical and philosophical realities: studies undertaken solely within the boundaries of the United States of America cannot be extrapolated to the desert as a whole; desert and grassland are not mutually exclusive vegetative conditions; the spatial arrangements, habits and relative densities of species can be more important than simple presence or absence in the definition of a biotic unit.

A brief survey of the pertinent literature indicates that the Chihuahuan Desert has been studied primarily in piecemeal fashion using political (usually state) boundaries to delimit areas of analysis. The vegetative survey of MARROQUIN, BORJA, VELAZQUEZ & LA CRUZ (1964) most closely approaches the ideal of unified analysis of the entire region, being confined only to the south of the U.S.-Mexican boundary (thus including about 80% of the desert). Other broad coverage accounts of Mexican Chihuahuan vegetation are those of JOHNSTON (1941, 1943-1944), the map of Mexican vegetation produced by LEOPOLD (1950), the more specific study of GARCIA, SOTO & MIRANDA (1960) on *Larrea* and climatic parameters, and SHREVE (1942a, 1942b) on grasslands.

There are numerous studies on specific sections of the Chihuahuan Desert within the United States with a particular emphasis on grassland/scrub relationships. Selected references for the filter barrier region of southeast Arizona are BLUMER (1909), DARROW (1944), MARSHALL (1957), HASTINGS (1959), and LOWE (1964). The Chihuahuan Desert in New Mexico is treated by SHREVE (1942b), GARDNER (1951), and BRANSCOMB (1958). The desert in west Texas is described in some detail by WARNOCK (1970) as well as BRAY (1905) and CORRELL & JOHNSTON (1970). Of the Mexican States in which the desert is contained, Chihuahua is the best studied. Major references include BRAND (1936), LE SUEUR (1945), GARDNER (1951), and HERNANDEZ & GONZALEZ (1959). The outstanding reference for Nuevo Leon is MULLER (1939) and for Coahuila, an article by the same author, MULLER (1947). RZEDOWSKI *et al.* (1957, 1958) provided extensive information for the states of Zacatecas and San Luis Potosi.

The subsequent discussion will define and describe the vegetative mosaic of the desert in terms of major aggregations of plant species I am employing the categorization of CLEMENTS (1936a,b) for climax vegetation, not so much as indication that all vegetation is demonstrably in climax, but because his detailed terminology lends itself so well to the problem before me. As I have altered the definitions of CLEMENTS' terms in the present application, I offer the three following definitions. Formation is the largest unit of distinction to be used here. It designates a major geographical-climatic unit characterized by a continuum of plants of similar habit and physiology. The formation has geographical, climatic, and some structural uniformity (in physical habit and arrangement of dominant species), but does not necessarily include the same dominant species throughout. In the CLEMENTS hierarchy of terms formation is followed by association. Like formation, association is a geographical as well as an ecological division, but the association can be characterized by a generalized set of dominant spcies. Faciation is a minor geographical and climatic variant of association. Faciations of the same association contain some dominant species in common, but one or more characteristic species may be missing or replaced. Often subdominant vegetation includes species not typical of the association as a whole.

The subsequent discussion is a presentation of the specific units of the Chihuahuan mosaic in accordance with the literature sources (supplemented by field work of the author), principles, and categories previously discussed. In the following accounts, species are generally listed in descending order of dominance and the lists are confined to vascular taxa. See Appendix I for photographic illustrations of the vegetation described here.

A. Chihuahuan Desert Vegetation

Southwestern Scrubland Formation

Chihuahuan Desert Scrubland Association

GENERAL DESCRIPTION

Well-spaced sclerophyllous shrubs, usually less than $2\frac{1}{2}$ meters in diameter, characterize the region. Succulent species, especially cylindro-opuntias, and low platy-opuntias are also common. Grasses are reduced but still edaphically important. High-water table often produces a more dense and developed thorn scrub (*Prospis-Acacia*). There is an extensive *Yucca filifera* woodland in the southern or Salado faciation of the desert. These tree yuccas may reach 15 meters in height or more. Substrates occur in a complex variety and their edaphic influences will be described subsequently. Generally, the scrubland is floored with an alluvial gravel (largely limestone) with a layer of uneven thickness (usually less than $\frac{1}{2}$ meter). This alluvial location is the most widespread and may be used as a most typical expression of this association. (Numerous other edaphic faciations exist that would unnecessarily complicate the definition here).

Elevation: 500-1800 meters (in the Pecos and Mapimian faciations in the association does not extend above 1500 meters).

Precipitation: 75 or less to 400 mm annually; 65-80% of rainfall in June-September.

Temperature: average extreme maximum $40°C$; average extreme minimum $8°C$.

CHARACTERISTIC SPECIES OF CHIHUAHUAN DESERT SCRUB

I. Species Ubiquitous in or Endemic to the Chihuahuan Desert:
    1. *Flourensia cernua*
    2. *Parthenium incanum*
    3. *Agave lechigiulla*
    4. *Euphorbia antisyphylitica*
II. Other Dominant Species:
    1. *Larrea divaricata*
    2   *Fouquieria splendens*
    3   *Jatropha dioica*
    4. *Koeberlina spinosa*
    5. *Condolia ericoides*
    6. *Yucca* (esp. *elata* and *fillifera*)
    7. *Dasylirion scariosa*
    8. *Hectia* sp.
    9. *Rhus microphylla*
    10. *Acacia neovernicosa*
    11. *Nama* sp.
    12. *Lycium* sp.
III. Important Cacti:
    1. Cylindro and platy-opuntias
    2. *Coryphantha* sp.
    3. *Echinocactus* sp.
    4. *Echinocereus* sp.
IV. Major Grasses:
    1. *Hilaria mutica*
    2. *Erioneuron pilosun*
    3. *Bouteloua ramosa*
    4. *Sporobolus* sp.

Edaphic Associations of the Chihuahuan Desert Scrubland
1. Alluvium Association

Characteristic Species:	1. *Larrea tridentata*
Scrub:	2. *Flourensia cernua*
	3. *Parthenium incanum*
	4. *Jatropha dioica*
	5. *Koeberlina spinosa*
	6. *Prosopis* sp.
Other Major Components	1. *Lycium* sp.
	2. *Cornalia ericoides*
	3. *Acacia* (*constricta*, etc.)
	4. *Rhus microphylla*
	5. *Microrhamnus ericoides*
	6. *Nama* sp.

EDAPHIC FACTOR: The alluvial association is essentially the vegetation of the well-drained slope and open basin. It is most typically present along the gravelly alluvial fans and bajadas of the desert basin and range system. Slopes of less than ten degrees usually support this vegetation up to the angle of repose where the alluvium drops away revealing a solid substrate. The vegetation often continues across basin floor excepting only high water table, alkalines, gypsum, or organic soil. The alluvial association is the most widespread aspect of the Chihuahuan Desert vegetation.

FACIATIONS: The southern faciation, the Saladan, is the most mesic, a fact reflected by the presence of a *Yucca filifera* woodland. Furthermore, there are locally pocketed cactus forests (discussed in the igneous soil faciation). Further evidence of increasing mesic conditions are seen in the higher densities and diversity of herbaceous composites, as well as the herb, *Bouvardia* and the shrub *Mimosa*. Pecos and Mapimi aspects of this vegetation are generally similar to each other. They both support small local erect Yucca stands (*Y. elata* and *Y. torreyi*). Grassland tracts, especially on slopes, are more extensive in the northern Pecos faciation.

2. Karst Associations:

Characteristic Species:	
Scrub:	1. *Buddleia marrubifolia*
	2. *Parthenium argentatum*
Succulent:	1. *Agave lechiguilla*
	2. *Hectia* sp.
	3. *Fouquieria splendens*
	4. *Euphorbia antisyphylitica*
	5. *Dasylirion* sp.
	6. *Agave striata*
	7. *Opuntia stenopetela*

EDAPHIC FACTOR: A residual karst topography typifies much of the Chihuahuan Desert. It is characterized by a landscape of limestone ridges, streamless valleys, and complex underground drainage systems resulting in caverns and springs (see Physiography Chapter). Limestone, carried in solution, pervades the entire region with $CaCO_3$. The most common source of this substrate is simply a limestone outcropping. Alluvial vegetation is also here.

FACIATIONS: Pronounced faciations of this association are not easily discernible, at least in terms of dominants. Karst topography is especially typical of the eastern Mapimian and the Saladan faciations. Local differences occur in the species of *Agave, Hectia, Dasylirion*, and *Opuntia* participating in this location.

3. Igneous Substrate Association:

Characteristic Species:	
Scrub:	1. *Lophocereus* sp.
	2. *Cassia* sp.
	3. *Acacia schaffueri*
	4. *Prosopis* sp.
	5. *Celtis pallida*
Succulents:	1. *Agave filifera*
	2. *Myrtillacactus geometricans*
	3. *Lemaireocereus* sp.
	4. *Opuntia robusta*
	5. *Opuntia leucotricha*
	6. *Opuntia streptacantha*
	7. *Yucca carnerosana*

Major Grasses:	1. *Bouteloua* sp.
	2. *Panicum* sp.
Herbs:	1. *Dichondra* sp.
	2. *Evolvulus alsinoides*
	3. *Alternanthera repens*
	4. *Savhitalia ocymoides*

EDAPHIC FACTOR: This vegetation is indicative of a substrate of igneous derivation. It includes granite, rhyolite, basalt, and andesite outcroppings as well as basaltic clay and granitic loam.

FACIATIONS: The most striking faciation of igneous vegetation is to be found in the southern periphery of the Salado region. Here dense cactus forests of *Opuntia, Myrtillacactus* and *Lemarieocereus* develop in localized pockets, probably as relicts of the more continuous stands south and east of the desert. A reduced and largely succulent vegetation occurs in the more northern and arid Mapimian and Pecos faciations.

4. Ephemeral, Mesic Association:

Characteristic Species:	
Scrub:	1. *Prosopis* sp.
	2. *Acacia constricta*
	3. *Celtis pallida*
	4. *Zizipus obhisifulia*
	5. *Rhus (choriophylla)*
	6. *Berberis trifoliata*
	7. *Chilopsis linearis*
	8. *Porlieria angustifolia*
Major Grasses:	1. *Hilaria mutica*
	2. *Bouteloua (eriopoda* and *gracilis)*

EDAPHIC FACTOR: Water is the edaphic factor and its influence is the result of one of two physiographical situations. Locally mesic situations have been created by the cutting and channel action of desert arroyos or by the development of a collecting basin near the water table in closed drainage systems. Alternately, fault block produced ranges and karst landscapes often result in localized seepages and springs, often high up on mountainous slopes. Permanent surface water, however, is not a requirement of this vegetation. Within certain elevational limits, a uniform vegetation tends to take opportunity of both situations. The *Hilaria* grassland (see faciation) illustrates the one important exception. A true riparian woodland occurs only in association with permanent water sources within the desert and will be discussed elsewhere. The vegetation discussed here is essentially a dense thorn scrub dominated by *Prosopis-Acacia*. It is highly predictable in its occurrence and lines virtually every major arroyo and spring. This vegetation expands outward and upward into better drainage soils when mesic climatic conditions on the desert periphery develop a mesquite-grassland to replace true desert scrub.

FACIATIONS: For the most part there are no conspicuous geographical modifications of the general condition. However, the Pecos and Mapimi faciations of the desert sustain a very different mesic vegetation in the silt aggregating floors of their drier valleys. In these closed basins feeble runoff carries silt to the floor of the basin rather than forming the arroyos, canyons, or the alkaline playas of stronger runoffs. As a result the valley floor may be carpeted with a tough bunched grass, called tobosa or wire grass (*Hilaria mutica*). This grass, often in association with *Bouteloua* and the shrubs *Flourensia cernua* and *Prosopis glandulosa* is sustained by the rich silty soil of these basins. Other non-grasses include *Florestina tripteris Viguiera*, and *Xanthocephalum gymnospermoides*. This grassland is no prairie relict but a vigorous and predictable component of some of the most arid and pristine aspects of the Chihuahuan Desert (see SHREVE, 1942b; and MULLER, 1940).

5. The Alkaline Association:

Characteristic Species:
Scrub:
    1. *Atriplex* sp.
    2. *Allenrolfea* sp.
    3. *Distichlis* sp.
    4. *Monanthochloe* sp.
Major Grasses:
    1. *Sporobolus* sp.
More Peripheral Scrub:
    1. *Prosopis glandulosa*
    2. *Acacia* sp.
    3. *Lycium* sp.

EDAPHIC FACTOR: This is a conspicuous and familiar aspect of most North American desert vegetation. These plants, especially the first four, are physiologically adapted to living in alkaline substrates. This halophytic scrub grass generally presents a low profile, individuals usually not attaining a half meter in height (*Atriplex* excepted). Its association with alkaline conditions is most apparent in playas or dry lakes. However, any closed basin, especially with strong runoff from adjacent slopes, may develop alkaline flats. Moderate alkaline conditions (often combined with high water table) may induce the *Prosopis-Acacia* development.

FACIATIONS: While highly localized halophytic endemics do, occur no major regional faciations can be discerned.

6. Gypsum Association:
Characteristic Species:
Scrub:
    1. *Dicranocarpus parviflorus*
    2. *Nerisyrenia*
    3. *Drymaria*
    4. *Sartwellia*
    5. *Nama*
    6. *Phaecelia gypsogenia*
Major Grasses:
    1. *Muhlenbergia villiflora*

EDAPHIC FACTOR: The factor involved here is, in a word, gypsum ($CaSO_4$). It occurs most frequently as an aggregation on the floors of closed basins, either in the form of dunes or as 'yesos', flats of chalky, crunchy (and hollow-sounding) white soil. JOHNSTON (1941) discussed gypsophilous vegetation of northern Mexico in some depth and defended the distinction between gypsum and alkaline edaphic vegetation with considerable force.

FACIATIONS: With a high number (and apparently high relative percentage) of regional endemics, gypsophilous vegetation provides exceptional illustration of three major faciations of the desert. While the list offered at the opening of this discussion cites the more ubiquitous species, each of the faciations can be characterized by the additional forms in Table 6. The table indicates through the first three species in each column a rather dramatic pattern of replacement between the faciations. Furthermore, isolated gypsum localities within each of the faciations demonstrate considerable homogeneity on comparison. It should also be noted that the Mapimi faciation, which in subsequent biogeographical discussion is viewed as the oldest and most stable of the desert subdivisions, is seen here to be the most richly endowed with an endemic gypsophilous vegetation. While 'yeso' substrates are fairly widespread, they are most evident in the southern and eastern portions of the desert (Mapimi and Salado). Gypsum dunes occur in both Pecos and Mapimi faciations. These dunes are often associated with major pluvial lake beds.

Table 6. Characteristic Plants of Major Faciations in the Gypsoph Vegetation of the Chihuahuan Desert.

Faciation:	Pecos	Mapimi	Saladan
Species:	1. *Nerisyrenia linearifolia*	1. *Nerisyrenia castilloni*	1. *Nerisyrenia gracilis*
	2. *Sartwellia flaveriae*	2. *Sartwellia mexicana*	2. *Sartwellia humilis*
	3. *Nama carnosum*	3. *Nama stewartii*	3. *Nama hispidum*
	4. *Sporobolus* sp.	4. *Sporobolus nealleyi*	4. *Flaveria anomala*
		5. *Fouquieria shrevei*	
		6. *Euphorbia astyla*	
		7. *Drymaria elata*	
		8. *Atriplex reptans*	
		9. *Selinocarpus purpusianus*	
		10. *Selinocarpus palmeri*	
		11. *Petalonyx crenatus*	
		12. *Thelesperma scabridulum*	
		13. *Thelesperma ramosius*	

7. Silaceous Dune Association:

Characteristic Species:
Scrub:

1. *Prosopis glandulosa*
2. *Artemesia filifolia*
3. *Dalea scoparia*
4. *Yucca elata*
5. *Poliomintha incana*
6. *Pentstemon* sp.
7. *Oenothera pallida*

Major Grasses:

1. *Sporobolus contractus*

EDAPHIC FACTOR: Sand, approximately 94% silica, and piled in unstable dunes up to 5 meters in height is the determining substrate of this association. These dunes are often associated with playas or with the Rio Grande but are not restricted to such associations. LE SUEUR (1945) described a serial zonation of dune vegetation reflecting immediate substrate conditions, but indicated that this zonation was not a demonstrable process of succession nor dune stabilization, at least in northern Chihuahua.

FACIATIONS: No silaceous dunes have been reported from the Salado faciation. The dunes of Pecos faciation have been described in some detail and the preceding list is based on information based on these dunes in northern Chihuahua. The following summarizes the observation of LE SUEUR (1945) on their vegetative zonation:

Zonation of Vegetation in Dunes of Northern Chihuahua (from LE SUEUR, 1945).

Dune Structure	Characteristic Vegetation
Moving Dunes	1. *Prosopis glandulosa*
	2. *Yucca elata*
	3. *Penstemon ambiguus*
	4. *Oenothera pallida*
Backwash	1. *Artemesia filifolia*
	2. *Senecio ridelli*
	3. *Artemisia dracunculoides*
	4. *Sporobolus cyptandrus* (and other species)
Blowouts	1. *Bouteloua barbata*
1st Series Stable Dunes	1. *Atriplex canescens*
	2. *Atriplex obovata*
2nd Series Stable Dunes	1. *Sphaeralcea incana*
	2. *Wislizenia refracta*
	3. *Atriplex* sp.
Aged Dunes	1. *Chloris virgata*
	2. *Setaria macrostachya*
	3. *Panicum* sp.
	4. *Bouteloua* sp. grassland

The dunes associated with the playes of the Rio Nazas drainage in the Mapimi faciation appear to have higher percentages of organic material and include both siliceous and gypsiferous substrates. These dunes are lower, usually under a single meter in height (though barchan dunes 5 meters high occur near Viesca, Coahuila). The most unstable dunes support only *Prosopis* with *Drymaria* in the blowouts. However, more peripherally, *Larrea*, cylindro-opuntias, numerous herbaceous composites, and the grass *Bouteloua* sp. are common.

B. Non-Desert Relict and Border Vegetation

1. Mesquite Scrub Grassland Ecotone (between Juniper-Pinyon, Grassland, and Desert Formations)

GENERAL DESCRIPTION: This is essentially an unstable grassland combined with mesquite thorn scrub that virtually rings the periphery of the entire Chihuahuan Desert. At middle altitudes of the desert, this ecotone is highly compressed to less than 2000 meters across and stratified onto the foothills of the Sierra Madres. However, to the north and south, expecially along the northwestern filter barrier, climate and topography have conspired to produce plains dominated by the constantly shifting aspects of this ecotone. It reappears as ecological islands upon plateaus within the desert proper.

The ecotone may be viewed as an arid *Bouteloua gracilis* grassland which is replaced by thorn scrub when rocky outcroppings, slope, water erosion, or overgrazing reduce organic soil. These short grass plains (less than a meter high) are also modified locally into woodland or savanna situations by the presence of standing yuccas, platy-opuntias and oaks about three to five meters tall.

Overgrazing by cattle in the last century has inflicted a dramatic shift upon the ecotone favoring the thorn scrub (*Prosopis-Mimosa-Acacia*) aspect.

However, these shifts should not be so surprising in an ecotonal situation, and they only involve the shifting of relative densities of species already intrinsic to the region, not major invasions. Unfortunately, most of the so-called Chihuahuan Desert within the United States is, in fact, a portion of the ecotone. As a result some American investigators (i.e., WEAVER & CLEMENTS, 1938) have gone so far as to conclude that the entire Chihuahuan Desert is nothing but an overgrazed desert grassland. LOWE (1964) stated clearly and emphatically why this conclusion is without basis. Several points are worthy of restatement here. Firstly, the ecotone itself is not part of the Chihuahuan Desert (a thorn scrub does participate but only as an edaphic mesic growth). Secondly, mesquite-grassland is an unstable ecotone and not a climax association. Finally, and related to the last point, shifts between grassland and scrub are reversible (see LOWE, 1964; WARNOCK, 1970; BUFFINGTON & HERBEL, 1965) and have probably been reversed with every major and many minor shifts in evaporation/precipitation ratios throughout the Quarternary. LE SUEUR (1945) indicated that scrub vegetation might have been favored more by the seed-eating habits of the indigenous kangaroo rat (genus *Dipodomys*) than by the overgrazing of livestock.

While soils are generally of limestone origin, this vegetation is also supported by granitic loams and red volcanic clays, especially in the Salado faciation.
Elevation: 1500-2000 meters; however, in the Salado faciation, it does not appear until about 1700 meters and may extend to 2200 meters.
Precipitation: 375-450 mm annually; 60% or more of this occurring July through September.
Temperature: average extreme maximum: 40° C; average extreme minimum: −8.0° C.

Characteristic Species:

Scrub:	1. *Prosopis glandulosa* (and others)
	2. *Acacia constricta*
	3. *Mimosa* sp.
	4. *Ephedra* sp.
	5. Platy and Cylindro-Opuntias
	6. *Haplopappus* sp.
Major Grasses:	1. *Bouteloua gracilis*
	2. *Bouteloua curtipendula*
	3. *Andropogon* sp.
	4. *Eragrostis* sp.
	5. *Aristida* sp.

EDAPHIC FACTORS: The mesquite thorn scrub of the mesic edaphic association described previously is almost indistinguishable in content from the scrub aspect of the mesquite-grassland ecotone. However, in the desert proper this thorn scrub is restricted to mesic areas and positions of high water table, while in the non-desert ecotone, this same scrub extends far up the well-drained slopes of alluvial fans and bajadas, replacing the *Larrea-Flourensia* scrub of desert alluvium association.

FACIATIONS: The characteristic plants previously listed serve best to illustrate the vegetation of the Trans-Pecos and Mapimi faciations. However, in the high plateaus and southern periphery of the Salado region occurs a distinct faciation described by SHREVE (1942b) as "cactus-*Acacia* grassland". It is higher, cooler in terms of extreme summer heat, and possibly more mesic than the northern faciation and can be characterized as follows:

Cactus-*Acacia*-Grassland Faciation
Trees:
1. *Opuntia durangensis*
2. *Opuntia streptoacantha*
3. *Acacia schaffneri*
4. *Yucca carnerosana*

Scrub:
1. *Celtis pallida*
2. *Acacia paucispina*
3. *Opuntia imbricata*

Major Grasses:
1. *Bouteloua* sp. (*gracilis*)
2. *Panicum* sp.

Another aspect, possibly part of this faciation, is illustrated by the local occurence of grassland in association with improved soil and level topography in eastern Nuevo Leon (MULLER, 1939). Here it appears to alternate with a chaparral scrub and is strongly influenced by it. Its major components are as follows:

Major Grasses:
1. *Bouteloua gracilis*
2. *Buchloe dactyloides*
3. *Stipa* (*tenuissima* and *mucronata*)
4. *Aristida wrighti*
5. *Andropogon barbinodis*
6. *Muhlenbergia globrata*

Scrub:
1. *Aplopappus spinulosus*
2. *Verbesina stricta*
3. *Hymenopappus flavescens*
4. *Verbena canescens*
5. *Castelleja lanata*
6. the genera *Dyssodia, Salvia, Convolvulus, Erigonum, Penstemon*, and *Erigeron* are also present.

2. Chaparral Formation

a. Chaparral

GENERAL DESCRIPTION: A dense sclerophyllous scrub, usually situated on fairly steep well-drained slopes, forms the chaparral vegetation of the Mexican central plateau. Scrub is usually less than 1.5 meters high. It may be dotted with unbranched standing yuccas and short junipers up to approximately three meters in height. Plants are typically woody, leatherleafed shrubs, but an understory of herbs and grasses is virtually always developed to some appreciable degree. Soil is generally very thin (less than 30 cm), with high rock and gravel content. The density of the vegetation and the trend toward a slightly more broad-leafed plant habit help distinguish the chaparral from the true desert scrub.

Elevation: Chaparral on the Mexican Plateau has been reported from about 1500 to 2500 meters, but its most predictable occurrence centers at about 1700 to 2000 meters on both sides of the Chihuahuan Desert between latitudes 24 and 26 north. Relictual montane islands of chaparral occur above 1500 meters in the Mapimi region and 1700 meters in the Salado.

Precipitation: 750 to 1100 mm annually. Relict stands and those in the cooler northwestern periphery of the desert may persist at an annual precipitation of 440 mm (LE SUEUR, 1945). Over 60% from July through September.

Temperature: 11 to 16.5° C mean annual temperature (MULLER, 1939, for Nuevo Leon). Extremes − 9°C to 40°C (SHREVE, 1945, for Chihuahua).

Characteristic Species:

Scrub:
1. *Rhus* sp. (*trilobata, virens*)
2. *Ceanothus* sp.
3. *Cercocarpus*
4. *Garrya* sp.
5. *Agave* sp. (*falcata*)
6. *Nolina* sp.
7. *Arctostaphylos pungens*
8. *Mahonia* sp.
9. *Berberis trifoliata*
10. *Condalia ericoides*
11. Other locally common genera include: *Mimosa, Haplopappus, Verbesina, Castillja, Salvia, Convolvulus, Eriogonum,* and *Hesperaloe*

Trees:
1. *Quercus* sp.
2. *Yucca* sp.
3. *Juniperus mexicana*
4. *Pinus cembroides*

EDAPHIC FACTORS: Slope and face of exposure are crucial to the presence of chaparral, especially in its peripheral and relict positions relative to the Chihuahuan Desert. It is best developed on appreciable inclines, above five degrees, and locally gives way to true desert scrub, or more often to a grassland just previously described (see Mesquite-Grassland Ecotone).

Along the central eastern edge of the desert, especially in Nuevo Leon, shifts in vegetative associations are sudden and dramatic. Chaparral is well developed on south and west exposures above 1700 meters and below 1900 meters. On the north and east exposures of these same foothills of the Sierra Madre Oriental, greater exposure to summer rain clouds passing from the southeast results in modified pinyon-juniper woodland (often a parkland) or even a montane pine forest.

FACIATIONS: Chaparral consists of dominants of the same genera, at least for the most part, on both Occidental and Oriental ridges of the Sierra Madres, but they are usually distinct and possibly sibling species. An illustrative sample is offered below:

Complementary Sets of Major Shrubs and Trees from the Two Ridges of the Sierra Madre.

Sierra Madre Occidental	Sierra Madre Oriental
(Chihuahua, LE SUEUR, 1945)	(Nuevo Leon, MULLER, 1939)
1. *Quercus santaclarensis*	1. *Quercus cordifolia*
2. *Arbutus arizonicas*	2. *Arbutus xalapensis*
3. *Cercocarpus breviflorus*	3. *Cercocarpus majadensis*
4. *Ceanothus buxifolius*	4. *Ceanothus lanuginosus*

b. Sotol-Yucca Woodland

GENERAL DESCRIPTION: A narrow woodland strip consisting of unbranched tree yuccas and sotols stands the upper edges of the Chihuahuan Desert scrub and the lower periphery of the chaparral and pinyon-juniper woodland. Scrub, cacti, and grasses are generally below one meter in height; yuccas average between three and five meters, and the standing sotols are approximately three meters tall. Vegetation forms a cover of about 60 to 70% with the remainder exposed soil. Topsoil is very shallow, under 30 cm deep, and usually contains a high percentage of rock and gravel. Sotols (*Dasylirion*) grow on the steeper slopes (above five degrees) in association with limestone outcroppings. The yuccas tend to replace sotol as the slope levels and a gravelly soil develops. The scrub understory includes considerable mixtures of chaparral and desert scrub components, with *Larrea* and *Flourensia* still very much in evidence on the lower slopes.

Elevation: 1500 to 1800 meters.

Precipitation: 400 to 700 mm annually (estimated); more than 60% in July through September.

Characteristic Species:

Scrub:
1. *Dasylirion*
2. *Croton* sp.
3. *Euphorbia* sp.
4. *Jatropha dioica*
5. *Parthenium argentatum*
6 *Erigonum* sp.
7. *Nama* sp.
8. *Hesperaloe* sp.
9. *Hectia* sp.
10. *Agave lecheguilla* (also *A. falcata*, *A. striata*, *A. asperrima*)
11. *Coldenia* sp.

Trees:
1. *Dasylirion longissimum*
2. *Yucca carnerosana*

Cacti: (for Nuevo Leon)
1. *Echinocereus* sp.
2. *Ferocactus pringlei*
3. *Thelocactus conothelos*
4. *Thelocactus fossulatus*
5. Cylindro and platy-opuntias

Major Grasses:
1. *Bouteloua* sp.
2. *Tridens* sp.
3. *Panicum* sp.
4. *Erigrostis* sp.

EDAPHIC FACTORS: The Sotol-Yucca Woodland occurs on well-drained slopes of both steep and moderate incline. Limestone outcropping and a derived alluvium are ordinarily present.

FACIATIONS: The woodland as it has been defined here is restricted to the Salado region, and it is particularly developed as montane relict and peripheral desert vegetation. Relictual montane pockets of chaparral in the Mapimi region often have a corona of standing yuccas and sotols about 1.5 to 3 meters tall.

c. Piedmont

GENERAL DESCRIPTION: This is essentially an eastern or Tamaulipan equivalent of the chaparral lying on the western or Chihuahuan slopes of the Sierra Madre Oriental. It is a scrubland of moderately sclerophylous forms covering the eastern slopes of the Sierra Madre foothills. By definition this scrub ends when a majority of area is occupied by trees. The presence of occasional palm stands (*Brahea bella*) indicates local mesic conditions. It comes in contact with the Chihuahuan Desert only along its northeastern edge, specifically around the headwaters of Rio Sabinas and along the eastern slopes of the Sierra del Carmen, both in north central Coahuila. The piedmont scrub is completely separated by the Chihuahuan Desert by the montane forests of Sierra Madre Oriental which assume greater continuity and proportions to the south.
Elevation: (MULLER, 1939, Nuevo Leon) 500 to 1000 meters.
Locally tongues of piedmont scrub extend up south facing slopes to 1500 meters, but this vegetation is primarily lowland and its advances toward the Chihuahuan Desert to its northwest are reflective of the lower elevations in this part of Mexican Plateau.
Precipitation: 500-800 mm annually.

Characteristic Species:
Scrub:
1. *Diospyros texana*
2. *Bumelia lanuginosa*
3. *Sophora secundiflora*
4. *Bauhinia lunarioides*
5. *Rhus virens*
6. *Acacia farnesiana*
7. *Leucaena* sp.
8. *Cordia boissieri*
9. *Agave (americana, amenacea)*
10. Platy and Cylindro-opuntias
11. *Celtis reticulata*

Trees:
1. *Quercus fusiformis*
2. *Brahea bella*
3. *Juglans microcarpa*

FACIATIONS: No significant faciations have been described. The piedmont scrub of Coahuila appears to be a xeric and depauperate version of the same association in Nuevo Leon. MULLER (1947) suggested that a strong affinity may exist between this association and the oak scrub of the Edwards Plateau and other montane scrubs of central and western Texas.

3.Deciduous Woodland Formation

a. Desert Riparian Woodland Association

GENERAL DESCRIPTION: This woodland association often takes the form of narrow gallery forest. Trees are often over 20 meters in height, and canopy may be locally quite dense. This vegetation is at its best development along the banks of the few major rivers transversing the Chihuahuan Desert. It is especially developed in relatively level areas in relatively good organic soil beyond the cobble and sand bars of the flood plains so often associated with these rivers. An understory of broad-leaf shrubs and herbs grows in sheltering canopy and leaf mulch generated by the gallery forest. Understory and scrub vegetation usually are under five meters.

The desert riparian woodland is best viewed as a branched and peripherally disjunct system of intrusions from the continuous lowland deciduous forests of Austro-riparian North America. Edaphic conditions produce a variety of familiar aquatic associations which deviate radically from the woodland just described. These will be treated separately under 'edaphic factors'. Included here also is a brief account of vegetation entirely restricted to permanent surface water, regardless of the presence or absence of a woodland.

Elevation: Throughout the Chihuahuan Desert, 500-2000 meters, generally present and distinct below 1800 meters.

Precipitation: Variable: desert riparian woodland is locally well-developed along permanent river courses even in exceptionally arid portions of the desert.

Temperature: Variable.

Characteristic Species:

Trees:	1. *Platanus*
	2. *Salix* sp. (*nigra*)
	3. *Populus wislizeni*
	4. Other genera include: *Fraxinus, Tamarix, Carya, Arundo*
	5. *Taxodium mucronatum*
Scrub:	1. *Baccharis glutinosus*
	2. *Chilopsis linearis*
	3. *Rhus microphylla*
	4. *Prosopis glandulosa*
Aquatic Vegetation:	1. *Typha* sp.
	2. *Nitella* sp.
	3. *Elecharis* sp.
	4. *Chara* sp.
	5. *Myriophyllum* sp.
	6. *Potamogeton* sp.
	7. *Cosmos* sp.
	8. *Spirodela* sp.

EDAPHIC FACTORS: Common to all the situations encompassed here is the presence of permanent (or nearly permanent) surface water. However, the physical arrangement of this water in both quantitative and qualitative terms has radical consequences on the character. MINKLEY (1969) provides a concise but detailed description of edaphic expressions of aquatic vegetation in the marshes of Cuatro Cienegas Basin. Unfortunately, Cuatro Cienegas has a strong biogeographical affinity for Tamaulipan vegetation and fauna (with probably repeated connection to the lower Rio Grande drainage) and a number of endemic species as well. It provides a rather unrepresentative basis for extrapolation to other aquatic environments in the Chihuahuan Desert.

FACIATIONS: The general list of plants presented previously can be applied with modest modifications to all the major tributaries of the Rio Grande, namely the Pecos River, the Rio Conchos, the Rio Salado, and the Rio Sabinas. LE SUEUR (1945) noted the presence of *Typha latifolia, Heteranthera dubia* and a hydrophytic *Aster* along the Rio Conchos and their absence along the Rio Grande, as well as the absence of such Rio Grande genera as *Tamarix* and *Arundo*. The fish fauna of the Rio Nazas, further south, indicates that this also is a Rio Grande derivative (now part of closed Parras Basin drainage). A "gallery forest" of relictual Mexican Bald Cypress (*Taxodium*) grows along its higher banks. The Salado region is totally devoid of major rivers or standing permanent surface water. A few spring-fed perennial flows occur and these may be bordered by a riparian woodland of *Fraxinus, Populus*, and *Salix*. A dense shaded mesic undergrowth of herbs, grasses, ferns, and mosses often accompanies the tree stands. The Cuatro Cienegas Basin, previously noted, protrudes like a renal glomerulus of aquatic and largely Tamaulipan biota into the Chihuahuan Desert at the eastern edges of the Mapimi region. Its arteriole, The Rio Salado, is an off and on connection, which has contributed to the isolation and speciation of some of its flora and fauna. MINKLEY (1969) noted the presence of the lily (*Nymphaea*) in permanent standing water. The local marshes support *Eleocharis rostellata, Scirpus olynei, Spartina spartinae, Buccharis glutinosa, Cladium californicum*, and *Sporobolus airoides*. The major laguna maintains a vegetation including the genera *Fimbristylis, Cyperus, Andropogon, Typha, Potamogeton, Eleocharis* (*macrostachya*), and *Phragmites* (*communis*).

See LE SUEUR (1945) for an account of succession in aquatic vegetation (stream bottom and reed swamp) in Chihuahua.

4. Evergreen Woodland Formation

a. Pinyon-Juniper Woodland

GENERAL DESCRIPTION: The pinyon-juniper woodland is a slope or foothill vegetation comprised of low (usually under five meters) widely spaces pines, junipers and oaks with an undergrowth of grama grasses resulting in a parkland, or a low scrub (under one and one-half meters) very similar to chaparral and almost certainly derived from it. Steep slopes, good drainage, thin gravelly soils are under-bedded with limestone or igneous substrate and covered with a feeble and patchy leaf litter. The western, eastern, and northern faciations of this woodland involve many different species, especially in the composition of the undergrowth and in the representatives of the genus *Quercus*.

Relict stands of *Pinus cembroides*, many of which are no longer reseeding, occur throughout the higher ranges of the Mapimian region. Gnarled survivors often occur within chaparral slopes.

Elevation: In the Pecos and Mapimian regions this vegetation occurs peripherally and as a relict from 1500 to 1800 meters elevation. It occasionally approaches

2000 meters in locally isolated ranges, and in the northwest periphery of the desert (LE SUEUR, 1945) may drop down as low as 1000 meters on north facing slopes. In the Saladan region, pinyon-juniper woodland occurs from about 1800 meters up to 2500 meters (MULLER, 1939, 1947).

Precipitation: Variable, especially in regard to relict stands, but generally above 250 mm to above 500 mm annually.

Temperature: Extreme annual temperature ranges from a minimum of $-13°$ C to a maximum of $38°$ C. Mean annual temperature is approximately 15-$17°$ C.

Characteristic Species:

Trees:	1. *Pinus cembroides*
	2. *Quercus emoryi*
	3. *Quercus* sp. (*arizonica*)
	4. *Juniperus* sp.
Scrub:	1. *Salvia* sp.
	2. *Carrya ovata*
	3. *Rhus trilobata*
	4. *Vitis arizonica*
	5. *Cercocarpus* sp.
	6. *Ceanothus* sp.
	7. *Eriogonum* sp.
	8. *Rhanmus* sp.
	9. *Penstemon* sp.
	10. *Mimosa* sp.
	11. *Erigeron* sp.
	12. *Bouvardia* sp.
Grasses:	1. *Bouteloua gracilis*
	2. *Stipa* sp.
	3. *Muhlenbergia* sp.
	4. *Aristida* sp.
	5. *Eragrostris* sp.

EDAPHIC FACTORS: Riparian conditions with pinyon-juniper woodland provide conditions for the development of a more mesic tree vegetation. In eastern Coahuila and western Nuevo Leon, *Pinus arizonica*, *Acer brachypterum*, *Fraxinus cuspidata*, and *Arbutus xalapensis* constitute the main edaphic additions to the vegetation. These species represent invasions from adjacent vegetation associations; *Pinus arizonica* is derived from the higher montane pine forests while *Arbutus* may come from a more mesic piedmont scrub.

FACIATIONS: Three geographical faciations are distinguishable. In the northern periphery (Trans-Pecos region) this woodland, as described by LOV'E (1964), includes *Quercus oblongifolia*, *Juniperus deppeana* and *monosperma*, and *Cupressus arizonica*.

A faciation for the western foothills of the Sierra Madre Oriental (including much of relict vegetation on the ranges of central Coahuila) has been described by MULLER (1947). In addition the following species are generally characteristic of the woodland: *Juniperus pachyphloea*, *J. flaccida*, *Quercus gravesii*, *Q. hypoleucoides*, *Q. chisosensis laceyi*, *Q. sinuata* as well as the previously noted *Arbutus* and *Fraxinus*.

The pinyon-juniper woodland of the eastern slopes of the Sierra Madre Occidental include *Quercus grisea*, *Q. undulata*, *Q. dumosa*, *Q. reticulata*, *Juniperus monosperma*, *J. pachyphloea*, and *J. mexicana* as distinguishing components (at least in combination).

5. Coniferous Forest Formation

a. Montane Pine Forest

GENERAL DESCRIPTION: The montane pine forest consists of dense stands of tall (over 30 meters in height) coniferous trees. This association represents a radical departure from true desert conditions that can be found within the geographical confines of the Chihuahuan Desert. The evaporation/precipitation conditions required to maintain such a vegetation virtually eliminate the possibility of direct peripheral contact between Chihuahuan Desert and montane pine forest; however, topography and precipitation patterns conspire to produce a very narrow compression of vertical vegetational zones along the eastern wall of the desert. At some points (especially in southwestern Nuevo Leon) the transition from one extreme to the other can be accomplished in a 400 meter vertical climb. Furthermore, montane islands characterized by this vegetation occur in the more major mountain ranges of the Trans-Pecos region and the southeastern Arizona filter barrier. And lastly, a peninsular intrusion of this forest extends from Sierra Madre Oriental across the transversely placed ranges that sever the Mapimian from the Saladan regions.

These deep mesic forests develop on both igneous and limestone substrates (primarily the former in the northwestern portions of the desert). They are associated with a well-developed humus soil covered with a needle leaf litter. The extensive crowns and close spacing of the dominant trees results in restricted illumination and undergrowth development, the latter usually reaching eleborate expression along creek, cliffs, burns, and human disturbance where the overhead canopy is broken. High montane meadows also occur where forest opens to an unwooded valley or plain. The montane pine forest typically occurs on well-drained slopes, but it also covers valleys except where extensive marsh conditions persist.

Elevation: In the Pecos and Mapimian region this forest is restricted to elevations above 2000 meters. Along the eastern wall of the Saladan region local tongues of deep pine forest extend down to slightly below 1800 meters. The forest reaches its upper limits between 2500 and 2800 meters through the regions under consideration.

Precipitation: At least for northern Chihuahua, LE SUEUR (1945) suggests an average annual precipitation of 650 mm with 65% of this amount falling July through September. He also offers 480 to 800 mm annual precipitation as the range of values capable of sustaining the montane pine forest in that region. Along the southeastern edge of the desert, annual precipitation may shift by about 100 mm upward.

Temperature: Average annual temperature range from 9.5 to 13°C. These are estimated for the forests of the Sierra Madre Oriental.

Characteristic Species:

Trees:	1. *Pinus arizonica*
	2. *Pinus ayacahuite*
	3. *Populus tremuloides*
	4. *Cypressus arizonica*
	5. *Juniperus pachyphlaea*
Shrubs:	1. *Lonicera*
	2. *Mahonia* sp.

3. *Ceanothus* sp.
4. *Viburnum* sp.
5. *Rhamnus* sp.

Grasses: 1. *Piptochaetium fimbriatum*
2. *Stipa* sp.
3. *Muhlenbergia* sp.

EDAPHIC FACTORS: As mentioned previously, breaks in the forest canopy, regardless of the responsible agent, produce an enriched undergrowth. Also, valley floors with particular marshy conditions produce meadows of grasses and sedges (*Juncus* and *Cyperus*).

FACIATIONS: As with pinyon-juniper woodland, a western and eastern faciation can be described, but there is no northern distinct aspect. LOWE (1964) gave apt description to the northern faciation, at least as illustrated in the ranges of the Chihuahuan Desert filter barrier of southeastern Arizona. Here the predominant tree, *Pinus ponderosa*, is associated with *Pinus leiophylla*, *Quercus gambeli*, *Acer grandidentatum*, and *Alnus oblongifolia*. The predominant shrub, *Ceanothus fendleri* accompanies *Pachystima*, *Rosa*, *Symphoricarpos*, and *Ribes*. Predominant grasses are of the genus *Muhlenbergia*. At the lower edges of this forest, pinon-juniper and chaparral components are increasingly conspicuous. At the upper edge of these forests, they give way to spruce-fir forests, but only on highest peaks. LE SUEUR (1945) described an essentially identical ponderosa pine forest for northern Chihuahua.

MULLER (1939, 1947) described a Sierra Madre Oriental mesic forest along the eastern edge of the desert and including the Sierra del Carmen, the Sierra de la Madre, and Sierra de Parras of Coahuila. In these ranges *Pinus arizonica* and *P. teocote* predominate. Other distinctive trees include *Quercus gravesii*, *Q. hypoleucoides*, *Q. muehlenbergii* and *Acer brachypterum*. *Cypressus*, *Juniperus*, and *Arbutus* are conspicuous especially at lower elevations. Among the principal shrubs are *Ceanothus coeruleus* and *Lonicera pilosa*. The grasses *Andropogon scoparius*, *Bouteloua curtipendula*, *Elymus canadensis*, *Muhlenbergia emersleyi* are principal constituents of the ground cover (MULLER, 1947 -- Sierra del Pino, Coahuila).

b. Spruce-Fir Forest

GENERAL DESCRIPTION: This forest often forms a thin crest or cap to the montane forest surrounding it. In entire spectrum of vegetative association contained by or peripheral to the Chihuahuan Desert none is more alien, in both climate and contents, to the desert itself than the spruce-fir forest. It is a cold, wet, deep forest of huge trees (30 meters in height). These massive forests occur in the Chihuahuan region only along the very highest ridges bordering the desert (the Sierra Madres and fragmented ranges of the southern Rocky Mountains), and in the fragmented derivatives of the Sierra Madre Oriental (i.e., Sierra de Parras, Sierra de la Madre, Sierra del Carmen, and Sierra del Pinto) that in part characterize the Mapimian region. Soil is generally well-developed humus, gradually deteriorating at the highest elevations, and the rock under-bedding may be igneous, metamorphic, or sedimentary. They are poorly differentiated from the surrounding montane pine forests.

Elevation: 2300 to above 2500 meters (the latter on Cerro Potosi in Nuevo Leon).

Table 7. Categorization of Chihuahuan Desert Vegetation.

Formation[1]	Association	Faciation
True Desert: I. Southwestern Desert Scrub	A. Chihuahaan Desert Scrub	1. Trans-Pecos 2. Mapimi 3. Salado
	Edaphic Associations:[2] 1. Alluvium (*Larrea, Flourensia*) 2. Karst (*Agave-Fouquieria*) 3. Igneous Substrate (*Opuntia*) 4. Ephermeral Mesic and High Water Table (*Prosopis-Acacia-Hilaria*) 5. Alkaline Flat 6. Gypsum 7. Siliceous Dunes (*Prosopis-Ephedra-Arte-mesia*)	
These are non-desert relict and border vegetation:		
II. The Mesquite-Grassland Ecotone	A. Mesquite-Grassland	1. Trans-Pecos and Mapimi 2. Salado
III. Chaparral	A. Chaparral	1. Sierra Madre Occidental 2. Sierra Madre Oriental
	B. Sotol-Yucca Woodland C. Piedmont	
IV. Deciduous Woodland	A. Desert Riparian Woodland	1. Rio Grande Drainage 2. Cuatro Cienegas 3. Salado
V. Evergreen-	A. Pinyon-Juniper Woodland	1. Sierra Madre Occ. 2. Sierra Madre Or. 3. Northern Ranges
VI. Coniferous Forest	A. Montane Pine Forest	1. Sierra Madre Occ. 2. Sierra Madre Or. 3. Northwest Ranges
	B. Spruce-Fir Forests	

1. Formation classification modified from LOWE (1964).
2. In descending order of importance.

Precipitation: Generally about 750 to 1000 mm annually (mostly June to September), but probably considerably higher (up to 2000 mm) locally on certain high slopes. Light but regular winter snows occur (20 cm deep on Cerro Potosi, Nuevo Leon).
Temperature: Mean annual temperature, 4.5 - 13°C (MULLER, 1939). Mean annual extreme maximum and minimum are 30°C to −10°C.

Characteristic Species:
Trees:　　　　　　　　　　　1. *Pseudotsuga menziesi*
　　　　　　　　　　　　　　2. *Abies* sp.
　　　　　　　　　　　　　　3. *Pinus ayacahuitae*
　　　　　　　　　　　　　　4. *Populus tremuloides*
Scrub:　　　　　　　　　　　1. *Viola* sp.
　　　　　　　　　　　　　　2. *Geranium* sp.

	3. *Lupinus* sp.
	4. *Achillea* sp.
	5. *Castilleja* sp.
	6. *Senecio* sp.
Grasses:	1. *Poa* sp.
	2. *Festuca* sp.
	3. *Blepharonouron* sp.
	4. *Bromus*
	5. *Trisetum* sp.

EDAPHIC FACTORS: The exposure of the slope in relation to wind and precipitation effect the altitudinal limits of this forest in obvious ways, vegetational zones generally being depressed on the north slopes. Excessively wet situations may contribute to the formation of grass and sedge meadows.

GEOGRAPHICAL FACIATIONS: This forest is so fragmented and locally restricted in the major mountain systems of the Chihuahuan region that each occurrence of this vegetation might be treated as a faciation. The geographical distribution generally follows the outlines for the montane pine forest and possible similar faciations could be constructed. The spruce-fir forests of southeastern Arizona and adjacent New Mexico are endowed with *Picea englemani*, absent to the south. The Sierra de la Madera of central Coahuila (and possibly the Sierra de Parras as well) support the endemic fir, *Abies coahuilensis*. In the subalpine forests of the Cerro Potosi of Nuevo Leon, *Pinus hartwegii* was the dominant forest component, accompanied by *P. strobiformis*, *Abies vejari*, and *Pseudotsuga menziesii*.

More detailed accounts for particular regions are LOWE (1964) for southeastern Arizona; MULLER (1939), and BEAMAN & ANDERSEN (1966) for Nuevo Leon and Coahuila; LE SUEUR (1945) for the Sierra Madre Occidental in Chihuahua; and WARNOCK (1969) for the Chisos of Texas.

The categorization of the foregoing account are summarized in Table 7 and arranged in ascending vertical zonation in Table 8.

Table 8. Elevational Distribution of Chihuahuan Desert and Peripheral Vegetation.

Life Zone	Association	Elevation (in meters)	
		N. of 24 N. Lat.	S. of 24 N. Lat.
Upper Sonoran	1. Chihuahuan Desert Scrub		
	2. Desert Riparian Woodland	500-1500	1500-1800
Lower Transition	3. Mesquite Grassland	1400-2000	1700-2200
	4. Chaparral	1500-1900	1700-2000
	5. Sotol-Yucca Woodland	1500-1700 (often absent)	1600-1800
	6. Piedmont	500-1000 (local tongues to 1500, often absent)	500-1000
Upper Transition	7. Pinyon-Juniper Woodland	1500-1800	1800-2500
	8. Montane Pine Forest	2000-2800	2000-2800 (local tongues lower)
Canadian	9. Spruce Fir Forest	2300-3000 (mixed stands with montane pine forest)	2300-3500

VI. DEFINITIONS, TERMINOLOGY, AND THE BASE MAP OF THE CHIHUAHUAN DESERT

The Concept of a Desert

The word "desert" can convey a valuable biological concept when properly defined and applied. Unfortunately, it has two very limiting drawbacks. The first is that it is part of the common usage vocabulary, thus vulnerable to constant non-scientific interpretation. The second disadvantage is that the English language has been developed by a people having only a feeble knowledge of arid land environments. To these people, the term desert has had vague, foreign, and negative connotations. This was reflected in the primary definition of Webster's Unabridged English Dictionary (1960) which defined desert as simply an uninhabited place, an empty place, in the sense of biblical desert wilderness. This simplistic biblical meaning of desert, namely as "empty" place is of little practical utility. For an Anglo-Saxon people the concept of a desert has little real meaning and thus this simplistic definition could stand without test in Great Britain. In the language of arid land dwelling people, Hebrew being an excellent example, the concept of a desert demands a very explicit meaning and several refined terms are used to describe arid environments.

Webster's second definition of desert did have more biological relevance. Its major points established the desert as a region with scant or no rainfall, usually less than 10 inches annually, periodic heat and drought, and soil resulting from rock decomposition (physical) rather than rock decay (chemical). Vegetation was desiccation resistant in form. This second definition will effectively discriminate geographical regions and is biologically meaningful. With some elaboration and refinement, I shall employ it here.

I view deserts as ultimately climatic in a causal sense. Two climatic factors are ultimately responsible for the physical form and biological aspect of the desert biome. These factors are incident solar radiation and incident precipitation. The two in combination determine temperature, light exposure, erosional processes, hydrology, soil structure, and the biota. Solar radiation is often measured in Langleys and precipitation in millimeters. If cold deserts are included, the following two criteria will set apart deserts from all other biomes of the earth: an annual incident solar radiation of 450 Langleys or higher, and annual incident precipitation of 250-300 mm or less.

Since the warm deserts of North America have a highly uniform shared biota, differing strikingly from the colder or mesic regions adjacent to them, I prefer to discriminate warm deserts as those with an annual incident solar radiation in excess of 500 Langleys.

The definition of warm desert may be refined using the following secondary characteristics:

(1) P/T ratio under 20 (P/T = Precipitation in mm/Temperature in $^\circ$C).

(2) Soils are either Lithosols or Sierozems.

(3) Physiography of closed basin drainage, no locally supplied permanent surface water, ranges bordered by alluvial fans and bajadas, valleys covered by dunes and playas.

(4) Vegetation adapted to, and characteristic of, highly desiccating environments, namely sclerophyll scrub, short life cycle herbs and grasses, and succulent growth.

Severe ethnic (lingual) geographical, and professional incongruities still exist in the application of the term desert. It is a powerful biological concept, of tremendous value in both theoretical analysis and practical utility. Standardization of the term as a technical tool in biology should be an objective of biogeographers of this decade.

Biogeographical Terminology

The term "assemblage" simply indicates a list of particular species making up a contemporary ecological or spatial unit. Other ecological terms have already been discussed in the vegetation section. The biogeographical terminology of SAVAGE (1960 and 1966) has been employed here with some modification. Two basic types of classification are used here, spatial and temporal.

The spatial classification goes in descending order from region to province. My utilization of SAVAGE'S terminology is altered in two aspects. One, SAVAGE'S provinces are equivalent to my superprovinces — for quantitative reasons discussed in HAGMEIER & STULTS (1964). My definition of a biotic province corresponds qualitatively to that of DICE (1943), namely a geographical unit containing a uniformly dominant physiography, vegetation, and fauna, usually in the form of a single dominant climax community. Quantitatively, the province should have a terrestrial vertebrate fauna internally homogenous at the 60% level as measured by the coefficient of community. This follows HAGMEIER & STULTS' (1964) quantitative criteria for mammalian provinces. Further province boundaries should be determined by the points of greatest biotic change (index of faunal change) as measured by a grid of uniform squares across the entire area being analyzed.

My second deviation from SAVAGE'S spatial terminology is the addition of a new term, that of a "participant" fauna. Since a province should have a dominant community, the fauna that ecologically operates in that community is designated the participant fauna, and other species geographically present are considered relictual or ecologically restricted.

Temporal or historical biogeographical terms are essentially identical with those of SAVAGE (1960). The descending order of terms progresses from element to complex to component for herpetofaunas. The faunal elements and complexes are usually expressed by assemblages of genera or species groups. Faunal components are represented by lists of individual species. Parallel geofloras are equivalent to faunal elements.

Historical Résumé of the Chihuahuan Desert as a Biogeographical Unit

A vague southwestern "Chihuahua Region" was first named by HINDE (1843). Later COOPER (1859) recognized both a "Chihuahuan" and a "Coahuilan" region. DICE & BLOOM (1937) included both desert and semi-arid grasslands of Mexican Plateau in a Chihuahuan biotic province, but DICE (1943) confined the province to the Chihuahuan Desert proper in 1943. His definition of desert in terms of climate, drainage and vegetation conforms in all major points with my definition, upon which the base map is constructed. The DICE definition differed from mine primarily in the lesser scope and lack of quantification of his definition.

SMITH (1939, 1940) provided the first herpetofaunal biogeography for the region. His analysis, based on the distribution of species in the lizard genus *Sceloporus*, divided the Mexican Plateau across the Anticlinorium of Arteaga into northern and southern provinces. Otherwise his unit generally corresponded to the Mapimian subdivision of the Chihuahuan Desert. SMITH (1947) re-assessed the arrangement of Mexican biotic provinces using the entire herpetofauna. GOLD-MAN & MOORE (1946) used climate, vegetation, avian and mammalian faunas to assess the biotic provinces of Mexico -- essentially extending the units of DICE (1943) and using the same concept of province. Their definition differed from the base map only in that they extended the desert south into Aguascalientes and set upper elevational limits of the desert at 5000 feet (1650 m). BLAIR (1950) refined the application and definition of DICE'S provinces in Texas. My terminology for all geographical definition of province boundaries is virtually identical with that of BLAIR. His ecological definition did differ in some details, but none of these are contradictory to those used here. As a climatic character, he noted a moisture deficiency index of −40 to −60%. He also provided valuable reviews of indigenous biotas including the herpetofauna. Generally his treatment of the herpetofauna was compatible with patterns established in the ecological and spatial biogeographical sections developed here.

WEBB (1950) undertook the first quantitative faunal analysis of the region, based on mammal and snake faunas of Texas and Oklahoma. Based on the Coefficient of Community, calculated across a grid, isopleths of faunal homogeneity were constructed. Regions homogeneous at the 75% level were designated as communities. Using both mammals and snakes, WEBB was able to isolate the Chihuahuan Desert in Texas as the 'Trans-Pecos Community'. He discriminated between this region and the less uniform ecotones surrounding it. SAVAGE (1960) recognized a Chihuahuan Desert herpetofauna as a subprovince and historical component. HAGMEIER & STULTS (1964) in their quantitative analysis of North American mammal provinces discriminate the Chihuahuan Desert within United States borders as the "Mapimian Province". Their unit is essentially identical with the corresponding section of the base map. Other aspects of their work are discussed in sections on procedure and spatial analysis.

STUART (1964) used essentially the same provinces as GOLDMAN & MOORE (1945). He, too, recognized the Chihuahuan-Zacatecas Province, essentially the Chihuahuan Desert as defined here, but included the ecotonal mesquite-grassland border as well.

LOWE (1964) provided an excellent definition of the Chihuahuan Desert in terms of Arizona plant communities. His accounts offered excellent survey of pertinent literature, especially in regard to the character and climax of Chihuahuan Desert scrub.

GEHLBACH (1967) also defined the Chihuahuan Desert in terms of vegetation. He designated the limestone scrub of southwest Texas and adjacent Mexico as true Chihuahuan Desert due to its distinctive endemic species. He considered the creosote scrub of the contiguous alluvium to be a continuum with the Sonoran Desert. GEHLBACH'S approach does not lend itself to biogeographical application. Furthermore, the alluvial scrub does contain species which clearly distinguish it

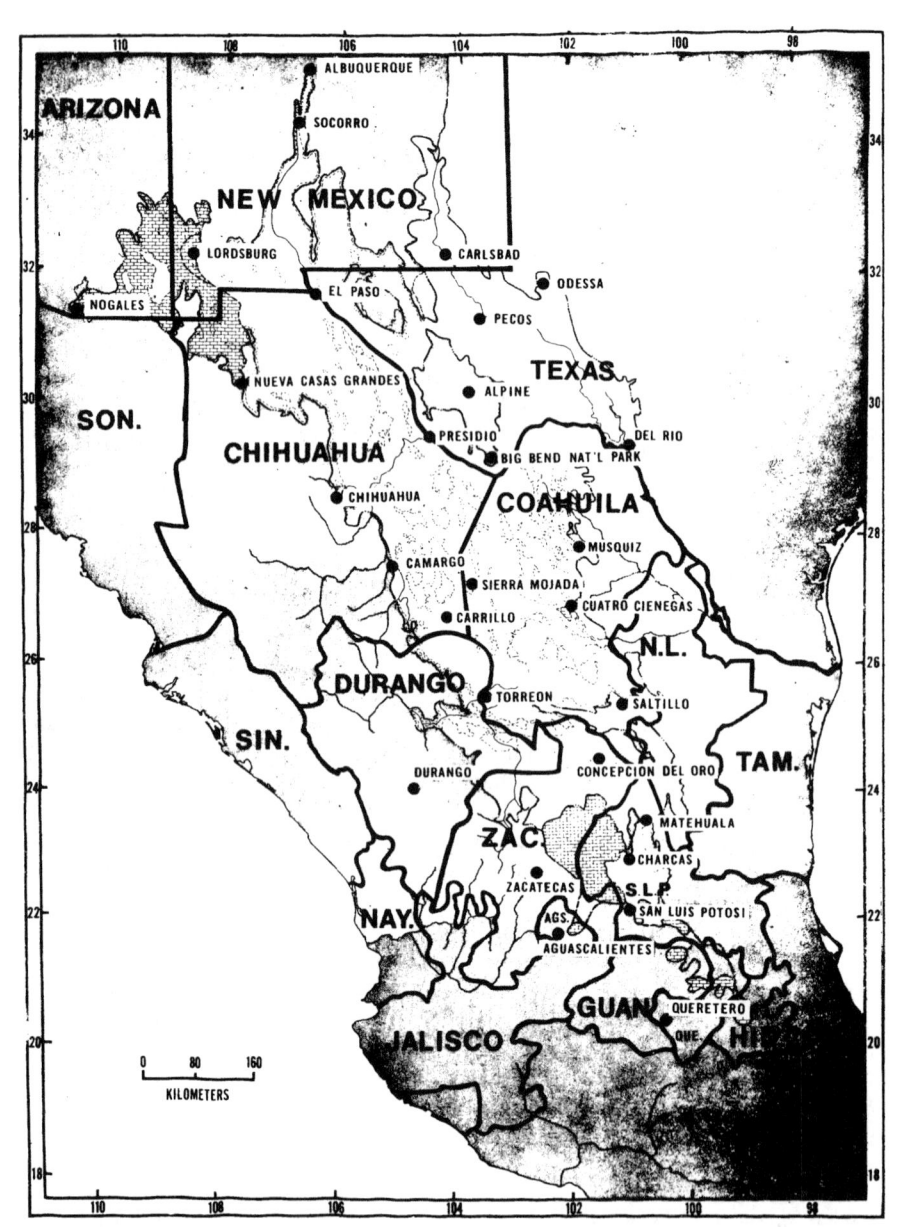

Map 2. Base Map of the Chihuahuan Desert in Relation to Political Borders and Cities. Key: Clear = Desert; Stippled = Non-Desert; Bricks = Filter-Barrier; Dashes = 1500 M Contour; Dots = 1800 M Contour.

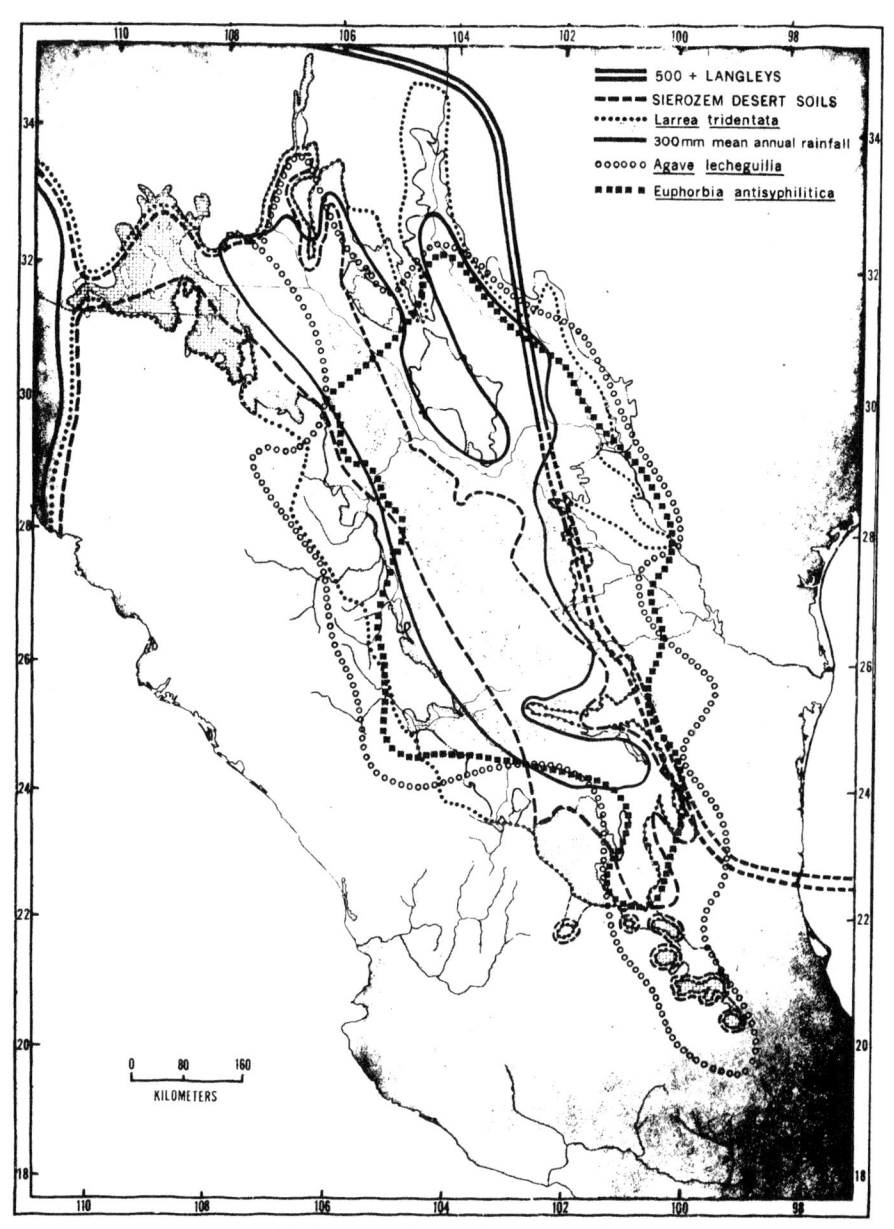

Map 3. Contours Used in Construction of Base Map.

Category	Range of Values or Indicator ($+$ /$-$)
I. Climate	
a. Average Annual Temperature	: 19 (16-22)°C
b. Average Annual Precipitation	: 75-300 mm (350 mm on periphery)
c. Precipitation/ Temperature Ratio (based on specific annual averages)	: 14.3 (7-21)
d. Solar Radiation (average annual)	: 500($+$) Langleys
II. Physiography	
a. Soil Group	: Sierozem — under 0.5 meters to soil horizon (alluvial soils also common with halomorphic and azonal soils locally important)
b. Elevation	: 600-1500 m north of 25th parallel; 1500-1800 m south of 25th parallel
c. Hydrographical Divisions	: all east of Continental Divide — except portions of Cochise Filter Barrier in Gila and Yaqui drainage — remainder ultimately Rio Grande or closed interior drainage systems — southern desert relicts in Rio Panuco drainage
d. General Topography	: Basin and range province with fault blocks generally parallel diagonals running north-west, karst landscape, many basins filling with aggregating alluvium
III. Vegetation	
a. Sclerophyll Scrub (all three indicators in combination only)	: 1. *Larrea tridentata* 2. *Flourensia cernua* 3. *Parthenium incanum*
b. Saxicolous Scrub ('crassicole')	: 1. *Agave lechiguilla* 2. *Parthenium argentatum* 3. *Euphorbia antisyphylitica* 4. *Dasylirion* sp. 5. *Fouquieria splendens* (east of Continental Divide)
c. Desert Grassland	: 1. *Bouteloua gracilis* characteristic but not endemic: 2. *Hilaria mutica* 3. *Tridens puchellus*

1. Roughly in descending order of importance.

from alluvial vegetation of Sonoran Desert, *Flourensia cernua* and *Parthenium argentatum*, being abundant examples of these distinguishing Chihuahuan species.

CORRELL & JOHNSTON (1970) designated the Chihuahuan Desert in Texas as the "Trans-Pecos, Mountains and Basins" vegetational area. This area corresponds well to the Texas portion of the base map, but the former unit has its eastern boundary often 100 kilometers west of that marked on the base map. The vegetational area was given good definition in terms of climate, soil, and indicator plants. The definition is generally compatible and largely overlapping with that employed here.

To summarize, the Chihuahuan province of DICE (1943), the Chihuahuan-Zacatecas of GOLDMAN & MOORE (1945), the Chihuahuan and Austral of SMITH

(1939). the Trans-Pecos community of WEBB (1950) and CORRELL & JOHN-STON (1970), and the Mapimian province of HAGMEIER & STULTS (1964) all correspond fundamentally to the unit described by the base map used here.

Definition of the Chihuahuan Desert and the Base Map

Geographical illustration in this analysis is expressed through a base map which is used throughout. The base map, Map 2, includes the major political boundaries and cities as practical points of reference. The criteria listed below establish an explicit definition of the Chihuahuan Desert and at the same time define the geographical parameters, mutually superimposed, from which the base map was constructed (See Map 3). Climate and physiography (including soil and drainage) were primary considerations, but the ranges of indicator plant species were utilized to resolve more detailed boundaries, being viewed as operational indicators of desert climates when direct data was lacking (see MULLER, 1937; GARCIA, SOTO & MIRANDA, 1961).

VII. THE HERPETOFAUNA

Systematic Problems in the Classification of Chihuahuan Herpetofauna

The informational bedrock of this analysis is the spatial distribution of well-defined biotic species. It does little good to provide detailed distributions of species which are of dubious validity or inadequate definition. Yet, I am forced by practical considerations and limited information to make somewhat arbitrary decisions about species' validity and identity.

It is both pretentious and unnecessary for this biogeographical investigation to provide a complete systematic species account. In its place I offer a review of selected systematic problems and a checklist in order to establish my position on major points of current controversy and also as a means of presenting new and relevant data. As a foundation to further discussion my positions regarding the nature and application of subspecies, species and genera must be established.

I view the current usage of the trinomial as a terminology of last resort (see DUNN *et al.*, 1943). It is a most arbitrary and abused technique of classification of intraspecific population variation which often obscures more evolutionary and genetic reality than it reveals. WILSON & BROWN (1953[1]) and VANZOLINI & WILLIAMS (1970) suggest the restricted application of subspecies to populations well differentiated morphologically and allopatric from their closest allies, in other words, candidates for incipient speciation. The term has some practical value when applied to allopatric semispecies and based on thorough comparative analysis of morphological and physiological characteristics. AUFFENBERG (1955) in his excellent review of the subspecific variation of the snake *Coluber constrictor* made clear the potential for chaos in defining subspecies on the basis of coloration or a random fragment of the total meristic characters. His analysis drew attention to two profound flaws in subspecific systematics. Firstly, some minor and superficial characters (such as coloration) or scale or vertebral number may vary in accordance with ontogenetic responses to different physical environmental parameters and thus obscure the real genetic affinities of different populations (FOX, 1948). The second major flaw in this system is that small sets of characters, be they morphometric, pigmentation, or physiological, may not vary in coordinated fashion geographically, either with each other, or with the vast array of unsampled characteristics. As a result, two workers, each manipulating different small clusters of characters might easily divide the populations of species into completely different geographical races. A third criticism of the subspecific unit might also be considered. It usually employs such a small and superficial set of criteria for definition that the probabilities of polytypic origin for a single subspecies is quite likely. Thus populations, either converging genetically under similar selective pressures, or responding similarly to ontogenetic influences might easily be lumped into a single subspecies. Ultimately, subspecies lacks a standard definition, even in the abstract. In real application, its use is chaotic.

For all of these reasons, the evolutionary biologist is forced to view subspecies

1. Systs Zoo. 2: 92-111.

as extremely unreliable measurement of phyletic relationship. HEYER & SAVAGE (1967) would have the term abolished altogether. MAYR (1970) defined the subspecies as "...an aggregate of phenotypically similar populations of a species inhabiting a geographic subdivision of the range of the species" (p. 210). While MAYR condoned the term as a convenient tool of taxonomy, he warned (1970, p. 211) "...the subspecies is an artifact rather than a unit of evolution." While I do recognize the taxonomic convenience of the term, especially when based on a broad spectrum of comparisons, I cannot find many instances where the term has been used responsibly. I am forced to operate on the assumption that most subspecies are highly arbitrary and very poorly defined. This is especially the case in the Chihuahuan Desert where huge geographical gaps in sampling data exist, and individual samples fail to reveal the spectrum of variation for a single deme. Biogeography is speculative enough without employing such a dubious unit, and I consequently reject the general use of subspecies in this analysis. Exceptions to this decision will be discussed on their individual merits.

The species is the chosen basic unit of interpretation in this study and is used in the sense of MAYR (1970) of a protected gene pool constituting a reproductive and ecological as well as genetic entity. Allopatric forms (sibling species or just races?) separated by the Continental Divide are, at best, tentative or presumed species. Still, when one deals with sympatric sexual species intrinsic to the Chihuahuan Desert, there is very little uncertainty regarding their specific status. I will address myself to the few specific species problems later in this chapter.

The Chihuahuan Desert herpetofauna does provide a more fundamental challenge to the species concept in the form of parthenospecies of the lizard genus *Cnemidophorus*. The all female populations include diploids, triploids, and tetraploids of possibly hybrid origin from sexual parental stocks. At least six forms occur within or are peripheral to the Chihuahuan Desert. Each parthenospecies consists of complexes of morphologically and genetic, similar clones, not demes, and no gene flow occurs normally within or between clones. These biological entities are in a narrow literal sense protected genetically, and ecological units, but their basic population structure, their inability to maintain a gene pool, gene flow, or variation through recombination, and their consequential lack of evolutionary adaptive potential eliminate them as biological species. Furthermore, morphological and karyotypic evidence indicate constituent clones of a given parthenogenetic form (see WRIGHT & LOWE, 1967) may not even be of monophyletic origin. MAYR (1970) suggested parathenogenetic forms be treated as sibling species in relation to their sexual parental stocks. I cannot reconcile this decision with MAYR'S own definition of a biological species. The parthenogenetic *Cnemidophorus* of the Chihuahuan Desert are ecologically and biogeographically important components of its herpetofauna. But they are systems of similarly derived hybrid clones, not species in an evolutionary biological sense. The biological species connotates strongly a dynamic process with a potential to respond to temporal environmental change. Acceptance of obligatory parthenogenetic species is a partial reversion to typological species classification, denying the dimensions of evolutionary potential and adaptive response.

On questions of generic classification, I concede that virtually all decisions are subjective, since rigid coherent criteria for a genus, especially biologically relevant criteria, are vague, comparative and difficult to apply uniformly. All definitions, especially those designed to be universally applicable, reveal the subjective nature of the unit (see DUNN *et al.,* 1943). MAYR, LINSLEY & USINGER (1953) provided such a definition which serves as well as its many alternatives, "... a species or group of species of presumably common phylogenetic origin, which is separated from other similar units by a decided gap." I view an ideal herpetological genus as a cluster of similar species, apparently of monophyletic origin, sharing a broad range of characters in common which distinguish them from other species clusters, but not being so distinctive as holding major differences in deep morphology (osteology or myology). While this statement only makes explicit values that remain subjective, I may be able to convey my opinion on effective usage of the genus category by an illustrative example. The lizard genus *Sceloporus* defines a cluster of species, distinctive from other iguanid lizards in number of characters. The group appears to be monophyletic whether judged on physiological, behavioral, morphological, or biogeographical grounds. Within the genus are several phyletic lines, distinctive in both superficial morphology and ecology, which indicate an adaptive radiation within a heliothermic insectivore adaptive zone. Yet assigning each of these clades with a generic title would do more to obscure their relationships and radiation than to elucidate them. Morphological or physiological attributes of the genus do not so distinguish it from other iguanid lizards that rank above the genus level would be warranted (i.e., no major osteological character differences). To express my view of the genus as a systematic tool more bluntly, I am a "lumper".

The subsequent list of positions will address itself to explicit decisions on the generic and specific level. Topics will be discussed in the order that they have been treated in the checklist.

A Statement of Systematic Positions

CLASS AMPHIBIA

ORDER SALIENTIA

FAMILY LEPTODACTYLIDAE

1. The status of *Syrrhopus marnocki, S. guttitalus*, and *S. gaigea*: LYNCH (1970) synonymized *gaigae* with *guttilatus*, a decision with which I concur. However, he fails to consider *guttilatus* as conspecific with *marnocki*, retaining the former on basis of shade of coloration alone. Morphologically and biogeographically the Edwards Plateau *marnocki* and the Sierra Madrean *guttilatus* should be conspecific. I have collected intermediate *Syrrhopus* in the Sierra de Agava east of Monclova, Coahuila (now deposited in Carnegie Museum). Allopatric populations of *Eleutherodactylus augusti, Gerrhonotus liocephalus* and *Crotalus lepidus* occur in both regions and offer no indications of speciation. The name having priority is *Syrrhophus marnocki*, and I consider both *gaigea* and *guttulatus* synonymous with it.
2. The genus *Hylactophryne:* Despite LYNCH (1968), I question the validity of dermal and slight osteological differences in the digits as generic characters, and maintain the use of *Eleutherodactylus* inclusive of *E. augusti.*

FAMILY BUFONIDAE

3. *Bufo speciosus* and *Bufo compactilis:* I consider both of these valid species, following BOGERT's (1960) report of their distinctive calls. SMITH (1947) ably lists the subtle but consistently distinguishing characters for the two forms. My examination of museum material indicates no intermediate individuals. Ecologically the species are very distinctive, *B. compactilis* being a montane forest-meadow form and *B. speciosus* occurring in lowland mesquite-grassland and peripheral Chihuahuan Desert situations.

FAMILY HYLIDAE

4. The status of *Hyla wrightorum:* I follow DUELLMAN (1970) in treating *Hyla wrightorum* as conspecific with *Hyla eximia*.

CLASS REPTILIA

ORDER TESTUDINATA

FAMILY KINOSTERNIDAE

5. *Kinosternon flavescens, K. integrum, K. scorpioides*, and *K. sonoriense* and *K. hirtipes* — a review of specific status: I have generally followed WERMUTH & MERTENS (1961) in recognizing *flavescens, hirtipes, scorpioides*, and *sonoriense* as valid species. I consider *integrum* a Mexican high plateau race of *scorpioides*, and *murrayi* as form (or even clinal variant) of *K. hirtipes* (see MOLL, WILLIAMS *et al.*, 1963). I consider this classification practical. It is based largely on currently published accounts. The ultimate biological relationships between the *scorpioides* complex, *hirtipes*, and *sonoriense* remain to be determined.

FAMILY EMYDIDAE

6. The synonymy of *Pseudemys* and *Chrysemys:* I feel that MCDOWELL (1964) has provided sufficient evidence to synonymize the genus *Pseudemys*. His arguments, anatomically based, were augmented by the detailed osteological and paleontological analysis of WEAVER & ROSE (1967). MOLL & LEGLER (1971) rejected these arguments, invoking the existence of distinctive coloration, pattern and soft anatomy in favor of maintaining *Pseudemys* as distinct from *Chrysemys*. I cannot accept color or pattern as valid basis of generic distinction, and especially in the case of the "*Pseudemys*" species where interspecific variation virtually outstrips the supposed generic differences (i.e., melanism). Concerning differences in soft anatomy, MOLL & LEGLER failed to make their objections specific.
7. The status of *Chrysemys concinna:* I follow RAUN & GEHLBACH (1972) in accepting *Chrysemys concinna* as valid species and not a form of *C. floridana*.
8. The status of *Chrysemys gaigea:* I concur with STEJNEGER & BARBOUR (1943) in their treatment of this form as a distinct species. The morphological evidence provided by WEAVER & ROSE (1967) coupled with the elimination of

alleged integrades with *C. scripta* (see CARR, 1952; LEGLER, 1960) justifies this conclusion, at least until more widespread sampling in the Rio Grande drainage of northeastern Coahuila and Reeves County, Texas (HARTWEG, 1938) has been undertaken.[1]

9. The specific assignment of *Chrysemys* population from Cuatro Cienegas Bolson, Coahuila, Mexico: LEGLER (1960) established the occurrence of integrades between the Cuatro Cienegas *Chrysemys* and *C. scripta* populations to the east. This contradicts the tentative suggestion of WEAVER & ROSE (1967) that the Cuatro Cienegas form *taylori*, is morphologically closest to the *gaigae* group of the genus. If one accepts the use of trinomials, then *Chrysemys scripta taylori* is valid.

FAMILY TESTUDINIDAE

10. The status of certain species in the genus *Gopherus:* There are four living species of *Gopherus*. This is well established by both morphological and behavioral evidence (AUFFENBERG, 1969; BRAMBLE, personal communication). The conspecific treatment of the living forms afforded by WERMUTH & MERTENS (1961) was undefended in its presentation. Captive *G. berlandieri* and *G. agassizi* have been crossed successfully (at least for a single generation), but this is hardly a convincing basis for conspecificity in itself. Some fossil species may be conspecific with living forms, especially the large *Gopherus flavomarginatus*. The shell proportions and absolute size of the mid-Pleistocene tortoise of west Texas, *Gopherus huecoenis* (STRAIN, 1966) indicate possible conspecific status with *G. flavomarginatus* extant about 500 kilometers south of the fossil locality.

FAMILY TRIONYCHIDAE

11. The status of *Trionyx ater:* I consider the form *Trionyx ater* at best a race, or more probably, a morph of *Trionyx spinifer*. WEBB (1966) conceded that juveniles of the two forms are virtually indistinguishable, but argues the specific validity of *ater* primarily on the sympatric occurrence of this form with *spinifer*. WEBB (1962) restated this argument even while acknowledging (p. 531): "On the basis of morphological criteria, I suspect that *ater* and *emoryi* are genetically compatible." The melanistic coloration and absence of septal ridges (in the single known male) which distinguish *ater* could best be explained as the result of genetic polymorphism or even ontogeny with or without a genetic basis. SMITH, NIXON & MINTON (1949) described a captive, starved, *Trionyx spinifer* from Kansas that both lacked septal ridges and underwent ontogenetic melanism, in effect becoming an *ater* morph. Melanism is common to the other aquatic turtles of Cuatro Cienegas and should not be considered conclusive proof of speciation.

ORDER SQUAMATA

FAMILY GEKKONIDAE

1. DEGENHARDT & CHRISTIANSEN (1974) indicated integrades with *C. scripta* in the Pecos River of New Mexico, near the Texas border. This fact reverses my earlier conclusions.

12. The status of *Coleonyx reticulatus:* I accept *Coleonyx reticulatus* as a valid species. Its distinctive squamation, size, and apparent sympatry with *C. brevis* have been reaffirmed by SEIFERT & MURPHY (1972). SEIFERT (personal communication) further affirmed its extended distribution across more than one hundred kilometers of Brewster and Presidio Counties, Texas. WEBB (1974) even further expanded its range into Durango. Extended surface activity by these lizards largely restricted to heavy summer precipitation indicates a behavior distinct from *C. brevis.* As an alternative the presence of a pleiotrophic atavistic gene or gene complex within local *brevis* populations is still not impossible. If this is the explanation of *C. reticulatus*, its large size and tubercular surface would be consistent with the primitive condition of a tropical scrub *Coleonyx*, presumably like the ancestors of *brevis* (KLAUBER, 1945; KLUGE, 1975). This form might be expected throughout the Mapimian subprovince.

FAMILY IGUANIDAE

13. The generic relationships between *Callisaurus, Cophosaurus, Holbrookia*, and *Uma*: SAVAGE (manuscript) indicated no strong osteological evidence supporting the recognition of these species gruops as separate genera. Superficial characters, such as the scaling over of the external tympanum are unstable in other iguanid genera (i.e., *Phrynosoma*). EARLE (1962) did suggest some differences in deep anatomy. A much more extensive array of characters, osteological, behavioral, and proportional, bind these forms together as a highly distinctive cluster of arid southwestern iguanid species. Most important, the members of all four genera interact to produce an intricate pattern of geographical and ecological exclusion, much like closely related members of the iguanid genus *Sceloporus*. These exclusive interactions, coupled with a set of major morphological characteristics in common (such as counter-sunk lower jaw) argue strongly for congeneric treatment of these four groups.

14. The validity of *Uma paraphygas: Uma paraphygas* is probably not a valid species. The behavior observations of CARPENTER (1967) indicated the absence of pre-copulatory isolating mechanisms, and in fact documented a mating between captive *U. exsul* and *paraphygas*. None of the defining characters cited by WILLIAMS, CHRAPLIWY & SMITH (1959) are completely stable, even in additional samples from the type locality of *U. paraphygas*. The character most dramatically distinguishing the two forms, the presence of paired belly bars on each side, breaks down into a spectrum of intermediates between paired bars and the *exsul* condition of single bars in about one-third of the specimens examined by me (from Texas Memorial Museum). *U. paraphygas* extends its distribution in a 75 kilometer arc from the type locality east to Laguna del Rey in Coahuila. Furthermore, intermediates and *paraphygas*-like individuals have been found between previously known populations of the two forms on the Coahuila-Chihuahua line at Laguna del Rey. However, the remaining geographical gap between the two has not been closed by sampling in the Valle de Acatita of southwestern Coahuila, despite the presence of dunes. The strong polymorphism of the populations of *Uma* from the type locality of *paraphygas* might indicate that it is a zone of secondary contact between more uniformly two barred populations to the northeast (as yet undiscovered) and typically one barred *exsul* to the southeast. It is also possible that populations expressing the regular appearance of paired bars,

once occuring to the north, have been obliterated by deteriorating Pleistocene or Post-glacial pluvial climates. Most probably these are Pliocene relict populations in which even such basic characters as belly bars have never been genetically fixed. In any case, no known population is uniformly distinct from *Uma exsul.*

FAMILY SCINCIDAE

15. The status of relationships between *Eumeces brevilineatus, callicephalus,* and *tetragrammus:* The specific validity of these forms is currently under reevaluation by LIEB (personal communication). Their specific status is doubtful, but no alteration in current systematic treatment will be offered here.

16. The status and interrelationships of *Eumeces brevirostris* and *Eumeces dicei:* I accept the arguments and evidence of DIXON (1969) in treating *dicei* as a race of *Eumeces brevirostris.*

17. *Leiolopisma laterale, L. silvicolum* and *L. caudaequinae* — generic terminology and specific status: I follow MARTIN (1958) in using the generic term *Leiolopisma*[1] and accept his arguments favoring the treatment of *caudaequinae* as a race of *silvicolum.* I also agree with his speculation regarding the possible conspecificity of *L. laterale* with this form. Geographically, a *silvicolum*-like organism in the Sierra del Carman is separated by less than 150 kilometers from a *L. laterale* population at Munquiz, both in Coahuila (McCOY, personal communication). McCOY (personal communication) reported a *L. laterale* derived form, distinctive at racial or specific level, occurs at Cuatro Cienegas, Coahuila.

FAMILY TEIIDAE

18. The parthenogenetic forms of *Cnemidophorus* and their status: As discussed previously, I consider these forms hybrid clones, not valid biological species. The specific designations applied to these clones are convenient, however, in indicating particular morphological, ecological, and genetic units. As a compromise, I use these labels in quotes. The peripheral and Chihuahuan forms involved are: *"Cnemidophorus dixoni", "Cnemidophorus exanguis", "Cnemidophorus flagellicaudus", "Cnemidophorus neomexicanus", "Cnemidophorus tesselatus", "Cnemidophorus uniparns",* and *"Cnemidophorus velox".* Table 9 summarizes the clonal hybrid origins of some of these forms.

Table 9. A Summary of Known Origins of Hybrid Parthenogenetic *Cnemidophorus* Clones Occurring in the Chihuahuan Desert (From LOWE & WRIGHT, 1966; WRIGHT & LOWE, 1967).

Parthenospecies	Parental Origins of Hybrid Clones	Chromosome Complement
1. *"C. neomexicanus"*	*C. inornatus* X *C. tigris*	2N
2. *"C. uniparens"*	*(C. inornatus)* X *C. gularis*	3N
Pattern Class: C, D, E, F	*C. tigris* X	
	C. septemvittatus	2N
	(C. gularis)	
A, B	*C. tigris* X	
	C. septemvittatus X	3N
	C. sexlineatus	

1. Greer (1965) would make the genus *Scincella.*

19. The status and interrelationships between *Cnemidophorus scalaris* and *Cnemidophorus gularis:* The form *scalaris (C. septemvittatus)* is generally restricted to the Mexican Plateau, but is only feebly distinct morphologically from *C. gularis* of the Gulf Coast Plain. The two clearly integrade along the Rio Salado east of Cuatro Cienegas, Coahuila (WALKER, 1966) and in the Cumbres de Monterrey on the Coahuila-Nuevo Leon state line. No integradation has yet been reported from Trans-Pecos Texas, but the range of *scalaris* in that reticulation of grassland and desert scrub is highly restricted. Both forms have geographically patchy distributions. As a result, little local gene flow may occur between west Texas populations of either form. DIXON, LIEB & KETCHERSID (1971) argued that contact areas between the two forms are few and that changes in meristic counts and ontogenetic color patterns are abrupt over short geographical spans (15 to 35 kilometers). This argument is no basis for concluding specific distinction for the two lizards. The *scalaris* form is cradled within the Mexican Plateau by the ridges of the Sierra Madre Mountains, and as a result these ridges form an effective barrier to *scalaris-gularis* contact. The only two major gaps in this barrier that have been effectively sampled have both produced integrades over narrow contact zones. These few contact zones had probably been obliterated by cool mesic conditions during glacial and pluvial maxima of the Quaternary Epoch. The result would be temporary isolation between plateau and Tamaulipan Plain forms, followed by secondary contact in an ameliorated contemporary climate. Secondary contact in narrow gaps would be expected to produce abrupt and restricted zones of integradation (see ZWEIFEL, 1962 for an example in *C. tigris*). Furthermore, in Queretero where the two forms show dramatically different meristic characteristics in close geographical proximity, the intervening region is endowed with abrupt changes in elevation and micro-environmental conditions. Abrupt morphological changes, clinal or otherwise should hardly be unexpected (see TRUEB, 1968, for *Hyla lancasteri*; MARTIN, 1958, for *Gerrhonotus liocephalus*). No evidence supports sympatric occurrence or the existence of isolating mechanisms.

In conclusion, I find that form *scalaris* must be treated as a race or complexes of races conspecific with *Cnemidophorus gularis*, in accord with WALKER (1966).

20. The generic validity of *Barisia:* TIHEN (1949) split the genus *Gerrhonotus* into generic units, *Abronia, Coloptychon, Gerrhonotus, Barisia*, and *Elgaria*. STEBBINS (1958) reviewed this classification and reappraised the validity and stability of several of the characters employed, concluding that *Elgaria* was without adequate basis and that *Barisia* was only differentiated to the level of a sub-genus or species group within *Gerrhonotus*. *Barisia* are distinguished by their general pigmentation and pattern (involving both dorsal and ventral integument), their shorter tails, blunter snouts, and generally smaller size relative to *Gerrhonotus*, their viviparity in contrast to oviparity of *Gerrhonotus*, and their ecological association with higher or more northern or coastal coniferous forests, rather than the oak-pinyon associations of *Gerrhonotus*. Other characters used by TIHEN (1949) included head squamation, tooth number, and cranial proportions. STEBBINS's (1949) reevaluation argued against specific status for *Barisia* in pointing out that its characters are so superficial and plastic (i.e., viviparity) that very minor genetic shifts, perhaps in response to higher cooler habitats, could repeatedly and independently derive a *Barisia* morph several times from a general Gerrhonotine stock. Even accepting the two groups are valid phyletic clades, the differences

between them are superficial and many of them could be produced by slight alterations in allometric growth patterns. Furthermore, California *Gerrhonotus multicarinatus* appear to be almost intermediate between the two groups in most of the characters mentioned. Whether this is due to phylogeny or convergence, it still weighs heavily against generic recognition for *Barisia*. The species groups are in no way more distinctive than those of the genus *Sceloporus* or *Cnemidophorus*, and I consider them congeneric, the name *Gerrhonotus* having priority.

21. The specific validity of *Gerrhonotus lugoi:* I tentatively accept *Gerrhonotus lugoi* as a valid species, but I wish also to present an alternative evaluation which cannot as yet be eliminated. The alternative is that *lugoi* simply represents the lower elevational extreme in a present or past (glacial or pluvial) clinal gradient of *Gerrhonotus leiocephalus* populations. MARTIN (1958) specifically observed that *G. leiocephalus* samples in the Sierra Madre Oriental have lower dorsal scale rows averaging about 50 at 1500 meters elevation, but increasing to about 60 dorsal rows below 900 meters. Significantly, McCoy (1970) reported essentially the same trend in dorsal counts when comparing the low Cuatro Cienegas Bolson (800 meters elevation) with populations of *G. leiocephalus* (1500 meters and above). If similar clinal trends can be applied to the other characters distinguishing *lugoi* from the obviously allied *G. leiocephalus*, the former must be assumed to be simply a lowland morph of the latter species. It seems more doubtful that head scute configuration and absence of keeling can be so easily accounted for, but the possibility does exist.

FAMILY XANTUSIDAE

22. The status of *Xantusia extorris:* This form appears clearly to be a disjunct race of *Xantusia vigilis* on both morphological (WEBB, 1970) and karyotypic grounds (BEZY, 1972). Whether their phyletic affinities lie with the peninsular California race (*X. v. gilberti*) as suggested on meristic counts by WEBB (1965) or with Arizona populations (karyotypic evidence of BEZY, 1972) is an intriguing and unresolved biogeographical question.

23. The specific status and affinities of *Xantusia henshawi bolsoni:* A morphological comparison with *Xantusia arizonae* coupled with karyotypic comparisons (BEZY, 1972 and personal communication) provide convincing evidence that *bolsoni* is conspecific with that species, not *Xantusia henshawi*. Karyotypically it is virtually identical with *X. arizonae* (BEZY, 1972). The femoral pore counts which presently distinguish *bolsoni* seem hardly a basis for specific recognition, especially considering the small sample size upon which the original description was based. I concur with SAVAGE (personal communication) in retaining *X. arizonae* as a valid species. BEZY (1967) establishes karyotypic affinities between *X. arizonae* and Arizona populations of *X. vigilis*. However, I interpret his examples of integrades as saxicolous *Xantusia vigilis* populations which have converged slightly in size, proportions, and pigmentation with adjacent (but not demonstrably integrading) *X. arizonae*.

SUBORDER SERPENTES

FAMILY COLUBRIDAE

24. The generic validity of *Gyalopion* and *Ficimia:* I do not accept the genus *Gyalopion* (SMITH & TAYLOR, 1941; HARDY, 1970) as valid. I consider the value of the characters employed in its definition and their stability to be dubious on the generic level. DIXON & LIEB (personal communication) have reported to me a specimen from the Chihuahuan Desert relict region around Toliman, Queretero, which was subsequently identified as a natural intergeneric hybrid (by L.M. HARDY) between *Gyalopion canum* and *Ficimia olivacea*. To me this animal indicates the feeble nature of the supposed generic characters and in fact places the specific status of *canum* in doubt. While hybridization is at present based on a single specimen, it is noteworthy that the Queretero region (DIXON, KETCHERSID & LIEB, 1972a) is the only known area where the two forms are sympatric. *Ficimia* is the generic name having priority, and I use it here as inclusive of *Gyalopion*.

25. The status of the species *Hypsiglena torquata* with particular reference to the form *onchrorhyncha:* I accept the arguments of HARDY & MCDIARMID (1969) for treating *onchrorhyncha* as a northwest Mexican pattern morph of *Hypsiglena torquata*. The only other valid species is *Hypsiglena tanzeri*. Such dramatic differences in patterns and scale counts between the two known specimens of this species and *H. torquata* 58 kilometers away circumstantially justify its recognition (clinal variation with *H. torquata* elsewhere is not comparable). But knowing that polymorphism does express itself dramatically in this genus, I am forced to make my recognition tentative. The short geographical gap between the forms involves a tortuous topography and abrupt shifts in climate and habitat. The possibility that *tanzeri* is an ecological race or morph, or a component in some extreme elevational clinal variation cannot yet be eliminated. It also bears a neck pattern similar to the insular *Eridiphas slevini* but fails to conform in squamation.

26. The status of *Pituophis deppei* and its relationship to *Pituophis melanoleucus:* I believe that *Pituophis deppei* is conspecific with *Pituophis melanoleucus*, and I have treated the former as synonymous with *melanoleucus* in all further discussion. CONANT (1965) suggested the possibility of either hybridization or integration between the two forms in central Zacatecas and San Luis Potosi. Both racial forms of *deppei, d. deppei* and *d. jani*, which come into contact with *melanoleucus* have formed unquestionable integrades with the latter. These integrades are unstable in head and neck markings, often with two prefrontal scales, and one supralabial in contact with orbit, or the reverse, and are in some case asymmetrical (i.e., two prefrontals on one side, one on the other). Of approximately forty *Pituophis* I have examined, about 30% were very much intermediate in pattern and squamation, between the two supposed species. These individuals came from an extensive region of overlap covering about 300 square kilometer tetragon across the Mexican Plateau between the cities of Torreon and Saltillo on the north, and Zacatecas and San Luis Potosi in the south (previously noted by CONANT, 1965). This region essentially corresponds to the Salado division of the Chihuahuan Desert. Intermediates were especially clustered around Matahuala in San Luis Potosi, Dr. Arroyo in Nuevo Leon, and Fresnillo in Zacatecas. Nowhere are *Pituophis deppei* and *melanoleucus* known to occur sympatrically in the absence of integrades. My field observations in these three Mexican states indicate no clear habitat segregation between these two strikingly similar snakes (both occur in the desert proper and the peripheral mesquite grassland). DOWLING (1958) suggested on the basis of vertebral morphology that all North American forms are conspecific. While further

sampling in the south central Mexican Plateau would be desirable, I find no evidence supporting the maintenance of *deppei* as a species.

DUELLMAN (1960) concludes that *Pituophis lineaticollis* is a species distinct from *P. deppei*, and sympatric with that form. His analysis, based on trunk and subcaudal scale counts, would make *lineaticollis* the only other valid species of the genus. It is intriguing that this last form resembles strikingly *Elaphe subocularis* in having a single pair of prefrontals, and extremely distinctive dorsal patterns consisting of neck longitudinal stripes (black) and posterior trunk blotches (brown/black) on a yellowish ground color. *E. subocularis* furthermore has *Pituophis*-like hemipenis (DOWLING, 1957). *P. lineaticollis* is completely allopatric to *E. subocularis*, the former being a montane species of southern Sierra Madres south into the highlands of Guatemala, while the latter is restricted to the Chihuahuan Desert north of the 25th parallel. In the 500 kilometer interval between the two forms, plateau populations of *Pituophis* appears to undergo a series of character displacements as they proceed northward approaching populations of *E. subocularis*. In the northward shift from *P. lineaticollis* to *deppei* the *subocularis*-like pattern gives way to a completely blotched dorsal pattern (excepting the head which often remains unicolor). Further north, as the *deppei* morphs integrade with *melanoleucus*, the head becomes blotched and barred with increasing frequency, the *Elaphe*-like paired prefrontals become four, and two labials in the orbit are replaced by one. It is this highly modified *melanoleucus* which is in actual sympatric contact with *E. subocularis*. The biogeographical significance of this apparent character displacement will be discussed later.

27. The specific status of *Tantilla atriceps*: On the basis of available evidence, I agree with MCDIARMID (1968) in recognizing the specific validity of *Tantilla yaquia* and *Tantilla atriceps*. His arguments, based on the stability of distinguishing pattern and meristic characters from individuals in virtual sympatry are a convincing rebuttal to the conclusions of TANNER (1966) that the two forms are geographical races of *Tantilla planiceps*. TANNER did make the valid assertion that these forms are closely related, and the possibility of intergradation between *Tantilla planiceps* in Utah and northern Arizona *Tantilla atriceps* cannot be eliminated. Still, such integradation has not been established, and the conditions described by MCDIARMID (1968) in the supposed contact area of southeastern Arizona make its existence dubious.

28. The specific status of *Tantilla cucullata, diabola* and *rubra*: I accept the conclusion of SMITH & WERLER (1969) that *diabola* is a geographical variant of *Tantilla rubra*, as is *cucullata* (DEGENHARDT *et al.*, 1975).

I also concur with the suggestion of RAUN & GEHLBACH (1972) that *Tantilla rubra* should also include *cucullata* as a race or coloration morph. These forms are similar in size, squamation and most aspects of their pigmentation. The primary distinction of *cucullata*, its solid black head cap, could be interpreted as clinal terminal in the northward reduction of a light head marking (post-ocular and neck band) of *Tantilla rubra* progressing from the Sierra Madre Oriental. Subcaudal scale count, the other distinguishing character, also increases to the northwest (FOUQUETTE & POTTER, 1961). *Diabola* is intermediate in pigmentation, squamation and geography. In all three the black of the neck marks terminates about the fourth dorsal posterior to the parietals. It is noteworthy that MINTON (1959) did not compare *rubra* and *cucullata* in his erection of the latter species. Virtually sympatric examples of the two forms occur in the Chisos Mountains, Brewster

County, Texas. There squamation is virtually identical and only the head cap differs (polymorphism?) (DEGENHART & MILSTEAD, 1959).

29. The status of *Trimorphodon vilkinsoni:* I tentatively concur with the evaluation of GEHLBACH (1971) that *vilkinsoni* is conspecific with *Trimorphodon biscutatus*. Several specimens provide additional evidence favoring genetic contact between this form and *lambda* race of the Sonoran Desert. One individual (NMU) strikingly like typical *lambda* in the number and split nature of the dorsal body blotches was collected approximately 30 kilometers south of Chihuahua, Chihuahua, Mexico deep within the range of typical *vilkinsoni*. Another individual collected in pine woodland near the city of Durango, Durango, appeared to be intermediate in dorsal blotch spacing between the Sonoran and Chihuahuan forms.

30. The relationship of *Sonora episcopa* and *Sonora semiannulata:* These two forms are largely complementary in geographical occurrence (both recorded from Presidio County, Texas, and about 30 kilometers NE of San Pedro, Coahuila) but clear instances of sympatry in areas of peripheral contact have not been established. Pattern differences are unstable in distinguishing between the two snakes, a situation complicated by polymorphism in both species. Only ventral and subcaudal counts reliably determine the identity of an individual, and even here sexual dimorphism must be accounted for in the diagnosis using the latter. Furthermore the occurrence of *S. episcopa* in the center of *S. semiannulata* range (San Pedro, Cuahuila) may indicate polymorphism. I severely doubt the specific validity of the two forms, but lack sufficient information on populations in areas of peripheral contact to make a final resolution of their status.

31. The status of *Salvadora hexalepis* with particular reference to form *deserticola:* I accept the classification of SCHMIDT (1940) in its treatment of *deserticola* as a race of *Salvadora hexalepis*. BOGERT & DEGENHARDT (1961) treated this form as distinct species, but provide no justification for their decision. The distinguishing characters (position of lateral dark stripe, condition of loreal, and number of labials in contact with loreal), provide no greater differences than those used in recognizing other geographical races of *Salvadora hexalepis*.

32. The status of *Salvadora grahamiae* with particular reference to the form *lineata:* I accept the treatment of WRIGHT & WRIGHT (1957) of *lineata* as geographical variant of the *Salvadora grahamiae*. Integrades are known from montane central Durango and northern Zacatecas.

A Checklist of Extant Amphibian and Reptilian Species Occurring Within the Geographical Chihuahuan Desert

CLASS AMPHIBIA

ORDER CAUDATA

FAMILY AMBYSTOMATIDAE

    1. *Ambystoma tigrinum* (GREEN), 1825 – Tiger Salamander

ORDER SALIENTIA

FAMILY PELOBATIDAE

1. *Scaphiopus bombifrons* (COPE), 1863 — Plains Spadefoot Toad
2. *Scaphiopus couchi* BAIRD, 1854 — Couch's Spadefoot Toad
3. *Scaphiopus hammondi* BAIRD, 1839 — Western Spadefoot Toad

FAMILY LEPTODACTYLIDAE

4. *Eleutherodactylus augusti** (DUGES), 1879 — Barking Frog
5. *Syrrhophus marnocki** COPE, 1878 — Texas Cliff Frog

FAMILY HYLIDAE

6. *Acris crepitans* BAIRD, 1854 — Cricket Frog
7. *Hyla arenicolor* COPE, 1866 — Canyon Tree Frog
8. *Pseudacris triseriata* (WIED), 1838 — Western Chorus Frog

FAMILY BUFONIDAE

9. *Bufo cognatus* SAY, 1823 — Great Plains Toad
10. *Bufo debilis* GIRARD, 1854 — Green Toad
11. *Bufo marinus** (LINNAEUS), 1758 — Giant Toad
12. *Bufo punctatus* BAIRD & GIRARD, 1852 — Red Spotted Toad
13. *Bufo speciosus* GIRARD, 1854 — Texas Toad
14. *Bufo valliceps** WIEGMANN, 1833 — Gulf Coast Toad
15. *Bufo woodhousei* GIRARD, 1854 — Woodhouse's Toad

FAMILY RANIDAE

16. *Rana catesbeiana* SHAW, 1802 — Bullfrog
17. *Rana pipiens* SHREBER, 1782 — Leopard Frog

FAMILY MICROHYLIDAE

18. *Gastrophryne olivaces* (HALLOWELL), 1856 — Great Plains Narrow-Mouthed Toad

CLASS REPTILIA

ORDER TESTUDINATA

FAMILY CHELYDRIDAE

1. *Chelydra serpentina* (LINNAEUS), 1758 — Common Snapping Turtle

FAMILY KINOSTERNIDAE

2. *Kinosternon flavescens* (AGASSIZ), 1857 — Yellow Mud Turtle
3. *Kinosternon hirtipes* WAGLER, 1830 — Mexican Mud Turtle
4. *Kinosternon sonoriense* LE CONTE, 1854 — Sonoran Mud Turtle

FAMILY EMYDIDAE

5. *Chrysemys concinna* (LE CONTE), 1830 — River Cooter
6. *Chrysemys gaigae* (HARTWEG), 1938 — Big Bend Slider
7. *Chrysemys picta* (SCHNEIDER), 1783 — Painted Turtle
8. *Chrysemys scripta* (SCHOEPFF), 1792 — Pond Slider
9. *Terrapene ornata* (AGASSIZ), 1857 — Western Box Turtle

FAMILY TESTUDINIDAE

10. *Gopherus flavomarginatus* LEGLER, 1959 — Bolson Tortoise

FAMILY TRIONYCHIDAE

    11. *Trionyx spiniferus* (LE SEUER), 1827 – Spiny Softshell

ORDER SQUAMATA

SUBORDER LACERTILIA

FAMILY GEKKONIDAE

    1. *Coleonyx brevis* STEJNEGER, 1893 – Texas Banded Gecko
    2. *Coleonyx reticulatus* DAVIS & DIXON, 1958 – Big Bend Gecko

FAMILY IGUANIDAE

    3. *Cophosaurus texanus* (TROSCHEL), 1852 – Great Earless Lizard
    4. *Crotaphytus collaris* SAY, 1823 – Collared Lizard
    5. *Crotaphytus wislizeni* (BAIRD & GIRARD), 1852 – Leopard Lizard
    6. *Holbrookia maculata* GIRARD, 1851 – Lesser Earless Lizard
    7. *Phrynosoma cornutum* (HARLAN), 1825 – Texas Horned Lizard
    8. *Phrynosoma douglassi** (BELL), 1829 – Short-Horned Lizard
    9. *Phrynosoma orbiculare** (LINNAEUS), 1789 – None
    10. *Phrynosoma modestum* GIRARD, 1852 – Round Tailed Horned Lizard
    11. *Sceloporus cautus* SMITH, 1938 – None
    12. *Sceloporus goldmani** SMITH, 1937 – Goldman's Bunch-Grass Lizard
    13. *Sceloporus graciosus** BAIRD & GIRARD, 1852 – Sagebrush Lizard
    14. *Sceloporus grammicus* WIEGMANN, 1828 – Mesquite Lizard
    15. *Sceloporus jarrovi* COPE, 1875 – Yarrow's Spiny Lizard
    16. *Sceloporus maculosus* SMITH, 1934 – None
    17. *Sceloporus magister* HALLOWELL, 1854 – Desert Spiny Lizard
    18. *Sceloporus merriami* STEJNEGER, 1904 – Canyon Lizard
    19. *Sceloporus olivaceus** SMITH, 1934 – Texas Spiny Lizard
    20. *Sceloporus ornatus* BAIRD, 1858, 1859 – Ornate Spiny Lizard
    21. *Sceloporus parvus* SMITH, 1934 – None
    22. *Sceloporus spinosus* WIEGMANN, 1828 – None
    23. *Sceloporus torquatus* WIEGMANN, 1828 – None
    24. *Sceloporus undulatus* (LATREILLE), 1802 – Eastern Fence Lizard
    25. *Uma exsul* SCHMIDT, 1944 – None
    26. *Urosaurus ornatus* (BAIRD & GIRARD), 1852 – Tree Lizard
    27. *Uta stansburiana* BAIRD & GIRARD, 1852 – Side-Blotched Lizard

FAMILY SCINCIDAE

    28. *Eumeces bilineatus* COPE, 1880 – Short-Lined Skink
    29. *Eumeces brevirostris* (GUNTHER), 1860 – None
    30. *Eumeces multivirgatus* (HALLOWELL) 1857 – Many-Lined Skink
    31. *Eumeces obsoletus* (BAIRD & GIRARD), 1852 – Great Plains Skink
    32. *Leiolopisma laterale* (SAY), 1823 – Ground Skink

FAMILY TEIIDAE

    33. *"Cnemidophorus exanguis"* LOWE, 1956 – Chihuahua Whiptail
    34. *Cnemidophorus gularis** BAIRD & GIRARD, 1852 – Texas Spotted Whiptail
    35. *Cnemidophorus inornatus* BAIRD, 1858 – Little Striped Whiptail
    36. *"Cnemidophorus neomexicanus"* LOWE & ZWEIFEL, 1952 – New Mexican Whiptail
    37. *"Cnemidophorus tesselatus"* (SAY), 1823[1] – Checkered Whiptail

1 *"Cnemidophorus dixoni"* SCUDDAY, 1974 is here treated as synonomous with this form, it is no more distinct than A,B,C, types of C. tesselatus.

38. *Cnemidophorus tigris* BAIRD & GIRARD, 1852 – Western Whiptail
39. *"Cnemidophorus uniparens"* WRIGHT & LOWE, 1965 – Desert Grassland Whiptail

FAMILY ANGUIDAE

40. *Gerrhonotus liocephalus* WIEGMANN, 1828 – Texas Alligator Lizard
41. *Gerrhonotus lugoi* MCCOY, 1971 – None

FAMILY XANTUSIDAE

42. *Xantusia arizonae* KLAUBER, 1931 – Night Lizard
43. *Xantusia vigilis* BAIRD, 1858 – Desert Night Lizard

SUBORDER SERPENTES

FAMILY LEPTOTYPHLOPIDAE

1. *Leptotyphlops dulcis* (BAIRD & GIRARD), 1853 – Texas Blind Snake
2. *Leptotyphlops humilis* (BAIRD & GIRARD), 1853 – Western Blind Snake

FAMILY COLUBRIDAE

3. *Arizona elegans* KENNICOTT, 1859 – Glossy Snake
4. *Coluber constrictor* LINNAEUS, 1758 – Racer
5. *Diadophis punctatus* (LINNAEUS), 1766 – Ringneck Snake
6. *Elaphe guttata* (LINNANEUS), 1766 – Corn Snake
7. *Elaphe obsoleta* (SAY), 1823 – Common Rat Snake
8. *Elaphe subocularis* (BROWN), 1901 – Trans-Pecos Rat Snake
9. *Ficimia cana* (COPE), 1860 – Western Hook-Nose Snake
10. *Heterodon nasicus* BAIRD & GIRARD, 1852 – Western Hognose Snake
11. *Hypsiglena torquata* (GUNTHER), 1860 – Night Snake
12. *Lampropeltis getulus* (LINNANEUS), 1766 – Common Kingsnake
13. *Lampropeltis mexicana* (GARMAN), 1883 – Gray-Banded Kingsnake
14. *Lampropeltis triangulum** (LACEPEDE), 1788 – Milk Snake
15. *Masticophis flagellum* (SHAW), 1802 – Coachwhip
16. *Masticophis taeniatus* (HALLOWELL), 1852 – Striped Whipsnake
17. *Natrix erythrogaster* (FOSTER), 1771 – Plain-Bellied Water Snake
18. *Natrix rhombifera** (HALLOWELL), 1852 – Diamond-Backed Water Snake
19. *Opehodrys aestivus** (LINNAEUS), 1776 – Rough Green Snake
20. *Pituophis melanoleucus* (DAUDIN), 1803 – Bullsnake
21. *Rhinocheilus lecontei* BAIRD & GIRARD, 1853 – Long-Nosed Snake
22. *Salvadora hexalepis* (COPE), 1866 – Desert Patch-Nosed Snake
23. *Salvadora grahamiae* BAIRD & GIRARD, 1853 – Mountain Patch-Nosed Snake
24. *Sonora episcopa* (KENNICOTT), 1859 – Great Plains Ground Snake
25. *Sonora semiannulata* BAIRD & GIRARD, 1853 – Western Ground Snake
26. *Tantilla atriceps* (GUNTHER), 1895 – Mexican Black-Headed Snake
27. *Tantilla nigriceps* KENNICOTT, 1860 – Plains Black-Headed Snake
28. *Tantilla rubra* COPE, 1876 – Big Bend Black-Headed Snake
29. *Tantilla wilcoxi** STEJNEGER, 1902 – Huachuca Night Snake
30. *Thamnophis cyrtopsis** (KENNICOTT), 1860 – Black-Necked Garter Snake
31. *Thamnophis eques** (RUESS), 1834 – Mexican Garter Snake
32. *Thamnophis marcianus* (BAIRD & GIRARD), 1853 – Checked Garter Snake
33. *Thamnophis melanogaster* (PETERS), 1864 – None
34. *Thamnophis proximus** (SAY), 1823 – Western Ribbon Snake
35. *Thamnophis rufipunctatus* (COPE), 1875 – Narrow-Headed Garter Snake
36. *Thamnophis sirtalis** (LINNAEUS), 1758 – Common Garter Snake
37. *Trimorphodon biscutatus* (DUMERIL & BIBRON), 1854 – Lyre Snake

FAMILY ELAPIDAE

38. *Micrurus fulvius** (LINNAEUS), 1766 – Coral Snake

FAMILY VIPERIDAE

39. *Agkistrodon contortrix* (LINNAEUS), 1766 – Copperhead
40. *Sistrurus catenatus* (RAFINESQUE), 1818 – Massasauga
41. *Crotalus atrox* BAIRD & GIRARD, 1853 – Western Diamond-Back Rattlesnake
42. *Crotalus lepidus* (KENNICOTT), 1861 – Rock Rattlesnake
43. *Crotalus molossus* BAIRD & GIRARD, 1853 – Black-Tailed Rattlesnake
44. *Crotalus scutulatus* (KENNICOTT), 1861 – Mojave Rattlesnake
45. *Crotalus viridis* (RAFINESQUE), 1818 – Western Rattlesnake

*restricted to no more than three documented desert localities

Addenda to the Checklist of Amphibian and Reptile Species Occurring Within the Geographical Chihuahuan Desert

Peripheral Species Reported from Localities Immediately Adjacent to the Chihuahuan Desert

CLASS AMPHIBIA

ORDER CAUDATA

FAMILY PLETHODONTIDAE

1. *Aneides hardyi* (TAYLOR), 1941 – Sacramento Mountains Salamander
2. *Chiropterotriton prisca* RABB, 1956 – none
3. *Eurycea neotenes* BISHOP & WRIGHT, 1937 – Texas Salamander
4. *Pseudoeurycea galeanae* (TAYLOR), 1941 – None

ORDER SALIENTIA

FAMILY BUFONIDAE

5. *Bufo alvarius* GIRARD, 1859 – Colorado River Toad
6. *Bufo compactilis* WIEGMANN, 1833 – none

FAMILY LEPTODACTYLIDAE

7. *Syrrhophus cystignathoides* (COPE), 1877 – Rio Grande Frog
8. *Syrrhophus dennis* LYNCH, 1970 – None
9. *Syrrhophus verrucipes* COPE, 1885 – None

FAMILY HYLIDAE

10. *Hyla eximia* BAIRD, 1854 – Arizona Tree Frog

FAMILY RANIDAE

11. *Rana montezumae* BAIRD, 1854 – Montezuma's Frog
12. *Rana tarahumarae* BOULENGER, 1917 – Tarahumara Frog

CLASS REPTILIA

ORDER TESTUDINATA

FAMILY KINOSTERNIDAE

1. *Kinosternon scorpioides* (LINNAEUS), 1766 – Scorpion Mud Turtle

FAMILY EMYDIDAE

2. *Terrapene coahuila* SCHMIDT & OWENS, 1944 – Cuatro Cienegas Box Turtle

FAMILY TESTUDINIDAE

3. *Gopherus agassizi* (COOPER), 1863 – Desert Tortoise
4. *Gopherus berlandieri* (AGASSIZ), 1863 – Berlandier's Tortoise

ORDER SQUAMATA

SUBORDER LACERTILIA

FAMILY GEKKONIDAE

1. *Coleonyx variegatus* (BAIRD), 1858 – Banded Gecko
2. *Hemidactylus turcicus* (LINNAEUS), 1756 – Mediterranean Gecko

FAMILY IGUANIDAE

3. *Callisaurus draconoides* BLAINVILLE, 1835 – Zebra-Tailed Lizard
4. *Crotaphytus reticulatus* BAIRD, 1858 – Reticulate Collared Lizard
5. *Holbrookia lacerata* COPE, 1880 – Spot-Tailed Earless Lizard
6. *Phrynosoma solare* GRAY, 1845 – Regal Horned Lizard
7. *Sceloporus clarki* BAIRD & GIRARD, 1852 – Clark's Spiny Lizard
8. *Sceloporus couchi* BAIRD, 1858 – Couch's Spiny Lizard
9. *Sceloporus cyanogenys* COPE, 1885 – Blue Spiny Lizard
10. *Sceloporus exsul* DIXON, KETCHERSID & LIEB, 1972 – None
11. *Sceloporus scalaris* WIEGMANN, 1828 – Bunch Grass Lizard
12. *Sceloporus variabilis* WIEGMANN, 1834 – Rose-Bellied Spiny Lizard
13. *Sceloporus virgatus* SMITH, 1938 – Striped Plateau Lizard

FAMILY SCINCIDAE

14. *Eumeces callicephalus* BOCOURT, 1879 – Mountain Skink
15. *Eumeces tetragrammus* (BAIRD), 1858 – Four-Lined Skink
16. *Leiolopisma silvicolum* (TAYLOR), 1937 – Sierra Madre Ground Skink

FAMILY TEIIDAE

17. *Cnemidophorus burti* TAYLOR, 1938 – Giant Whiptail
18. *"Cnemidophorus flagellicaudus"* LOWE & WRIGHT, 1964 – None
19. *"Cnemidophorus velox"* LOWE, 1955 – Plateau Whiptail

FAMILY ANGUIDAE

20. *Gerrhonotus imbricatus* (WIEGMANN), 1828 – Imbricate Alligator Lizard
21. *Gerrhonotus kingi* (GRAY), 1838 – Arizona Alligator Lizard

FAMILY HELODERMATIDAE

22. *Heloderma suspectum* COPE, 1869 – Gila Monster

SUBORDER SERPENTES

FAMILY COLUBRIDAE

1. *Chilomeniscus cinctus* COPE, 1861 – Banded Sand Snake
2. *Drymarchon corais* (BOIE), 1827 – Indigo Snake
3. *Drymobius margaritiferus* (SCHLEGEL), 1837 – Speckled Racer
4. *Elaphe triaspis* (COPE), 1866 – Green Rat Snake
5. *Hypsiglena tanzeri* DIXON & LIEB, 1972 – None
6. *Lampropeltis pyromelana* (COPE) 1866 – Arizona Mountain King Snake
7. *Masticophis bilineatus* JAN, 1863 – Sonora Whipsnake
8. *Opheodrys vernalis* (HARLAN), 1828 – Smooth Green Snake
9. *Rhadinaea crassa* SMITH, 1882 – None
10. *Storeria occipitomacuata* (STORER), 1839 – Red-Bellied Snake
11. *Tantilla gracilis* BAIRD & GIRARD, 1853 – Flat-Headed Snake
12. *Thamnophis exsul* ROSSMAN, 1969 – None
13. *Tropidoclonion lineatum* (HALLOWELL), 1856 – Lined Snake

FAMILY ELAPIDAE

14. *Micruroides euryxanthus* (KENNICOTT), 1860 – Arizona Coral Snake

FAMILY VIPERIDAE

15. *Crotalus pricei* VAN DENBURGH, 1895 – Twin-Spotted Rattlesnake
16. *Crotalus willardi* MEEK, 1905 – Ridge-Nosed Rattlesnake

VIII. THE ECOLOGICAL AFFINITIES OF THE CHIHUAHUAN HERPETOFAUNA

Ecological Affinities and Faunal Assemblages in a Biogeographical Unit

In the previous chapter, a checklist of 170 species of amphibians and reptiles occurring within 25 kilometers of the geographical Chihuahuan Desert was provided. Based on mapped distributions, all of these forms either do, or could potentially exist in the desert. But this checklist of a potential herpetofauna fails to answer one most fundamental question, namely: "What is the participant herpetofauna of Chihuahuan Desert ecosystems?" Maps are not enough in this case to construct biogeographically a meaningful assemblage of organisms. DICE (1943) suggested provinces should ideally delimit a single climax community. The geographical desert is riddled with deciduous riparian woodland, marshes, springs and caverns, and montane islands. Each of these units shelters depauperate faunas alien to, and intolerant of, the physical environment of the desert floor. Thus before an explicit biogeographical analysis can commence, this necessary, if sometimes coarse, set of ecological categorizations must be made.

The categorizations set forth in Table 10 are based largely on: CONANT (1975); STEBBINS (1966); LITTLEMAN & KELLER (1938); ANDERSON (personal communication); BOGERT & DEGENHARDT (1961); MINTON (1958); DIXON, KETCHERSID & LIEB (1972a), ANDERSON & LIDEKKER (1963); and a survey of species specific reports from the literature. In addition, museum locality records were checked against field notes and photographs taken from a six month, 15,000 kilometer field survey undertaken by the author during July through September 1971 and 1972.

Table 10 illustrates the difference between the potential and the real Chihuahuan Desert herpetofauna. Only thirty-four percent of the total species recorded for the general geographical area actually occur in Chihuahuan Desert plant associations with any degree of consistency. This low percentage may be exaggerated by inadequate sampling from some portions of the desert relative to a better

Table 10. An Ecological Categorization of the Herpetofauna Occurring in the Geographical Proximity[1] of the Chihuahuan Desert.

Taxa	# of Species	Relicts	Sonoran and Tamaulipan Peripherals	Riparian and Aquatic	Chihuahuan Desert Scrub
Caudata	5	80%	0	0	20%
Salientia	26	50%	20%	3%	27%
Testudinata	15	7%	20%	59%	14%
Lacertilia	63	33%	19%	5%	43%
Serpentes	61	41%	11%	15%	33%
Total	170	37%	16%	13%	34%

1. Within 25 kilometers of the boundaries of the Chihuahuan Desert proper, as set by the base map used in this text.

populated and more accessible periphery. Other species may have been eliminated as relicts because their secretive microhabitat selection obscures their true distribution (i.e., the crevice dwelling frogs, genus *Syrrhophus*). Even granting these possible distortions, I believe the small percentage of true desert forms reflects the impact of harsh physical parameters, most obviously the lack of water. The Creosote scrub association, dominant in desert surface area, is also impoverished in terms of primary productivity. CHEW & CHEW (1965) (in Cochise County, Arizona) assessed efficiency of a Chihuahuan Desert *Larrea* community at 0.03-0.06% of the total annual solar energy converted to net annual primary productivity. This is about 1/30 the efficiency of an English forest. In addition, absolute productivity was always less than more mesic systems. KIESTER (1970) noted a general correlation between North American herpetofaunal species densities and increasingly mesic climates. Not surprisingly this correlation is most pronounced in aquatic faunas. Only 14% of the regional turtles actually occur within the desert environment. But even among the lizards, a generally terrestrial and heliothermic group, only 43% of the regional forms inhabited the desert.

The remainder of this chapter will be devoted to a resume of herpetofaunal assemblages correlated with major plant associations, both within and immediately peripheral to the desert. An ecologically based checklist of Chihuahuan Desert herpetofauna will be presented, excluding the peripheral and relictual components of the geographical list. It shall attempt to define assemblages, and assess their worth quantitatively using Coefficients of Community (C.C.) (JACCARD, 1902) and the Coefficient of Similarity (S.C.). Major ecological refugia and displacement patterns evident in the habitat distribution of relict non-desert species will also be categorized. Finally, three selected genera of particular diversity and biogeographical importance in the Mexican Plateau, *Cnemidophorus, Crotalus,* and *Sceloporus,* will be examined in detail for specific intrageneric patterns of habitat segregation.

Correlations Between Plant Associations and the Habitat Distribution

Indicative species will be listed by class for plant association or cluster of association. Species will often be listed for more than one category when they are a conspicuous part of the fauna. They will also be rated for the degree of their affinity for a particular association by the following characters, noted parenthetically after the species name:

saxicolous exclusively: (-sax)

Restricted to this one association habitat: (R)
(90% of total Chihuahuan records)

Characteristic, but not an obligatory
inhabitant of this association: (C)
(more than 50% of total Chihuahuan records)

Consistently present in this association throughout
its range, but no more so than to other associations: (P)
(less than 50% of total Chihuahuan records)
Restricted geographically to one or more faciations: *

86

A. Chihuahuan Desert Scrub Association:

COMMENT: The checklist that follows is the herpetofauna actually utilizing terrestrial desert habitats within the geographical Chihuahuan Desert. While all forms listed here are true desert species, at least in extensive portions of their range, not all permeate all geographical portions of the desert. The geographical faciations of this fauna will be discussed in the following chapter. Forms occurring in less than five desert localities and primarily associated with non-desert vegetation will be discussed separately in a subsequent section of this chapter. They are also designated by the symbol*.

Species are also listed by edaphic associations within the desert scrubland. However, several categories of edaphic associations have been clumped where preliminary inspection indicated separate treatment of the herpetofaunas unjustified.

Indicative Herpetofauna:

Amphibia:

Caudata:	1.	*Ambystoma tigrinum* (P)*
Salientia:	1.	*Scaphiopus couchi* (C/P)
	2.	*Scaphiopus hammondi* (P)
	3.	*Bufo cognatus* (P)
	4.	*Bufo debilis* (P)*
	5.	*Bufo punctatus* (P)
	6.	*Bufo speciosus* (P – may be relict)*
	7.	*Gastrophryne olivacea* (P – may be relict)*

Reptilia	1.	*Terrepene ornata* (P)*
Testudinata:	2.	*Gopherus flavomarginatus* (R)*

Squamata:	1.	*Coleonyx brevis* (C)*
Lacertilia:	2.	*Coleonyx reticulatus* (R)*
	3.	*Cophosaurus texanus* (C)
	4.	*Crotaphytus collaris* (P)
	5.	*Crotaphytus wislizeni* (P)
	6.	*Holbrookia maculata* (P)
	7.	*Phrynosoma cornutum* (P)
	8.	*Phrynosoma modestum* (R)
	9.	*Sceloporus cautus* (R)*
	10.	*Sceloporus grammicus* (P)*
	11.	*Sceloporus maculosus* (R)*
	12.	*Sceloporus magister* (P)*
	13.	*Sceloporus merriami* (R)*
	14.	*Sceloporus ornatus* (R)*
	15.	*Sceloporus poinsetti* (R)
	16.	*Sceloporus undulatus* (P)*
	17.	*Uma exsul* (R)*
	18.	*Uta stansburiana* (P)*
	19.	*Eumeces obsoletus* (P – or relict)*
	20.	*Cnemidophorus gularis* (P)
	21.	*Cnemidophorus inornatus* (R)
	22.	*"Cnemidophorus neomexicanus" (R)*
	23.	*"Cnemidophorus tesselatus" (C)*
	24.	*Cnemidophorus tigris* (P)*
	25.	*Gerrhonotus lugoi* (R)*

	26.	*Xantusia arizonae* (P)*
	27.	*Xantusia vigilis* (P)*
Serpentes:	1.	*Leptotyphlops dulcis* (C)
	2.	*Leptotyphlops humilis* (P)*
	3.	*Arizona elegans* (P)
	4.	*Elaphe subocularis* (R)*
	5.	*Ficimia cana* (R)
	6.	*Heterodon nasicus* (P)
	7.	*Hypsiglena torquata* (P)
	8.	*Lampropeltis getulus* (P)
	9.	*Masticophis flagellum* (P)
	10.	*Masticophis taeniatus* (C)
	11.	*Pituophis melanoleucus* (P)
	12.	*Rhinocheilus lecontei* (P)
	13.	*Salvadora hexalepis* (P)*
	14.	*Sonora semiannulata* (P)*
	15.	*Tantilla atriceps* (R)
	16.	*Thamnophis marcianus* (P)*
	17.	*Trimorphodon biscutatus* (P)*
	18.	*Crotalus atrox* (P)
	19.	*Crotalus molossus* (P)
	20.	*Crotalus scutulatus* (P)

1. Edaphic Associations within Chihuahuan Desert Scrub

a. Alluvium Edaphic Association:

COMMENT: The physical evolution of the Mexican Plateau has generated a topography where alluvium is the predominant aspect of its desert surfaces. The *Larrea-Parthenium-Flourensia* scrub association that characterizes these surfaces is therefore a predominant Chihuahuan Desert feature and so is its indicative herpetofauna.
Indicative Herpetofauna:

Amphibia:		
Caudata:	1.	*Ambystoma tigrinum* (P)
Salientia:	1.	*Scaphiopus couchi* (P)
	2.	*Scaphiopus hammondii*, (P)
	3.	*Bufo cognatus* (P)
	4.	*Bufo punctatus* (P)
Reptilia:		
Testudinata:	1.	*Terrepene ornata* (P)*
	2.	*Gopherus flavomarginatus* (P)*
Squamata:		
Lacertilia:	1.	*Coleonyx brevis* (P)*
	2.	*Coleonyx reticulatus* (P-?)*
	3.	*Cophosaurus texanus* (C)
	4.	*Phrynosoma cornutum* (P)
	5.	*Phrynosoma modestum* (P)
	6.	*Sceloporus cautus* (P)*
	7.	*Sceloporus grammicus* (P – restricted to Yucca and Dasylorion woodland)*
	8.	*Sceloporus magister* (P)*
	9.	*Sceloporus undulatus* (P)*

88

		10.	*Uta stansburiana* (P)*
		11.	*Cnemidophorus gularis* (P)
		12.	*Cnemidophorus inornatus* (P – low slopes only)
		13.	*"Cnemidophorus tesselatus"* (P)*
		14.	*Cnemidophorus tigris* (P)*
		15.	*Xantusia vigilis* (R – Yucca woodland)*
Serpentes:		1.	*Arizona elegans* (P – lower slopes)
		2.	*Elaphe subocularis* (P)*
		3.	*Masticophis flagellum* (P)
		4.	*Hypsiglena torquata* (P)
		5.	*Rhinocheilus lecontei* (P)
		6.	*Sonora semiannulata* (P – usually rocky)
		7.	*Crotalus atrox* (P)

b. Barrial Edaphic Associations:

COMMENT: This category includes High Water Table, Alkaline and Gypsum Edaphic Associations. Thus it lumps together all desert floor associations excepting only desert riparian woodland and gypsum dunes. See Displacement and Restriction, Group 1 (Fig. 5) for discussion of relict faunas. In addition, occasional turtles (and the snake *Thamnophis cyrtopsis*) are reported from ephemeral rain pools here, usually of the genus *Chrysemys* or *Kinosternon*. They presumably reached such situations by overland wandering during summer rains, or by passive transport from mountains and arroyos through the agent of flash flooding.
Indicative Herpetofauna:

Amphibia:
Caudata:		1.	*Ambystoma tigrinum* (P)*
Salientia:		1.	*Scaphiopus couchi* (P)
		2.	*Scaphiopus hammondi* (C/P)
		3.	*Bufo cognatus* (P)
		4.	*Bufo debilis* (C)*
		5.	*Bufo punctatus* (P)
		6.	*Bufo speciosus* (P – possible relict)*
		7.	*Gastrophryne olivacea* (P – possible relict)*

Reptilia:
Testudinata:		1.	*Terrepene ornata* (P – possible relict)*
		2.	*Gopherus flavomarginatus* (C)*
Squamata:			
Lacertilia:		1.	*Coleonyx brevis* (P)*
		2.	*Cophosaurus texanus* (P)
		3.	*Crotaphytus wislizeni* (P)*
		4.	*Holbrookia maculata* (P)
		5.	*Phrynosoma cornutum* (P)
		6.	*Phrynosoma modestum* (P)
		7.	*Sceloporus cautus* (P)*
		8.	*Sceloporus magister* (P)*
		9.	*Sceloporus undulatus* (P)*
		10.	*Uta stansburiana* (P)*
		11.	*Cnemidophorus gularis* (C/P)
		12.	*Cnemidophorus inornatus* (C/P)
		13.	*"Cnemidophorus neomexicanus"* (P)*
		14.	*Cnemidophorus tigris* (P)*
Serpentes:		1.	*Leptotyphlops dulcis* (P – ?)
		2.	*Leptotyphlops humilis* (P – ?)*

3. *Arizona elegans* (P)
4. *Ficimia cana* (C)
5. *Heterodon nasicus* (C)
6. *Hypsiglena torquata* (P – rare)
7. *Lampropeltis getulus* (P)
8. *Masticophis flagellum* (P)
9. *Pituophis melanoleucus* (P/C)
10. *Rhinocheilus lecontei* (P)
11. *Salvadora hexalepis* (P)*
12. *Sonora semiannulata* (P – ?)*
13. *Thamnophis marcianus* (P/C)*
14. *Crotalus atrox* (P)
15. *Crotalus scutulatus* (C)

c. Arenicolous Edaphic Associations:

COMMENT: Both gypsum and silaceous dune vegetation associations are combined here. No truly arenicolous amphibians are known from the Chihuahuan Desert (*Ambystoma, Scaphiopus couchi* and *Bufo cognatus* are peripheral). Among the reptiles, no arenicolous species appears to discriminate between gypsum based and silaceous dune substrates. See Displacement and Restriction, Groups 1 and 2a (Fig. 5) for relict faunas. (Probably the tortoise, *Gopherus flavomarginatus* also occurs in some of these associations).
Indicative Species:

Reptilia:
  Squamata:
    Lacertilia:

1. *Coleonyx brevis* (P)*
2. *Crotaphytus wislizeni* (P/C)*
3. *Holbrookia maculata* (P)
4. *Sceloporus magister* (P)*
5. *Sceloporus undulatus* (P)*
6. *Uma exsul* (R)*
7. *Uta stansburiana* (P)*
8. *Cnemidophorus inornatus* (P)
9. *Cnemidophorus tigris* (P)*

    Serpentes:

1. *Arizona elegans* (P/C)
2. *Heterodon nasicus* (P/C)
3. *Masticophis flagellum* (P)
4. *Rhinocheilus lecontei* (P)
5. *Sonora semiannulata* (P)*
6. *Crotalus atrox* (P)

d. Saxicolous Edaphic Associations:

COMMENT: Igneous and limestone plant associations have been treated as components of a single unit here. Relict amphibians and skinks (*Eumeces brevilineatus, brevirostris,* and *obsoletus*) do favor limestone substrates, probably in response to its more extensive fissuring and (porous) water transporting and holding properties. Otherwise, Chihuahuan reptiles do not appear to discriminate between the two types of rocky outcroppings and talus. *Xantusia arizonae* in Durango (and *Sceloporus ornatus* occurring in limestone) may be an exception, being known presently only from weathered andesite outcroppings.
Indicative Species:

Reptilia:
Squamata:

Lacertilia:

1. *Coleonyx brevis* (P)*
2. *Coleonyx reticulatus* (R − ?)*
3. *Cophosaurus texanus* (P)
4. *Crotaphytus collaris* (R/C)
5. *Phrynosoma modestum* (C)
6. *Sceloporus cautus* (P)*
7. *Sceloporus maculosus* (R)*
8. *Sceloporus merriami* (R)*
9. *Sceloporus ornatus* (R)*
10. *Sceloporus poinsetti* (R)
11. *Sceloporus undulatus* (P)*
12. *Uta stansburiana* (P)*
13. *Eumeces obsoletus* (C)*
14. *Cnemidophorus gularis* (P)
15. *"Cnemidophorus tesselatus"* (P/C)*
16. *Gerrhonotus lugoi* (R − ?)
17. *Xantusia arizonae* (R)*

Serpentes:

1. *Leptotyphlops dulcis* (P)
2. *Leptotyphlops humilis* (P)*
3. *Elaphe subocularis* (R/C)*
4. *Hypsiglena torquata* (C)
5. *Masticophis taeniatus* (C)
6. *Sonora semiannulata* (C/P − ?)*
7. *Tantilla atriceps* (R/C)
8. *Trimorphodon biscutatus* (R/C)*
9. *Crotalus atrox* (P)
10. *Crotalus lepidus* (R − may be considered a relict)*
11. *Crotalus molossus* (R)

B. The Mesquite Grassland Ecotone:

COMMENT: The ecotonal nature of mesquite grassland is affirmed by the nature of its herpetofauna, a fauna characterized by edge effect, lack of ecological endemic species, and the conspicuous present of hybrid clones. The edge effect (ODUM, 1971) is manifest by a higher number of amphibian and reptilian species than in any other category of vegetation. Numerous species of otherwise very diverse associations, specifically pinyon-juniper woodland and Chihuahuan Desert scrub, occur in ecological and geographical sympatry here. The hybrid pathenogenetic clones of *Cnemidophorus* provide further evidence of habitat, heterogeneity and temporal instability (LOWE & WRIGHT, 1968). See subsequent section on Analysis of Ecologically Based Herpetofauna Assemblages for a quantified affirmation of these ecotonal conditions.

Indicative Species:

Amphibia:

Caudata:

1. *Ambystoma tigrinum* (P)*

Salientia:

1. *Scaphiopus bombifrons* (C)*
2. *Scaphiopus couchi* (P)
3. *Scaphiopus hammondi* (P)
4. *Eleutherodactylus augusti* (P)
5. *Hyla arenicolor* (P)
6. *Bufo cognatus* (P)
7. *Bufo debilis* (P)

	8.	*Bufo marinus* (P – Tamaulipan/thornscrub border)*
	9.	*Bufo punctatus* (P)
	10.	*Budo speciosus* (P)*
	11.	*Bufo valliceps* (P – mesic eastern relict)*
	12.	*Bufo woodhousei* (P)*
	13.	*Rana montezumae* (P)*
	14.	*Rana pipens* (P)
	15.	*Gastrophryne olivacea* (P)*

Reptilia:
   Testudinata:

1. *Kinosternon hirtipes* (P)*
2. *Chrysemys scripta* (P)*
3. *Terrepene ornata* (P)*
4. *Gopherus berlandieri* (P – Tamaulipan peripheral)*

   Squamata:
   Lacertilia:

1. *Coleonyx brevis* (P)*
2. *Cophosaurus texanus* (P)
3. *Hemidactylus turcicus* (P – introduced Tamaulipan peripheral)*
4. *Crotaphytus collaris* (P)
5. *Crotaphytus reticulatus* (P – Tamaulipan peripheral)*
6. *Holbrookia lacerata* (P – Tamaulipan peripheral)*
7. *Holbrookia maculata* (P)
8. *Phrynosoma cornutum* (P)
9. *Phrynosoma douglassi* (P – rest. to N)*
10. *Phrynosoma modestum* (P)
11. *Phrynosoma orbiculare* (P)*
12. *Sceloporus cautus* (P)*
13. *Sceloporus cyanogenys* (P)*
14. *Sceloporus goldmani* (P)*
15. *Sceloporus grammicus* (P)
16. *Sceloporus olivaceus* (P – rest. to E)*
17. *Sceloporus poinsetti* (P)
18. *Sceloporus scalaris* (P)*
19. *Sceloporus torquatus* (P)*
20. *Sceloporus undulatus* (P)*
21. *Sceloporus variabilis* (P – Tamaulipan peripheral)*
22. *Urosaurus ornatus* (P – rest. to N)*
23. *Eumeces multivirgatus* (P – rest. to N)*
24. *Eumeces obsoleus* (P)*
25. *Cnemidophorus burti* (P – rest. to NW filter barrier)*
26. "*Cnemidophorus exanguis*" (R – rest. to N)*
27. *Cnemidophorus gularis* (P)*
28. "*Cnemidophorus neomexicanus*" (R/C)*
29. "*Cnemidophorus sexlineatus*" (P – peripheral to NE)*
30. *Cnemidophorus tigris* (P)*
31. "*Cnemidophorus uniparens*" (R)*
32. *Heloderma suspectum* (P – filter barrier NW)*
33. "*Cnemidophorus tesselatus*" (P – rest. to N)*

   Serpentes:

1. *Leptotyphlops dulcis* (P)
2. *Leptotyphlops humilis* (P)*
3. *Arizona elegans* (P)

4. *Coluber constrictor* (P)*
5. *Diadophis punctatus* (P – rest. to riparian woodland)*
6. *Drymarchon corais* (P – peripheral to NE)*
7. *Elaphe guttata* (P)
8. *Elaphe obsoleta* (P – rest. to NE)*
9. *Ficimia cana* (P)
10. *Heterodon nasicus* (P)
11. *Hypsiglena torquata* (P)
12. *Lampropeltis getulus* (P)
13. *Lampropeltis triangulum* (P – rest. to NW)*
14. *Masticophis flagellum* (P)
15. *Masticophis taeniatus*
16. *Pituophis melanoleucus* (P)
17. *Rhinocheilus lecontei* (P)
18. *Salvadora hexalepis* (P)*
19. *Sonora episcopa* (P – rest. to NE)*
20. *Sonora semiannulata* (P)*
21. *Tantilla atriceps* (P)
22. *Tantilla gracilis* (P – rest. to NE)*
23. *Tantilla rubra* (P)
24. *Tantilla wilcoxi* (P)
25. *Thamnophis cyrtopsis* (P)
26. *Thamnophis marcianus* (P)*
27. *Opheodrys aestivus* (P – rest. to NE)*
28. *Trimorphodon biscutatus* (P)*
29. *Micruroides eurythanxus* (P – rest. to NW filter)*
30. *Sistrurus catenatus* (P)*
31. *Crotalus atrox* (P)
32. *Crotalus lepidus* (P – rest. to saxicolous substrate)
33. *Crotalus molossus* (P – rest. to saxicolous substrate)
34. *Crotalus scutulatus* (P/C)
35. *Crotalus viridis* (P)*

C. Chaparral Association:

COMMENT: This is essentially the edaphic association of better drained slopes. It alternates with a valley grassland vegetation. The conditions of drainage reduce the availability of water, resulting in a depauperate and essentially saxicolous amphibian fauna. Chaparral provides a harsh and relatively homogeneous habitat. Even microhabitats provide little in ground cover, detritus, and canopy, or water. The reptiles are represented by relatively few species, none of which are confined to this association. Only four species could even be considered characteristic.
Indicative Species:

Amphibia:
  Salientia:
1. *Eleutherodactylus augusti* (P-sax)
2. *Syrrhophus marnocki* (P-sax)*

Reptilia:
  Squamata:
    Lacertilia:
1. *Phrynosoma orbiculare* (P/C)*
2. *Sceloporus parvus* (C)*
3. *Sceloporus poinsetti* (P-sax)

	4.	*Sceloporus spinosus* (P)*
	5.	*Sceloporus undulatus* (P – ?)*
	6.	*Urosaurus ornatus* (P-sax)
	7.	*"Cnemidophorus exanguis"* (P)*
	8.	*Cnemidophorus gularis* (P)
	9.	*Cnemidophorus tigris* (P)*
Serpentes:	1.	*Leptotyphlops dulcis* (P)
	2.	*Hypsiglena torquata* (P)
	3.	*Lampropeltis mexicana* (P – ?)
	4.	*Masticophis bilineatus* (P)*
	5.	*Masticophis taeniatus* (C)
	6.	*Pituophis melanoleucus* (P)
	7.	*Salvadora grahamiae* (C/P)
	8.	*Tantilla atriceps* (C/P)
	9.	*Crotalus lepidus* (R-sax)
	10.	*Crotalus scutulatus* (P)

D. Desert Riparian Associations (Woodland & Marsh):

COMMENT: The riparian vegetation participates geographically in the Chihuahuan Desert, but it is best viewed biogeographically and ecologically as a system of peninsular and insular extensions of an Austro-riparian association. Much the same can be said of its restricted and distinctive herpetofauna. While numerous desert forms occupy outwashes and exposed shorelines, only forms that penetrate the water, marsh, or riparian woodland are included here. Abundant water and the efficiency and productivity of the lush vegetation provide the basis for a high herpetofaunal species density, as high as the much more expansive Chihuahuan scrub association.

Indicative Species:

Amphibia:		
Salientia:	1.	*Scaphiopus hammondi* (P)
	2.	*Eleutherodactylus augusti* (P – ?)
	3.	*Acris crepitans* (R)
	4.	*Bufo alvarius* (C/P)*
	5.	*Bufo cognatus* (P)
	6.	*Bufo debilis* (P)
	7.	*Bufo marinus* (P)*
	8.	*Bufo punctatus* (P)
	9.	*Bufo speciosus* (P – ?)
	10.	*Bufo valliceps* (C)*
	11.	*Bufo woodhousei* (R/C)*
	12.	*Rana catesbeiana* (R)*
	13.	*Rana pipiens* (R)*
	14.	*Gastrophryne olivacea* (P/C)*
Reptilia:		
Testudinata:	1.	*Chelydra serpentina* (R)*
	2.	*Kinosternon flavescens* (R/C)*
	3.	*Kinosternon hirtipes* (C)
	4.	*Kinosternon sonoriense* (C)*
	5.	*Chrysemys concinna* (R)
	6.	*Chrysemys gaigeae* (R)
	7.	*Chrysemys picta* (R)
	8.	*Chrysemys scripta* (R)
	9.	*Trionyx spiniferus* (R)*

Squamata		
Lacertilia	1.	*Hemidactylus turcicus* (P)*
	2.	*Sceloporus clarki* (P)*
	3.	*Sceloporus grammicus* (P)
	4.	*Sceloporus magister* (P – ?)*
	5.	*Sceloporus undulatus* (P)
	6.	*Urosaurus ornatus* (P)*
	7.	*Eumeces multivirgatus* (P)*
	8.	*Eumeces obsoletus* (P)*
	9.	*Eumeces tetragrammus* (P)*
	10	*Leiolopisma laterale* (P)*
	11.	*Leiolopisma silvicolum* (P)*
	12.	*"Cnemidophorus exanguis,"*
	13.	*Cnemidophorus gularis* (P)
	14.	*"Cnemidophorus neomexicanus"* (P)*
	15.	*"Cnemidophorus tesselatus"* (P)
	16.	*Cnemidophorus tigris* (P)
Serpentes:	1.	*Drymarchon corais* (P)*
	2.	*Elaphe guttata* (C/P)
	3.	*Lampropeltis getulus* (P)
	4.	*Lampropeltis triangulum* (P/C)*
	5.	*Masticophis bilineatus* (P)*
	6.	*Masticophis flagellum* (P)
	7.	*Masticophis taeniatus* (P)
	8.	*Natrix erythrogaster* (R)
	9.	*Natrix rhombifera* (R)*
	10.	*Thamnophis cyrtopsis* (R/C)
	11.	*Thamnophis eques* (R/C)
	12.	*Thamnophis marcianus* (P)*
	13.	*Thamnophis melanogaster* (R)*
	14.	*Thamnophis proximus* (R)*
	15.	*Thamnophis rufipunctatus* (R)*
	16.	*Thamnophis sirtalis* (R)
	17.	*Agkistrodon contortrix* (C)*
	18.	*Sistrurus catenatus* (C/P)*
	19.	*Crotalus atrox* (P)

E. Pinyon-Juniper Woodland Association:

COMMENT: This is a peripheral vegetation that also occurs as montane relict within the desert. Faunas intimately associated with it follow a similar pattern. This open, moderately arid woodland offers a fair diversity in terms of microhabitats, moisture, and canopy. Undergrowth is generally of a grassland form, giving way to chaparral on steeper slopes. Pinyon-juniper woodland association shows considerable content overlap in both flora and fauna in the contiguous grassland and chaparral associations.

Only species occurring in relict or peripheral woodland localities on the Mexican Plateau will be included here.

Indicative Species:

Amphibia:		
Caudata:	1.	*Ambystoma tigrinum* (P)*
	2.	*Eurycea neotenes* (P – Edwards Plateau)*
	3.	*Pseudoeurycea galeanae* (P)*
Salientia:	1.	*Scaphiopus hammondi* (P)
	2.	*Eleutherodactylus augusti* (P)*

	3.	*Syrrhophus cystignathoides* (P)*
	4.	*Syrrhophus dennisi*
	5.	*Syrrhophus marnocki* (R/C)*
	6.	*Syrrhophus verrucipes* (P)*
	7.	*Hyla arenicolor* (C)
	8.	*Hyla eximia* (P)*
	9.	*Bufo compactilis* (P)*
	10.	*Rana pipiens* (P)
	11.	*Rana tarahumarae* (P)*

Reptilia:
  Testudinata:

1.	*Kinosternon hirtipes* (P)
2.	*Kinosternon sonoriense* (P)*

Squamata:
  Lacertilia:

1.	*Phrynosoma douglassi* (P/C)*
2.	*Phrynosoma orbiculare* (C)*
3.	*Sceloporus clarki* (P/C)*
4.	*Sceloporus couchi* (P/C)*
5.	*Sceloporus cyanogenys* (P)*
6.	*Sceloporus grammicus* (P/C)*
7.	*Sceloporus jarrovi* (C/P)
8.	*Sceloporus merriami* (P – rare)*
9.	*Sceloporus parvus* (P/C)*
10.	*Sceloporus poinsetti* (P)*
11.	*Sceloporus scalaris* (P)
12.	*Sceloporus spinosus* (P)*
13.	*Sceloporus torquatus* (P)*
14.	*Sceloporus undulatus* (P/C)*
15.	*Sceloporus virgatus* (C)*
16.	*Urosaurus ornatus* (C)*
17.	*Eumeces brevilineatus* (P)*
18.	*Eumeces callicephalus* (P)*
19.	*Eumeces dicei* (P)
20.	*Eumeces multivirgatus* (P)*
21.	*Eucemes obsoletus* (P)*
22.	*Leiolopisma silvicolum* (P – ?)*
23.	*Cnemidophorus burti* (P)*
24.	*"Cnemidophorus exanguis"* (P)*
25.	*"Cnemidophorus flagellicaudus"* (P)*
26.	*"Cnemidophorus gularis"* (P)*
27.	*"Cnemidophorus velox"* (C)*
28.	*Gerrhonotus kingi* (C)*
29.	*Gerrhonotus liocephalus* (C)

  Serpentes:

1.	*Leptotyphlos dulcis* (P)
2.	*Diadophis punctatus* (C)
3.	*Elaphe triapsis* (C)
4.	*Lampropeltis pyromelana* (P/C)*
5.	*Lampropeltis triangulum* (P)*
6.	*Masticophis bilineatus* (P)*
7.	*Masticophis taeniatus* (P)*
8.	*Pituophis melanoleucus* (P)
9.	*Salvadora grahamiae* (P)
10.	*Tantilla atriceps* (P)
11.	*Tantilla rubra* (P)*
12.	*Tantilla wilcoxi* (C)*
13.	*Thamnophis cyrtopsis* (P)*
14.	*Thamnophis elegans* (P)*
15.	*Thamnophis eques* (P)
16.	*Trimorphodon biscutatus* (P)*
17.	*Tropidoclonion lineatum* (P)*
18.	*Crotalus lepidus* (C)

19.	*Crotalus molossus* (C/P)
20.	*Crotalus pricei* (P)*
21.	*Crotalus willardi* (P)*

F. Montane Coniferous Forest:

COMMENT: This association is the highest, coolest, and most mesic (excepting riparian) of the plant 'associations within or contiguous with the Chihuahuan Desert. It is essentially montane pine forest with depauperate vestiges of a spruce-fir association combined at its highest points. The surface area of the Mexican Plateau occupied by this vegetation is far smaller than that of most of the other categories being considered. This fact, coupled with cool and strongly seasonal nature of climate (regular winter frosts) make the montane forest the habitat of relatively few poikilothermous vertebrates.
Indicative Species:

Amphibia:
    Caudata:

	1.	*Ambystoma tigrinum* (P)*
	2.	*Aneides hardyi* (R)*
	3.	*Chiropterotriton prisca* (R)*
	4.	*Pseudoeurycea galeanae* (C)*

    Salientia:

	1.	*Hyla eximia* (C)*
	2.	*Pseudacris triseriata* (R)*
	3.	*Bufo compactilis* (R/C)*
	4.	*Rana pipiens* (P)

Reptilia:
    Squamata:
        Lacertilia:

	1.	*Phrynosoma douglassi* (P/C)*
	2.	*Phrynosoma orbiculare* (P/C)*
	3.	*Sceloporus goldmani* (C)*
	4.	*Sceloporus grammicus* (P/C)*
	5.	*Sceloporus jarrovi* (C)
	6.	*Sceloporus poinsetti* (P)*
	7.	*Sceloporus scalaris* (R/C)
	8.	*Sceloporus spinosus* (P – ?)*
	9.	*Sceloporus torquatus* (P)*
	10.	*Sceloporus virgatus* (P)*
	11.	*Eumeces brevilineatus* (P)*
	12.	*Eumeces callicephalus* (P)*
	13.	*Eumeces dicei* (P/C)*
	14.	*Eumeces multivirgatus* (P)*
	15.	*"Cnemidophorus velox"* (P)*
	16.	*Gerrhonotus imbricatus* (P/C)

    Serpentes:

	1.	*Diadophis punctatus* (P)*
	2.	*Lampropeltis pyromelana* (P)*
	3.	*Opheodrys vernalis* (P)*
	4.	*Rhadinaea crassa* (P)*
	5.	*Salvadora grahamiae* (P)*
	6.	*Storeria occiptomaculata* (R)*
	7.	*Thamnophis cyrtopsis* (P/C)*
	8.	*Thamnophis elegans* (R)*
	9.	*Thamnophis eques* (C)
	10.	*Crotalus pricei* (C)
	11.	*Crotalus willardi* (R/C)*

Analysis of Ecologically Based Herpetofauna Assemblages

Three tables have been prepared in order to quantify the degrees of distinctiveness achieved by the assemblages based on plant association presented in the previous section of this chapter.

Table 11. Absolute and Relative Frequencies of Ecologically Restricted Herpetofauna.

Association		Total Number	Number of Restricted Species	Percentage of Total Species Restricted
I.	Chihuahuan Desert Scrub			
	A. Edaphic Associations:			
	1. Alluvial	28	1	4%
	2. Barrial	39	0	0%
	3. Arenicolous	15	1	7%
	4. Saxicolous	28	11	39%
	B. General (all scrub associations)	57	15	26%
II.	Desert Riparian Vegetation	59	18	31%
III.	Mesquite Grassland	86	1[1]	1%
IV.	Pinyon-Juniper Woodland	65	1	1%
V.	Coniferous Forest	35	9	26%

1. a parthenospecies

The first, Table 11 lists the number and relative percentage of species endemic to a given association. It clearly documents the distinctiveness of the saxicolous Chihuahuan Desert herpetofauna, with thirty-nine percent of the indigenous forms being endemic. The microhabitats — cooler and often more mesic — provided by this edaphic substrate are probably crucial to its importance. GEHLBACH (1967) confirmed a similar distinctiveness for saxicolous Chihuahuan (karst) vegetation in the Guadalupe Mountains of New Mexico. Other edaphic associations of the desert scrub are relatively undistinguished, although taken as a unit twenty-six percent of the total scrub herpetofauna is restricted to that major association.

Of the five non-desert associations, only two have distinctive complements of ecological endemics. Thirty-one percent of the desert riparian herpetofauna is restricted to that association, as is twenty-six percent of coniferous forest fauna. Again, local, cool, mesic conditions and the physiological "wilting points" of the indigenous fauna are important to their restriction. Mesquite-grassland, itself most certainly an ecotone, is without unique faunal complement, save one parthenogenetic *Cnemidophorus* clone. This one exception of hybrid origin serves to further document the unstable ecotonal nature of the mesquite-grassland. The adjacent chaparral and pinyon-juniper woodland are also virtually devoid of ecological endemics. Furthermore, mesquite-grassland faunal diversity is thirty percent higher than pinyon-juniper woodland, forty percent higher than desert scrub, indicating an edge effect. All three associations are virtually local edaphic (soil dependent) interfaces of one another and they probably hold in common a general set of climatic requirements.

The second set, Table 12 as well as Figure 3, use Coefficients of Community and Similarity (C.C. & S.C.). They evaluate the degree of similarity between

Table 12. Resemblance Coefficients for Chihuahuan Herpetofauna Assemblages Correlated to Edaphic Desert Associations.
Coefficients of Community appear in upper right triangle; Coefficients of Similarity in lower left; circled figures indicate number of samples (N).

Edaphic Associations	Alluvial	Barrial	Arenicolous	Saxicolous
Alluvial	㉘	.52	.34	.33
Barrial	.82	㊴	.35	.22
Arenicolous	.73	.93	⑬	.13
Saxicolous	.39	.43	.33	㉘

edaphic associations within the Chihuahuan Desert association. Three levels of similarity can be discriminated, a fifty percent level, a thirty-three percent level, and a twenty-five percent level. At the fifty percent level only the alluvial and barrial assemblages can be clumped. At the thirty-three percent level alluvial, dune and barrial associations can be clumped together, as can saxicolous and alluvial associations. Below the twenty-five percent level only saxicolous versus barrial and saxicolous versus dune associations remain distinctive. The generally low C.C. for the saxicolous faunal assemblages again affirms the observations of GEHLBACH (1967) for vegetation, and presence of distinctive microhabitats otherwise absent from the desert. S.C. patterns are similar but indicate closer clumping of dune and barrial assemblages (see Figure 3).

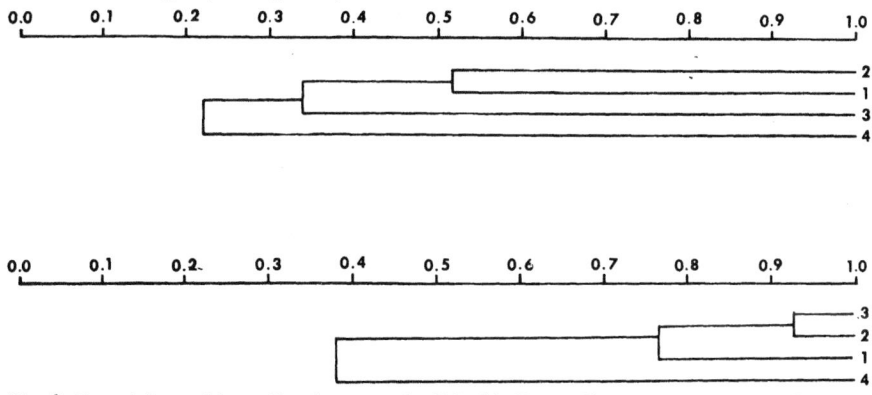

Fig. 3. Faunal Resemblance Dendrograms for Edaphic Desert Herpetofaunas. Upper Dendrogram: Coefficients of Community; Lower Dendrogram: Coefficients of Similarity; Key: 1. = Alluvial Herpetofauna; 2. = Barrial Herpetofauna; 3. = Dune Herpetofauna; 4. = Saxicolous Herpetofauna.

The third and final set of tables, Table 13 in this series, provides C.C. and S.C. comparisons for the major desert and contiguous or relict non-desert herpetofaunas of the Mexican Plateau. Several major conclusions can be drawn from this table. Firstly, some of the assemblages compared are more similar (higher C.C.) to one another than are some of the pairs of edaphic associations within Chihuahuan Desert scrub. Explicity, dune and saxicolous assemblages are more distinct from each other than eighty percent of the assemblages paired on Figure 4. Thus edaphic assemblages may constitute changes in faunas comparable in degree to those changes between desert and mesquite-grassland, or between pinyon-juniper woodland and coniferous forest.

Secondly, there is gradual and directional decline in similarities between faunas

Table 13. Resemblance Coefficients for the Vegetation Correlated with Assemblages of Chihuahuan Herpetofauna.
Coefficients of Community appear in upper right triangle; Coefficients of Similarity in lower left; circled figures indicate number of samples (N).

Associations	Chihuahuan Desert Scrub	Mesquite Grass-land	Desert Riparian Vegetation	Chaparral	Pinyon-Juniper Woodland	Montane Coniferous Forest
Chihuahuan Desert Scrub	⑤⑦	.47	.23	.15	.14	.04
Mesquite Grassland	.72	⑧⑥	.36	.10	.48	.16
Desert Riparian Vegetation	.37	.59	⑨⑨	.12	.14	.07
Chaparral	.62	.43	.43	㉑	.65	.35
Pinyon-Juniper Woodland	.26	.69	.25	.62	⑥⑤	.35
Montane Coniferous Forest	.05	.40	.17	.24	.71	㉟

as increasingly distant climatic and elevational associations are compared to desert scrub. By example, about half of the fauna of desert scrub and mesquite-grassland combined is shared by both. Scrub and desert riparian vegetation share about one-fourth. Scrub and pinyon-juniper woodland hold roughly a seventh of their faunas in common. At the extreme desert scrub and coniferous forest share approximately one species in twenty.

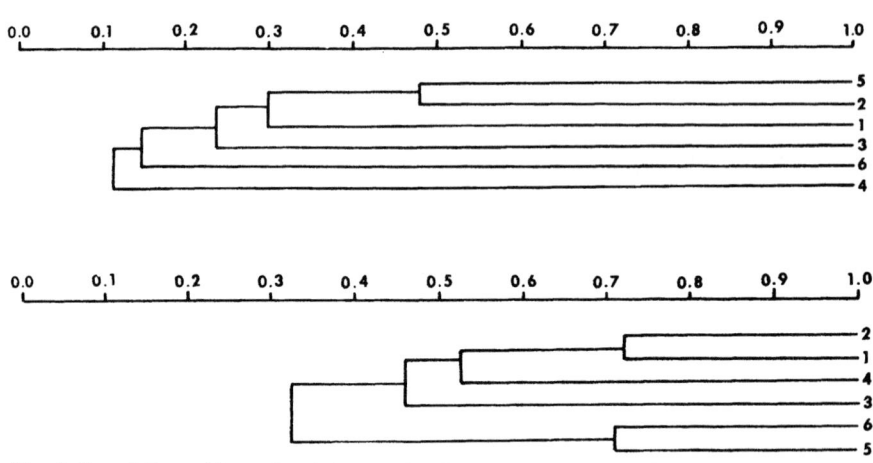

Fig. 4. Faunal Resemblance Dendrograms for the Herpetofaunas of Major Plant Associations; Upper Dendrogram: Coefficients of Community; Lower Dendrogram: Coefficients of Similarity; Key: 1. = Chihuahuan Desert Scrub Herpetofauna; 2. = Mesquite Grassland Ecotone Herpetofauna; 3. = Desert Riparian Vegetation Herpetofauna; 4. = Chaparral Herpetofauna; 5. = Pinyon Juniper Woodland Herpetofauna; 6. = Montane Coniferous Forest Herpetofauna.

Thirdly, the mesquite-grassland assemblage appears to be an essentially balanced ecotone between desert scrub and juniper woodland, with C.C. of almost fifty percent for each. It also holds a C.C. of about one-third with the desert riparian assemblage. With this added evidence, the arguments favoring treatment of the mesquite-grassland fauna as an ecotone may be summarized. This assemblage has higher simple species diversity than any other, by thirty percent minimally. It is devoid of ecological endemics, except for a single hybrid clonal form, tending to affirm its unstable composition. Furthermore, its affinities, as measured by C.C., indicate almost fifty percent species shared when combined with either adjacent association. Refer to the Vegetation account for additional purely botanical documentation.

A fourth general observation, based on this last table, is that no set of association based faunas share more than thirty-six percent of their species, excepting only the mesquite-grassland ecotone.

S.C. values indicate a simpler dendrogram of relationships, Figure 4, with arid adapted chaparral, desert, and grassland in one cluster and mesic woodland and forest in the other. In this arrangement riparian woodland is somewhat intermediate, but linked more closely to the arid group.

Habitat Displacement or Restriction

This section of the ecological account attempts to organize non-desert relict species of amphibians and reptiles into major units. Each unit provides a hypothetical model based on the distribution and ecology of the relict species. These models, Figure 5, attempt to assign geographical origins outside of the desert and link them to currently utilized refugia within the desert.

The purpose of this complex categorization is to develop a systematic picture of relict distribution patterns. Hopefully, a clarification of relict patterns will help resolve the characteristic desert herpetofauna from the remaining species. It will also provide a defined foundation for subsequent biogeographical discussions.

1. From Great Plain Grassland and Austro-Riparian Woodland to Chihuahuan Desert Lowlands: (Silaceous and Gypsum) Dune and Ephemeral Mesic Edaphic Associations (see Figure 5):

COMMENT: In some cases (i.e., *Ambystoma tigrinum, Heterodon nasicus*, and *Sistrurus catenatus*) there may simply be an *"in situ"* restriction of pluvial and riparian forms in the desiccating barrials of the desert valleys.

In other cases active displacement by competitive exclusion with better adapted desert scrubland forms is probably in effect. *Holbrookia maculata* may be displaced (to dunes and grassland) from open scrub desert by *Cophosaurus texana*. Also grassland species may be superior in locomotion in the soft shifting substrates of dunes and grassland, and thus survive there.
Examples:

Amphibia:
    Caudata: 1.   *Ambystoma tigrinum*
    Salientia: 1.   *Scaphiopus bombifrons*
                                     2.   *Bufo valliceps*
                                     3.   *Bufo woodhousei*
                                     4.   *Gastrophryne olivacea*

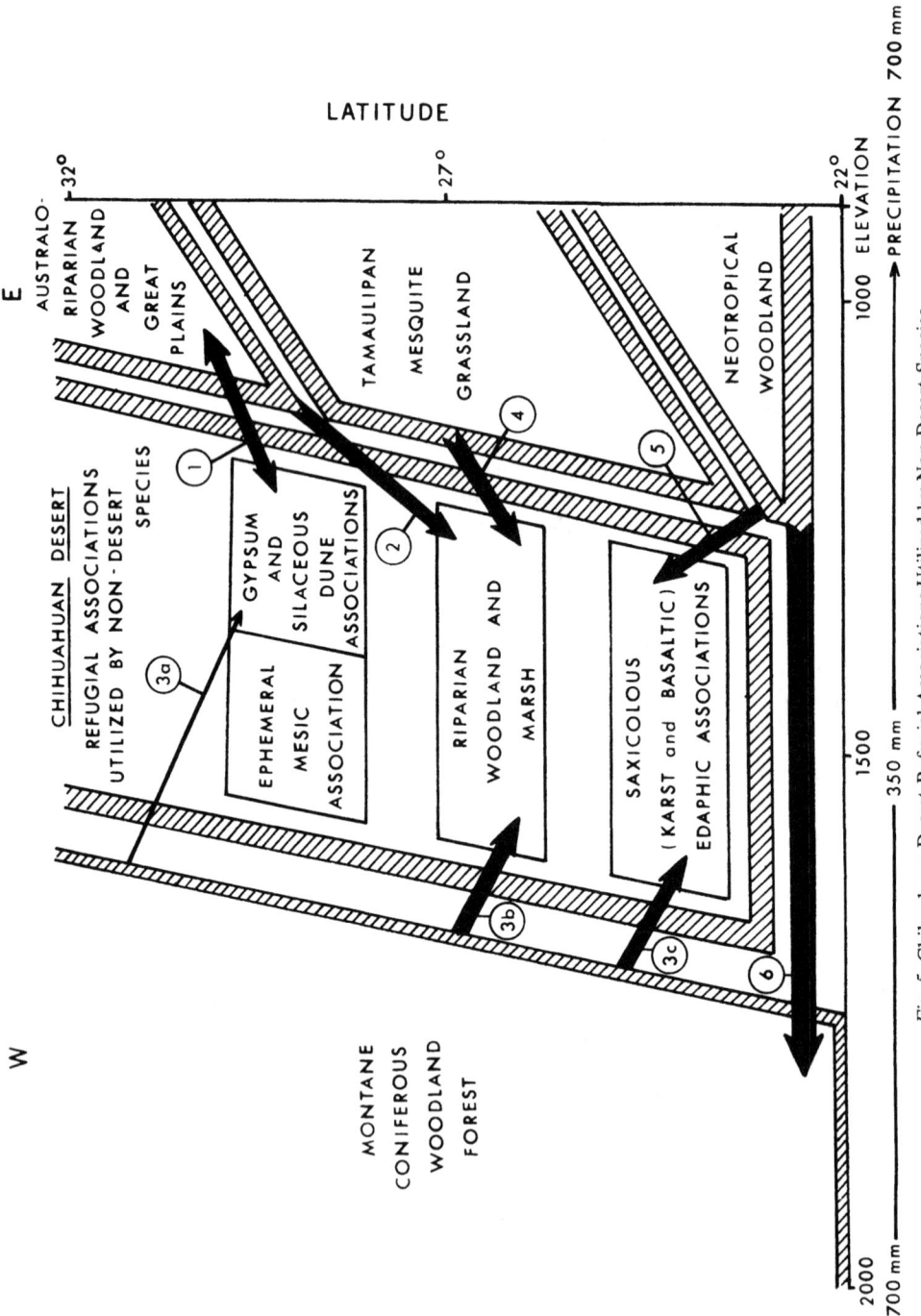

Fig. 5. Chihuahuan Desert Refugial Associations Utilized by Non-Desert Species.

Reptilia:
Testudinata:
1. *Terrepene ornata*
Squamata:
Lacertilia:
1. *Holbrookia maculata*
2. *Sceloporus undulatus*
3. *Eumeces obsoletus*
4. *Eumeces tetragrammus*
5. *Leiolopisma laterale*
Serpentes:
1. *Elaphe guttata*
2. *Elaphe obsoleta*
3. *Heterodon nasicus*
4. *Tantilla nigriceps*
5. *Sistrurus catenatus*
6. *Crotalus viridis*

2. From Great Plains and Austrabripanan Woodland to Desert Riparian Woodland and Marsh:

COMMENT: (See riparian woodland discussion in *Vegetation* Chapter). Along permanent stream courses well-developed woodlands of *Salix, Populus, Fraxinus*, and *Taxodium* – maintain homologous, if depauperate habitats with those of prairie and Gulf Coast Plain. Along the desert branches of the present Rio Grande system these riparian strips form mesic peninsulas projecting a lattice across the desert. Old derivatives of the Rio Grande, now closed basin drainage (the Rio Santa Maria and the Rio Nazas) are still essentially homologous without an endemic biota (except *T. coahuila*) on the specific level. This other situation probably involves *in situ* survival of relict populations rather than recent dispersion or displacement.

However, a dramatic example of habitat displacement is provided by the association of saxicolous *Agkistrodon contortrix* of the Big Bend National Park with riparian seep willow (*Baccharis glandulosa*) thickets (RO WAUER, personal communication). They are simply displaced by the saxicolous *Crotalus lepidus* and *C. molossus*.
Examples:

Amphibia:
Salientia:
1. *Bufo valliceps*
2. *Bufo woodhousei*
3. *Acris crepitans*
4. *Rana pipiens*

Reptilia:
Testudinata:
1. *Chelydra serpentina*
2. *Kinosternon flavescens*
3. *Chrysemys concinna*
4. *Chrysemys gaigeae*
5. *Chrysemys scripta*
6. *Terrepene coahuila*

Squamata:
Lacertilia:
1. *Sceloporus undulatus*
2. *Eumeces obsoletus*
3. *Eumeces tetragrammus*
Serpentes:
1. *Natrix erythrogaster*
2. *Natrix rhombifera*
3. *Thamnophis proximus*
4. *Agkistrodon contortrix*
5. *Sistrurus catenatus*

3a. From Montane Coniferous Woodland, Sagebrush, and Forest to Chihuahuan Desert Silaceous Dune Association:

COMMENT: Dunes involved here are peripheral to the desert in short grass prairie and may represent xeric sagebrush (*Artemisia tridentata*) relict islands in a more mesic grassland matrix. This lizard may be an unique reptilian relict of a cold-dry oriented biota.
Example:

Reptilia:
  Squamata:
    Lacertilia: 1.   *Sceloporus graciosus*

3b. From Montane Coniferous Woodland and Forest to Desert Riparian Woodland (and Marsh):

COMMENT: CONANT (1963) discussed this phenomenon ably in relation to snakes of genus *Thamnophis* in the Rio Nazas of Durango and Coahuila. In some cases the riparian descent of woodland species is so dramatic that actual dispersal along the river course seems more likely than simply relictual survival from more mesic times (e.g., *Thamnophis melanogaster*, a high montane form in the lower Rio Nazas). Beside the eastward incursions along the Rio Nazas and Rio Conchos (Chihuahua) there are also southward range extensions along the Rio Grande and Rio Santa Maria (the turtle *Chrysemys picta* and the snake *Thamnophis sirtalis*, in New Mexico and Chihuahua).
Examples:

Amphibia:
  Salientia: 1.   *Hyla arenicolor*
Reptilia:
  Testudinata: 1.   *Chrysemys picta*
                       2.   *Kinosternon sonoriense*
                       3.   *Kinosternon hirtipes*
  Squamata:
    Lacertilia: 1.   *Sceloporus grammicus*
                       2.   *Sceloporus undulatus*
                       3.   *Urosaurus ornatus*
                       4.   *Eumeces multivirgatus*
                       5.   *Eumeces obsoletus*
                       6.   *Gerrhonotus kingi*
    Serpentes: 1.   *Thamnophis cyrtopsis*
                       2.   *Thamnophis eques*
                       3.   *Thamnophis melanogaster*
                       4.   *Thamnophis rufipunctatus*
                       5.   *Thamnophis sirtalis*

3c. From Montane Coniferous Forest to Saxicolous Chihuahuan Desert Associations (Karst and Basalt Edaphic Associations):

COMMENT: Many of these forms (i.e., *Syrrhophus marnocki* and *Xantusia arizonae*) are obligatory saxicolous forms that simply survive *"in situ"* in rocky outcroppings which have been claimed geographically be expanding post-pluvial desert. Other component species, such as *Urosaurus ornatus* may have shifted (in part) to rocky perches and refuges in the absence of woodland.
Examples:

Amphibia:
    Salientia:
        1. *Eleutherodactylus augusti*
        2. *Syrrhophus marnocki*
Reptilia:
    Squamata:
        Lacertilia:
        1. *Eumeces brevilineatus*
        2. *Sceloporus jarrovi*
        3. *Urosaurus ornatus*
        Serpentes:
        1. *Lampropeltis mexicana*
        2. *Tantilla rubra*
        3. *Tantilla wilcoxi*
        4. *Crotalus lepidus*
        5. *Crotalus molossus*

4. From Tamaulipan Mesquite-Grassland to Riparian Woodland and Marsh:

COMMENT: This category overlaps with Group 3b. Subjective decisions made here consider total distribution of the species and its general habitat associations. In borderline cases highly aquatic forms (i.e., *Thamnophis proximus*) have been assigned to Austro-riparian woodland.
Examples:

Amphibia:
    Salientia:
        1. *Eleutherodactylus augusti* (saxicolous-Sierra Madre foothill)
        2. *Bufo marinus*
Reptilia:
    Testudinata:
        1. *Gopherus berlandieri*
    Squamata:
        Lacertilia:
        1. *Crotaphytus reticulatus*
        2. *Sceloporus couchi* (?)
        3. *Sceloporus olivaceus*
        4. *Eumeces tetragrammus*
        5. *Leilolopisma silvicolum*
        Serpentes:
        1. *Drymarchon corais*
        2. *Drymobius margaritiferus*
        3. *Lampropeltis triangulum*
        4. *Micrurus fulvius*

5. From Neotropical Woodland to Saxicolous (Karst and Basaltic) Edaphic Associations:

COMMENT: This category overlaps with 3c and b. Only a very few tropical woodland species actually occur as relict faunas in the Chihuahuan Desert proper, though a more considerable number inhabit southern relict pockets (for Queretero, see DIXON, KETSCHERSID & LIEB, 1972). However, the amphibian genera *Eleutherodactylus, Syrrhophus,* and the reptilian *Coleonyx* and *Leptotyphlops* should be considered saxicolous desert forms of clearly tropical silvicolous derivations.

Examples:

Reptilia:
    Squamata:
        Lacertilia: 1.   *Lepidophyma gaigae*

        Serpentes: 1.   *Trimorphodon biscutatus*

6. From Neotropical Woodland to Montane Coniferous Forest:

COMMENT: These species generally are completely absent from the desert proper. They are highly dependent on mesic micro-environments, so much so that this physiological need outstrips the high thermal preferences. Thus, they retreat above their "wilting points" despite declining average temperatures.

Examples:

Amphibia:
    Caudata: 1.   *Chiropterotriton prisca*
                                 2.   *Pseudoeurycea galeana*
Reptilia.   1.   *Kinosternon scorpioides*
    Testudinata:
        Squamata: 1.   *Eumeces brevirostris*
        Lacertilia: 2.   *Leiolopisma silvicolum*
                                 1.   *Elaphe triaspis*
        Serpentes: 2.   *Rhadinaea crassa*

Habitat Segregation and the Influence of Geographical Faciations

The tables and indicative species lists presented in previous sections are an attempt to correlate faunal distributions and plant associations. These correlations are restricted to the arid Mexican Plateau. Even so, they are subject to the physical and biotic differences between regional faciations. Three faciations discussed in relation to the Chihuahuan Desert vegetation will be used here, namely Trans-Pecos, the Mapimi, and the Salado. A more detailed examination of three representative groups, the lizard genera *Sceloporus* and *Cnemidophorus*, and the rattlesnakes, genus *Crotalus*, will serve to illustrate the nature and degree of faciation differences. In each case there is a shift in infrageneric species diversity along a latitudinal gradient, but the nature and impact of this shift differs in each instance.

The genus *Sceloporus* is centered by its distribution and species diversity in southern North America. The Mexican Plateau supports a wide diversity, which gradually increases with declining latitude to the tropics.

A total of twenty-two species occur within the desert in habitats contiguous with it, or occur exclusively as peripheral relicts. Of these species ten are present in some portion of the Trans-Pecos subdivisions. However, many are represented localized relict populations, or restricted to filter barrier situations. This division is huge geographically, constituting the northern half of the entire Chihuahuan Desert in surface area (750 kilometers along the 31st parallel, including filter barriers). The patchiness of its *Sceloporus* fauna is reflected in the fact that no

MAJOR ASSOCIATION:

ELEVATION:

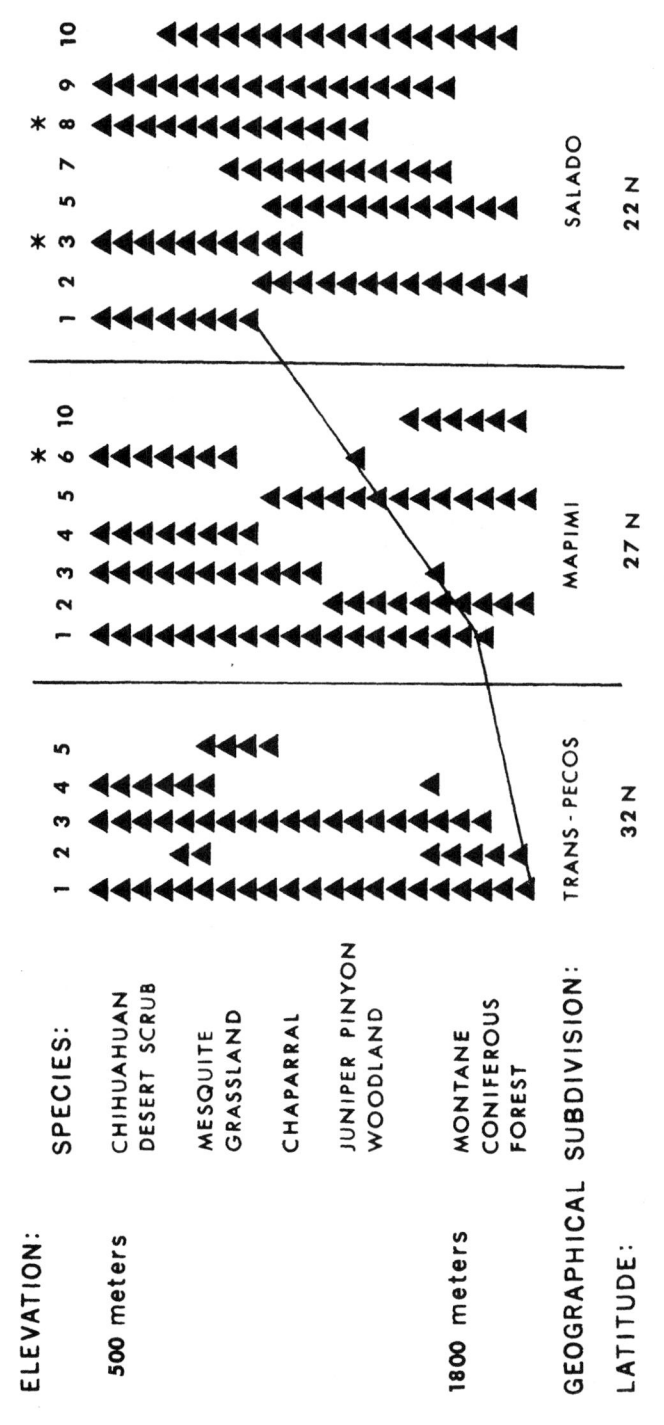

Fig. 6. Geographical Shifts in Vegetation Association of Saxicolous *Sceloporus* Species of the Chihuahuan Desert and the Contiguous Plateau; Legend: 1. *Sceloporus poinsetti*; 2. *Sceloporus jarrovi*; 3. *Sceloporus undulatus**; 4. *Sceloporus merriami**; 5. *Sceloporus cyanogenys*; 6. *Sceloporus maculosus**; 7. *Sceloporus parvus*; 8. *Sceloporus cautus**; 9. *Sceloporus spinosus*; 10. *Sceloporus torquatus*; Numbers used to label vertical distribution bars on the graph correspond to the species listed above. *3. *S. undulatus* and 8. *S. cautus* are essentially allopatric sibling species, as are 4. *S. merriami* and 6. *S. maculosus.*

107

more than three species occur sympatrically in the plant association. Furthermore, only three species are characteristically present throughout the Trans-Pecos.

The Mapimian region to the south shows a twenty percent increment in species diversity, with twelve present. As many as four species may occur in the same association sympatrically. Four species occur throughout the region.

The Salado region (and the southern relicts associated with it) comprise an area only about one-fourth the extent of the Trans-Pecos, yet it has forty percent increase in *Sceloporus* species. In the Salado region no more than three sympatric species occur in the same habitat. However, at least five species are widespread through the region geographically.

Figure 6 offers a comparison of association utilization and partitioning between saxicolous *Sceloporus* species. The vertical distribution bars of *S. poinsetti* have been joined together to emphasize the constriction in association distribution in the face of increasing species diversity to the south. At the northern edge of its range, this species is ubiquitous to all associations present. It may well be confined by physical limiting factors (i.e., temperature). At the southern end of its distribution, the correlation is obviously between increased diversity and reduced niche breadth — at least as measured in this one parameter. Here, competing species, especially *S. torquatus*, provide potential biotic limits. Thus, the species becomes nearly confined to desert scrub vegetation at the southern end of its distribution (see PIANKA, 1970 for limiting factors in *Cnemidophorus tigris*).

The situation demonstrated by *Sceloporus* and *S. poinsetti* in particular is a classical illustration of one often invoked model of increased tropical species diversity. The particular concept of increasingly narrow specialization and more complex habitat partitioning toward the tropics is almost a cliché in biogeographical speculations (see PIANKA, 1966). The next two examples emphasize the danger in over-generalizing this particular model to the exclusion of other (i.e., historical) phenomena.

The rattlesnakes, *Crotalus*, like *Sceloporus* are a genus centered in Sierra Madrean Plateau of Mexico, in terms of both diversity and geography (see KLAUBER, 1972).

The list of nine *Crotalus* species within or contiguous to the desert can reasonably be embellished with addition to two other Crotaline snakes, the copperhead, *Agkistrodon contortrix*, and the massasauga, *Sistrurus catenatus*. These last two pitvipers have considerable overlap with *Crotalus* in size, prey, habitat and microhabitat utilization. Figure 7 illustrates the relationship between geographical and habitat distribution of these forms on the arid plateau.

All nine species occur in the Trans-Pecos subdivision, inclusive of the Cochise County, Arizona filter barrier. Geographically eight of the nine species are sympatric in the vicinity of Animas Mountains of Hidalgo County, New Mexico (BOGERT & DEGENHARDT, 1961). Four of these eight may occur in the same general vegetation. Microgeographical complementarity between *C. scutulatus* and *C. viridis* may reduce the actual number to three. Only four of all nine forms are ubiquitous through most of the Trans-Pecos Desert: *Crotalus atrox, Crotalus molossus, Crotalus scutulatus,* and *Crotalus viridis.* In addition a fifth species, *Crotalus lepidus,* has an extensive, but ecologically restricted distribution across the montane and rocky portions of this region.

Crotaline snakes of the Mapimi subdivision drop to seven species from the Trans-Pecos high of nine, a twenty-two percent reduction in diversity. Rarely are

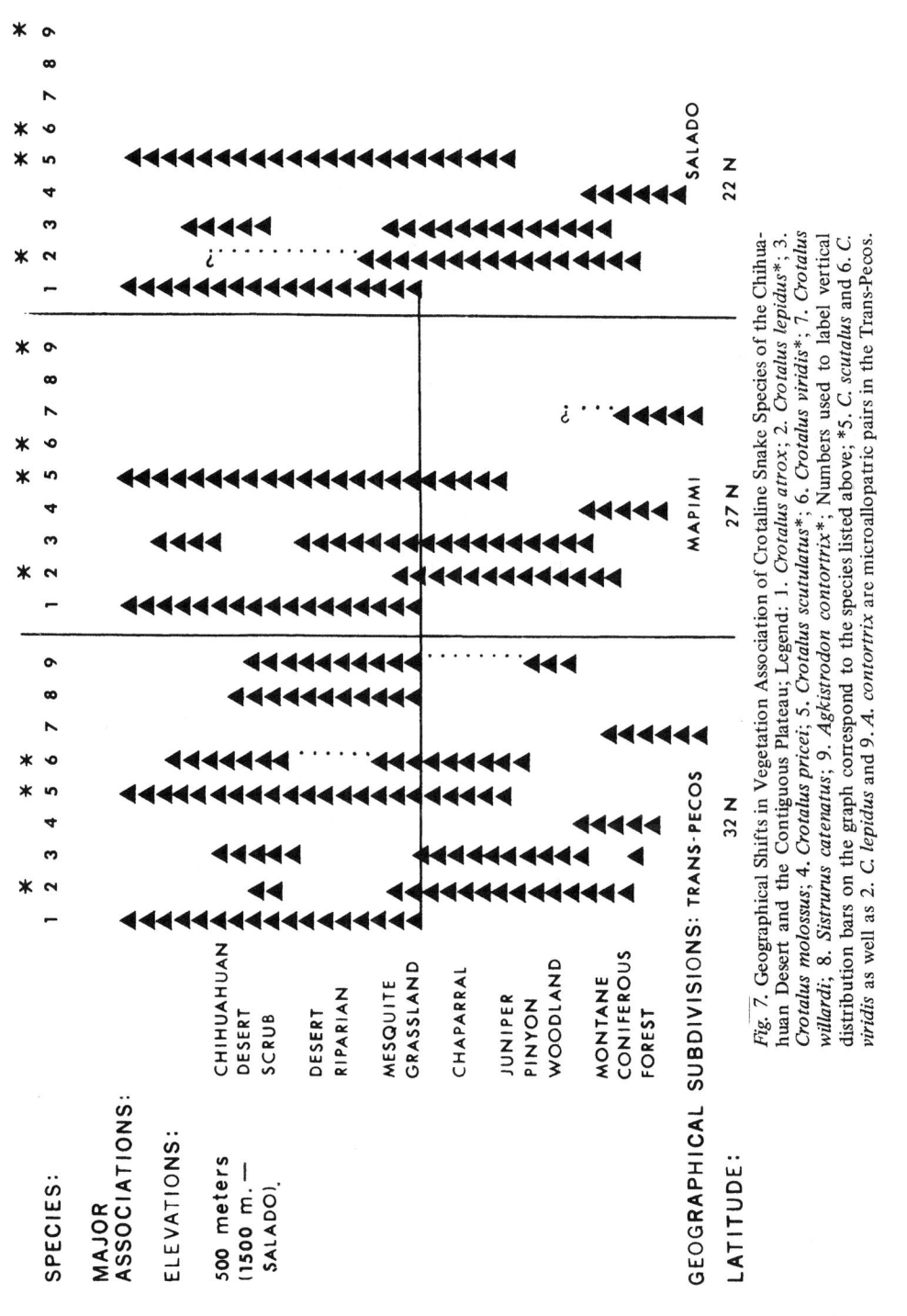

Fig. 7. Geographical Shifts in Vegetation Association of Crotaline Snake Species of the Chihuahuan Desert and the Contiguous Plateau; Legend: 1. *Crotalus atrox*; 2. *Crotalus lepidus**; 3. *Crotalus molossus*; 4. *Crotalus pricei*; 5. *Crotalus scutulatus**; 6. *Crotalus viridis**; 7. *Crotalus willardi*; 8. *Sistrurus catenatus*; 9. *Agkistrodon contortrix**; Numbers used to label vertical distribution bars on the graph correspond to the species listed above; *5. *C. scutulatus* and 6. *C. viridis* as well as 2. *C. lepidus* and 9. *A. contortrix* are microallopatric pairs in the Trans-Pecos.

more than two species ecologically and geographically sympatric. In the rocky foothills of eastern Durango (between Bermejillo and Nazas) four species occur in apparent geographical sympatry, though probably no more than three exist in the same edaphic association. Of the seven Mapimi crotalines, only *Crotalus atrox* and *scutulatus* appear to have continuous populations throughout suitable habitats. Two additional species, *C. lepidus* and *C. molossus*, have numerous and widely scattered saxicolous relict populations across the Mapimi region.

The Salado region sustains but five pit vipers. This represents a reduction to almost half of the Crotaline fauna of the Trans-Pecos. Even compared to the seven Mapimi forms, this indicates a thirty percent decline. A maximum of three species occur in geographical sympatry within the same association — and then only when local outcropping sustain pockets of *C. molossus* along with characteristic creosote scrub forms *C. atrox* and *C. scutulatus*. Only these three forms are widespread in the Salado region.

At this point it is worth noting that the Crotaline snakes essentially reverse the classical trend toward increased tropical species diversity. In contrast to *Sceloporus*, the species diversity in this group (even within *Crotalus* alone) declines toward the tropics. Furthermore, the breadth of vegetation distribution of most Crotalines (see *C. atrox* in Figure 7) are unaltered by changes in species diversity or changes in latitude. Only in one pair of species, *C. scutulatus* and *C. viridis*, does there occur a latitudinal shift in habitat utilization. When geographically overlapping the two forms are microallopatric in grassland and desert scrub pockets (see BOGERT & DEGENHARDT, 1961). This occurs only in the Trans-Pecos. The establishment of one species or the other seems to have no consistent ecological pattern. South of the Trans-Pecos only *C. scutulatus* occurs and its ecological distribution is much more predictable and continuous. Otherwise the only indications of partitioning or displacement are between *Crotalus lepidus* and *Agkistrodon contortrix*. The latter form appears to have become a secondary "moccasin", dwelling in riparian marshes and excluded from its typical wooded rock outcroppings by the *Crotalus*, at least in the Big Bend region.

The complexity and diversity of the Trans-Pecos Crotaline fauna cannot be dismissed as simple functions of its greater area relative to the two other subdivisions. Eight of the nine species occur within a sixty kilometer radius in the Animas Mountains (BOGERT & DEGENHARDT, prev. cit.). Furthermore, area inequities do not reverse the trend toward higher tropical diversity in *Sceloporus*.

Crotaline snakes are not unique among Chihuahuan genera in their ecological and geographical distribution pattern. The skinks, genus *Eumeces*, and the snake genus, *Tantilla*, conform to the same general pattern of highest diversity at northern ends of the desert.

The lizard genus *Cnemidophorus*, Figure 8, provides an accentuated version of this Crotaline pattern, with the added component of hybrid clones, often termed parthenospecies.

Again, the Trans-Pecos endowment of forms is not only the greatest in diversity, nine, but no additional forms contribute to the fauna further south. Of these nine as many as five are geographically sympatric. Four of these may occur in the same vegetation, especially along the riparian borders of the Rio Conchos, east of Julimes in Chihuahua (*C. gularis*, "*C. exanguis*", "*C. tesselatus*", *C. tigris*). Three forms occur throughout most of the Trans-Pecos Desert. These are as

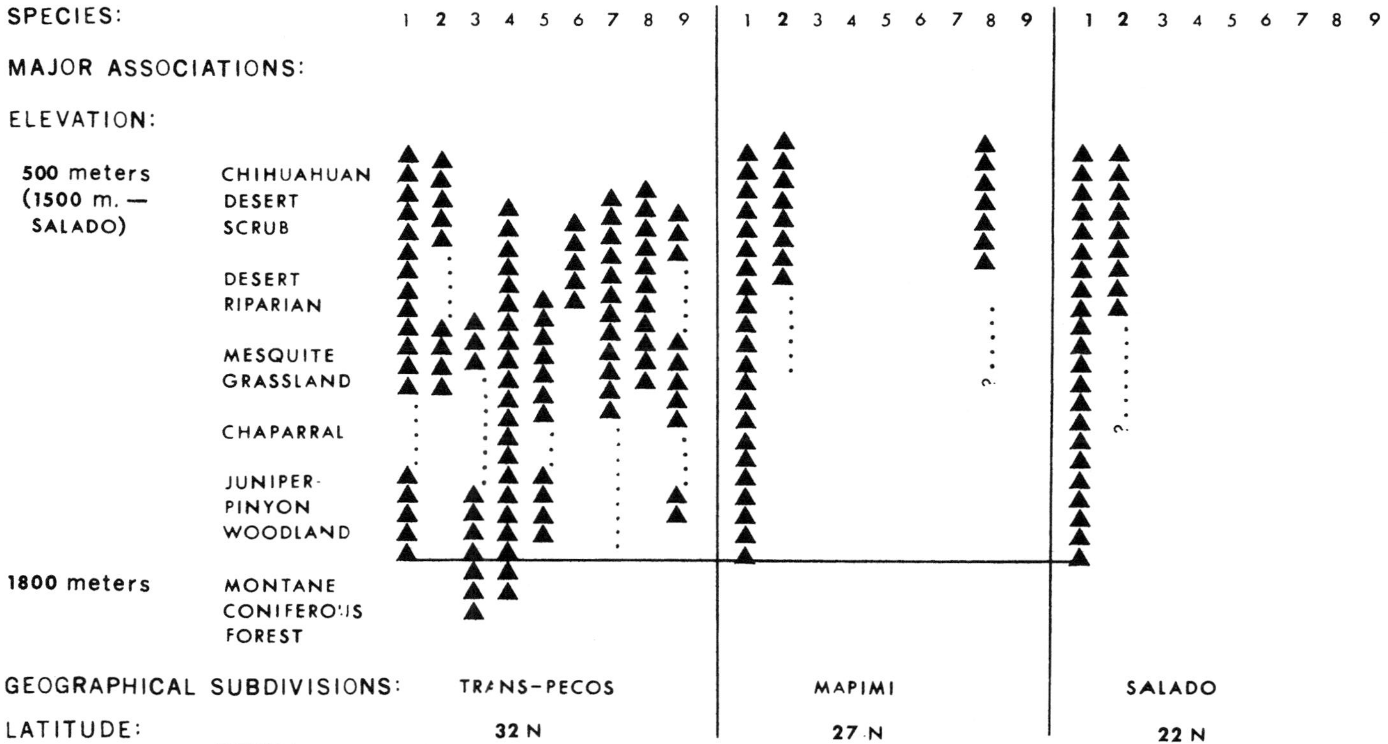

Fig. 8. Geographical Shifts in Vegetational Association of the Lizards Genus *Cnemidophorus* in the Chihuahuan Desert and the Continguous Plateau; Legend: 1. *Cnemidophorus gularis*; 2. *Cnemidophorus inornatus*; 3. *Cnemidophorus burti*; 4. *"Cnemidophorus exanguis"*; 5. *"Cnemidophorus flagellicaudus"*; 6. *"Cnemidophorus neomexicanus"*; 7. *"Cnemidophorus tesselatus"*; 8. *Cnemidophorus tigris*; 9. *"Cnemidophorus uniparens"*; Numbers used to label vertical distribution bars on the graph correspond to the species listed above.

follows: *C. inornatus, C. tigris, "C. tesselatus"*. In addition *C. gularis* is widespread in the eastern half of these regions. *"C. exanguis"* and *"C. uniparens"* are similary widespread in the west. It is noteworthy that five of these forms are in fact clonal hybrid populations, functioning as specific entities in an ecological sense only.

The three *Cnemidophorus* of the Mapimi region are all good sexual species. All three occur in geographical and occasional ecological sympatry. However, *C. gularis* predominates in more mesic settings, riparian woodland, mesquite arroyos, and upland slopes. *C. tigris* tends to replace it in more xeric microhabitats, especially dunes, alkaline flats, and other dry barrial floor associations. *C. inornatus* is so much smaller that it is not in active food competition with adults of either species. It coexists ecologically with both, but favors the lower and more xeric associations (except those inhabited by *C. tigris*). The deterioration in diversity relative to the Trans-Pecos fauna is a dramatic sixty-seven percent, from nine to three. Remember this decline is toward the tropics.

The Salado sustains but two (both sexual) species, a reduction of thirty-three percent of the Mapimi fauna, and a drop off of nearly eighty percent from the Trans-Pecos. No new species enter this region. No major alterations in habitat utilization, i.e., niche breadth, are indicated (see *C. gularis*).

To some extent, the Chihuahuan anurans *Bufo* and *Scaphiopus* conform to this pattern of sharp southern decline in diversity. The possible causes of these patterns, both ecologically and historically, will be investigated in later chapters on biogeography.

It should be re-empasized that the surprising condition demonstrated in the *Crotalus-Cnemidophorus* pattern cannot be assigned simply to the greater surface area of Trans-Pecos Desert. In both instances, the diversity of the smaller Cochise filter barrier at the western edge of the desert exceeds that of the Mapimi and Salado Deserts. Neither can it be claimed that these genera are northern or centered in distribution far beyond southern North America. All three genera, *Sceloporus, Crotalus* and *Cnemidophorus* are good Sierra Madrean Plateau stocks, or at least Mexican centered in diversity and distribution (*Cnemidophorus*). Nor do I believe that intergeneric competition provides an adequate explanation.

The possible causal agents molding these three patterns are both ecological and historical (paleogeographical and paleoecological). Some relevant factors are beyond the scope of this investigation (i.e., primary productivity of local ecosystems); others, especially historical, will be evaluated in the subsequent biogeographical analysis.

IX. A SPATIAL ANALYSIS OF THE CHIHUAHUAN HERPETOFAUNA

Objectives

The preceding sections have been directed toward the development of the operational concepts: the generation of ecological and geographical definition of the Chihuahuan Desert (culminating in the base map), the enumeration of the geographical Chihuahuan herpetofauna (i.e., the Checklist), and finally the refinement of the geographical herpetofauna into a participant Chihuahuan Desert herpetofauna by the use of ecological criteria (see previous chapter). The concept of a participant fauna as applied here may be defined as that faunal assemblage which actually participates in the dominant ecosystem of a particular geographical unit. For example, the 57 species of the Chihuahuan Desert Scrub Association constitute the participant herpetofauna of the Chihuahuan Desert. The importance of making this distinction is illustrated by the fact that only one-third of the 160 geographically present species contribute to the herpetofauna participating in the dominant ecosystem which occupies over 90% of the surface area of the geographical desert.

Sufficient foundation has now been established for a spatial biogeographical evaluation of the Chihuahuan Desert base map in terms of the regional herpetofauna. This section analyzes specific faunal distributions which, when quantified, ascertains the validity of the base map as a herpetofaunal primary area. Further evaluation will attempt to establish the status of this primary area quantitatively (using criteria of HAGMEIER & STULTS, 1964) as a faunal unit (i.e., province, subprovince, subregion, etc.), its internal homogeneity, and its affinities with other primary areas of equal status.

For the determination of the primary areas all terrestrial herpetofaunas in the region will be considered. However, for purposes of comparison within and between primary units, only the participant herpetofaunas of the dominant ecosystems of each of those units will be compared.

Parallel analyses of riparian and montane relict (participant) herpetofaunas within the geographical desert will also be evaluated. Essentially, these herpetofaunal distributions are being employed as biological litmus to test the resolution, content, and derivation of proposed biogeographical units, the Chihuahuan Desert in particular. The climate sensitivity and limited mobility of herpetofauna recommend them highly for this task.

From MORAFKA (1977), I will quote here in a abbreviated form: "the determination of primary areas seeks further to match the initial area generated against the broad set of species range maps developed in chapter three. Specifically, the limits of primary faunal areas in the general position of the Chihuahuan Desert, namely on the north Mexican Plateau, are determined here by establishing points of maximal faunal change. These points may be easily located by calculating the Index of Faunal Change (i.e., the percentage of the indigenous fauna whose distribution terminated in that unit of space) for a series of equal squares transecting the entire region under consideration. The squares with maximal IFC values fix the positions of maximal changes in the regional fauna. By joining these points, a primary faunal area may be enscribed and thus defined. If the primary area enscribed for herpetofauna corresponds to the Chihuahuan Desert base map constructed on other evidence, it would strongly affirm the general biogeographical relevance of the base map and its underlying definition".

Employing the strategy outlined above, I have constructed the following simplified method for tentatively resolving the primary area or areas within the approximate region of the Chihuahuan Desert.

The Mexican Plateau has been skewed across its greatest breadth by latitude 30 degrees north, and greatest length by longitude 104 degrees west. Along these lines transecting rows of 80 kilometer (per side) squares have been imposed on the base map covering the inner plateau and adjacent highlands (see Map 4). In addition, parallel and perpendicular transects have been drawn east about one-half degree above the 22nd parallel. These tangential transects were drawn in order to evaluate the transitional filter barriers depicted on the base map. These transects circumvent the use of grid over the entire Mexican Plateau. Though they are less thorough, two advantages favor their use. Firstly, they are much faster, and effectively sample the major axes of the region, and secondly, they are coincidently set across the best sampled portions of north Mexican Plateau (i.e., Cochise County, Big Bend, Sierra del Nido, etc.).

After the positioning of transecting squares was determined, the IFC values for those squares were calculated.

Using the IFC values from the squares on the grid transect, Map 5, I have tested the predictive ability of the initial base map (Map 3). If the base map accurately defines the biologically relevant borders of the desert, the discontinuities in herpetofaunal distributions should be maximal along these borders, intermediate across filter barriers, and minimal within the stable core of the desert. IFC values are a direct measure of the relative discontinuity of the herpetofaunas. Thus these values should correspond directly to borders, transitions, and the core region of the desert depicted on the base map. The Mann-Whitney test (a non-parametric rank-sum test) was employed to evaluate the null hypothesis that the IFC values of squares lying on the base map desert borders would be equal to those that were not. Acceptance of the null hypothesis would not decisively reject the existence of a biologically significant Chihuahuan Desert, only the specific definition used here, and only in its relation to herpetofaunal distribution patterns. Errors in choice of parameters, their limits, their mapped representation, or errors in faunal range maps could also cause an acceptance of the null hypothesis.

The results of this attempted correlation between the base map and the primary area may be summarized by the following quote from a more extensive discussion (MORAFKA, 1977):

"The IFC values on the crucifix indicate transitional faunal changes on the north and south of the Mexican Plateau with contrasting abrupt changes along the east and west slopes. The results are not surprising in light of topography, climatic patterns (especially precipitation) and vegetation. All four of these peripheral zones do generally correspond to the borders set for the initial area and could be joined to rationalize a single primary area identical to the Chihuahuan Desert base map of Map 2.

The second objective, already explained in the procedural section, does affirm the predictive value of the initial base map in setting the positions of maximal

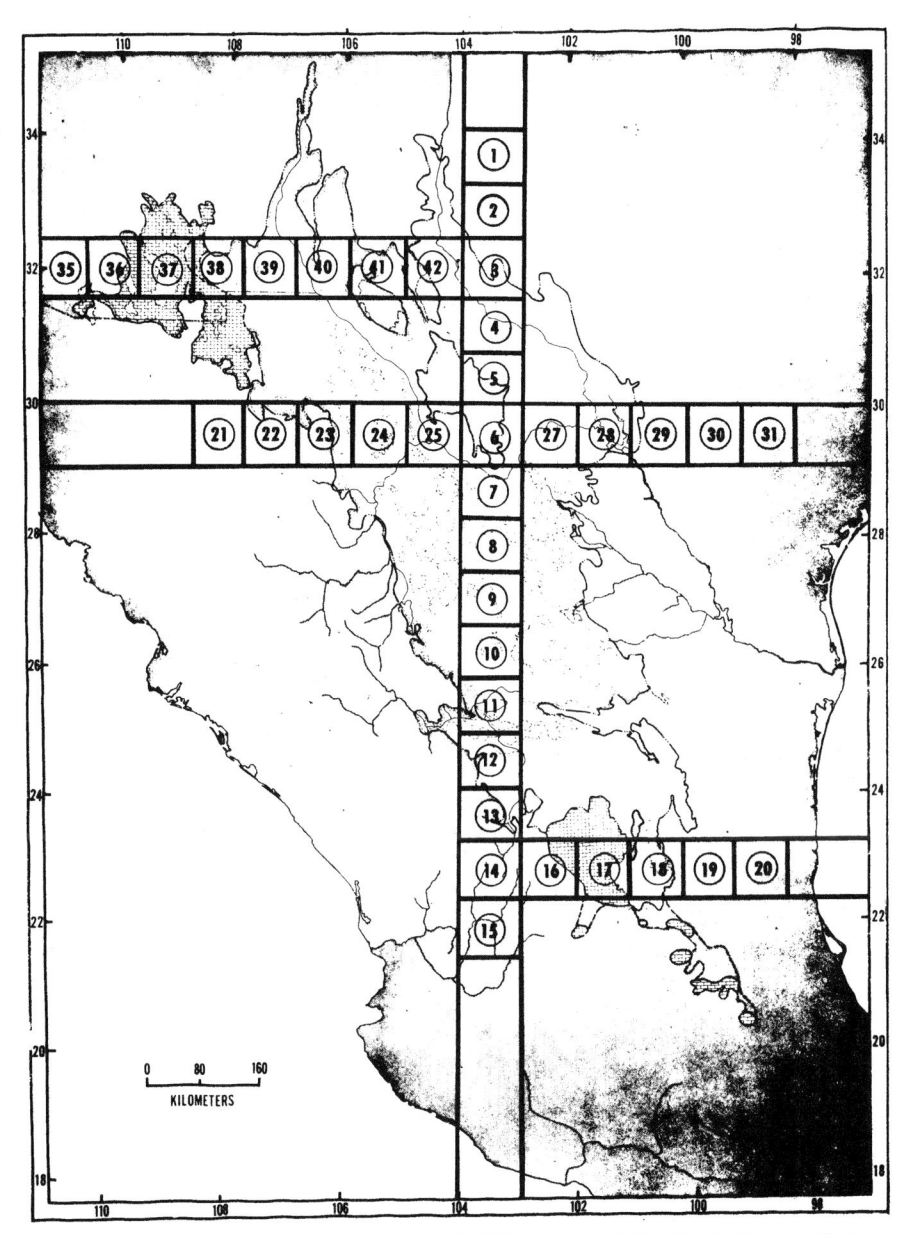

Map 4. Transecting Grid of the Longest and Widest Axes of the North Mexican Plateau. Explanation: each square is 6400 kms ^{2}; numbers serve only to identify squares of the grids.

faunal change. Specifically, the Mann-Whitney test rejects the null hypothesis at the $P \leqslant .05$ level of significance. The test does not effectively resolve, however, whether the initial base map inscribes one primary area or several."

Preliminary Evaluation of the Primary Area

The analytical procedure evaluates the significance of the Chihuahuan Desert as a herpetofaunal unit by three separate measures: one, correlation of the initial base map with natural faunal breaks just mentioned, two, the evaluation of internal and external affinities by means of faunal resemblance coefficients, and three, the degree of faunal differentiation as indicated by endemism.

Several additional evaluations of the primary area were also employed. After the primary area was established by the IFC test, its relative status as a biogeographical unit was tested against adjacent provinces using Coefficients of Community (C.C.) and Similarity of Coefficients (S.C.).

Table 14. Herpetofaunal Resemblance Between the Chihuahuan Desert and Adjacent Biotic Provinces.

Province	Coefficient of Community	Coefficient of Similarity
1. Kansan (Great Plains)	.32	.55
2. Balconian (Edwards Plateau)	.42	.59
3. Tamaulipan	.22	.37
4. Sierra Madre Oriental	.09	.18
5. Vera Cruzian	.04	.11
6. Trans-Volcanic	.09	.18
7. Sierra Madre Occidental	.07	.13
8. Sonoran	.32	.48
9. Navahonian	.12	.26

When C.C. values are used to test the Chihuahuan biotic province, they unequivocally affirm its validity at the province level of 0.65 as determined by HAGMEIER & STULTS (1964). The Chihuahuan province is compared to its adjacent vertebrate provinces, derived by intuitive and quantitative means (HAGMEIER, 1966; STUART, 1964) in Table 14. All nine contiguous faunal units proved to be distinct at the province level. The Edwards Plateau with a C.C. value of 0.42 was the only unit with which the Chihuahuan Desert could be combined at even the superprovince level. The Chihuahuan Desert shares smaller common herpetofaunas with the Sonoran Desert and the Kansas plains provinces, 0.32 in both comparisons.

The same comparisons were conducted using S.C. values, and while they emphasize the relative importance of shared faunas, the same general affinities and distinctions are expressed.

The Chihuahuan Desert herpetofauna was also compared to those of other North American deserts (adjacent and otherwise) using the same coefficients, as employed above. The matrices presented in Table 15 provide additional evidence for the provincial validity of the Chihuahuan herpetofauna, and for the internal homogeneity of the province. C.C. values between Chihuahuan and western deserts never exceeded 0.35. S.C. values again accentuated shared faunas but revealed the

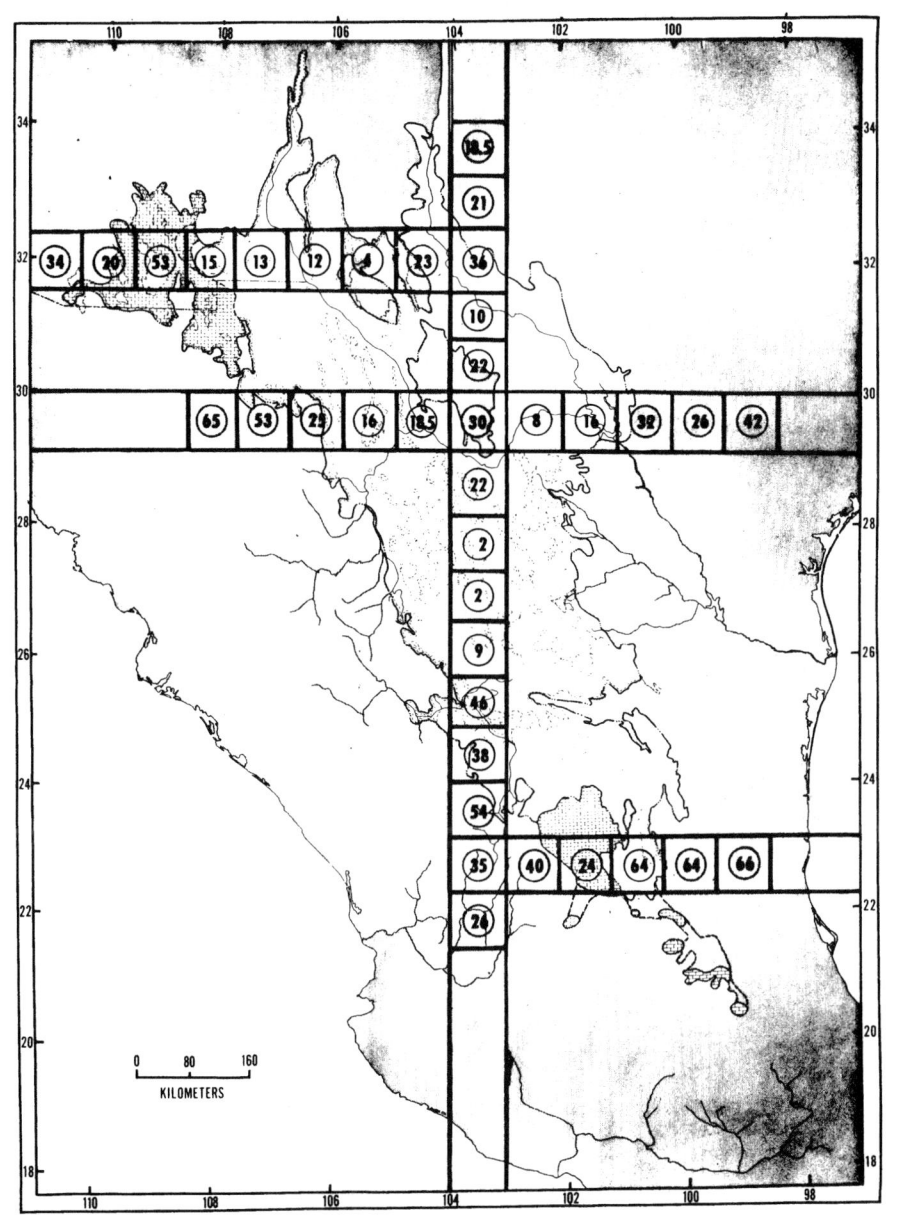

Map 5. Indices of Faunal (Herpetofaunal) Change for the Squares of the Mexican Plateau Grid. Key: numbers are percentage values for IFC.

117

same comparative pattern. Furthermore, the subdivisions of the Chihuahuan Desert clustered much more tightly than did the western deserts, generally at C.C. levels of 0.60. While these C.C. values are borderline between provincial and subprovincial levels of differentiation, S.C. values indicate much stronger affinities. The Saladan fauna is almost completely shared with Mapimian and Trans-Pecos (96%). Dendrograms on Figure 9 make explicit the clustered relationship between the three subdivisions and favors their categorical status as subprovinces.

The Chihuahuan province defined here is also evaluated for faunal endemism. The results are expressed in Tables 16 and 17. One-third of the entire participant herpetofauna has more than half its range contained within the defined Chihuahuan province. In addition, 79% of this third are endemic to the Chihuahuan Desert (isolated relict populations excluded). A ubiquitous desert and plains fauna is second in importance with 26%. A Sonoran province centered fauna (i.e., their distributions being more than 50% within the Sonoran Desert by area) is third with

Table 15. Matrix of Resemblance Coefficients for the Compared Herpetofaunas of North American Deserts.
Coefficients of Community appear in upper right triangle; Coefficients of Similarity in lower left; circled figures indicate number of samples (N).

	Trans-Pecos	Mapimian	Saladan	Sonoran	Peninsular	Mohavian
Trans-Pecos	㊼	.61	.52	.32	.18	.27
Mapimian	.75	㊽	.57	.35	.22	.34
Saladan	.96	.23	㉓	.12	.12	.18
Sonoran	.48	.46	.52	㊽	.36	.52
Peninsular	.31	.35	.30	.54	(51)	.49
Mohavian	.55	.61	.33	.87	.87	㉛

15%. Clearly, the Chihuahuan Province is not simply an arid overgrazed extension of the plains nor just an eastern aspect of the Sonoran Desert. Its largest herpetofaunal unit is Chihuahuan centered or endemic. This further affirms the biological significance of the base map and the unique nature of the indigenous herpetofauna.

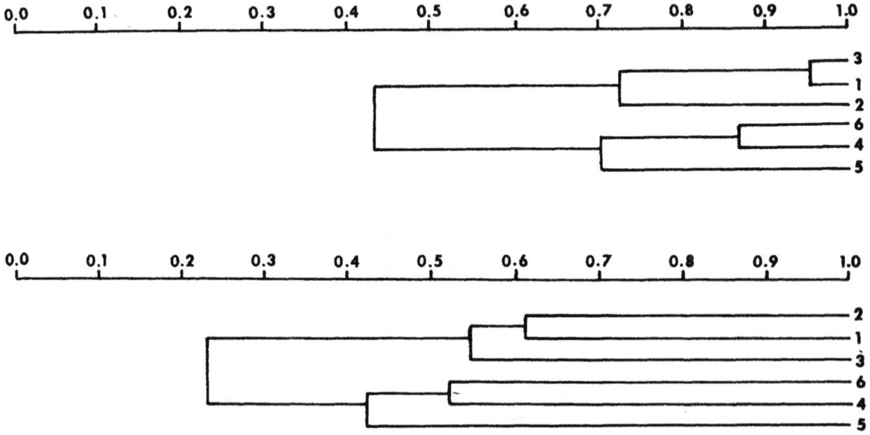

Fig. 9. Faunal Resemblance Dendrograms for the Herpetofaunas of North American Deserts; Upper Dendrogram: Coefficients of Similarity; Lower Dendrogram: Coefficients of Community; Key: 1. = Trans-Pecos Desert Herpetofauna; 2. = Mapimian Desert Herpetofauna; 3. = Saladan Desert Herpetofauna; 4. = Sonoran Desert Herpetofauna; 5. = Peninsular Desert Herpetofauna; 6. = Mohavian Desert Herpetofauna;

Table 16. A Distributional Classification of the Herpetofauna of the Chihuahuan Desert Scrub Association.

Nearctic Region:
A. Ubiquitous:
    Amphibia:
        Caudata: 1. *Ambystoma tigrinum*
    Reptilia:
        Lacertilia: 1. *Scleroporus undulatus*
        Serpentes: 1. *Lampropeltis getulus*
                          2. *Masticophis flagellum*
                          3. *Pituophis melanoleucus*

B. American Highland Sub-Region:
  I. Sierra Madre Super Province
      1. Ubiquitous:
          Reptilia:
             Lacertilia: 1. *Cnemidophorus gularis*
                          2. *Sceloporus grammicus*
             Serpentes: 1. *Crotalus molossus*
      2. Arizonian Province:
          Reptilia:
             Lacertilia: 1. *Xantusia arizonae*
  II. Desert and Plains Super Province
      1. Ubiquitous (or in a majority of provinces):
          Amphibia:
             Salientia: 1. *Scaphiopus couchi*
                          2. *Scaphiopus hammondi*
                          3. *Bufo cognatus*
                          4. *Bufo debilis*
                          5. *Bufo punctatus*
                          6. *Gastrophryne olivacea*
          Reptilia:
             Lacertilia: 1. *Crotaphytus collaris*
                          2. *Phrynosoma cornutum*
                          3. *Cnemidophorus gularis*
             Serpentes: 1. *Leptotyphlops dulcis*
                          2. *Arizona elegans*
                          3. *Hypsiglena torquata*
                          4. *Masticophis taeniatus*
                          5. *Rhinocheilus lecontei*
                          6. *Thamnophis marcianus*
                          7. *Crotalus atrox*
      2. Great Plains Province:
          Reptilia:
             Testudinata: 1. *Terrepene ornata*
             Lacertilia: 1. *Eumeces obsoletus*
                          2. *Holbrookia maculata*
             Serpentes: 1. *Heterodon nasicus*
      3. Chihuahuan Province:
          Reptilia:
             Testudinata: 1. *Gopherus flavomarginatus*
             Lacertilia: 1. *Coleonyx brevis*
                          2. *Coleonyx reticulatus*
                          3. *Cophosaurus texanus*
                          4. *Phrynosoma modestum*
                          5. *Sceloporus cautus*
                          6. *Sceloporus maculosus*
                          7. *Sceloporus merriami*

Table 16. A Distributional Classification of the Herpetofauna of the Chihuahuan Desert Scrub Association (Continued).

	8. *Sceloporus ornatus*
	9. *Scleoporus poinsetti*
	10. *Uma exsul**
	11. *Cnemidophorus inornatus*
	12. *"Cnemodophorus neomexicanus"*
	13. *"Cnemidophorus tesselatus"*
	14. *Gerrhonotus lugoi*
Serpentes:	1. *Elaphe subocularis**
	2. *Ficimia cana**
	3. *Tantilla atriceps**
	4. *Crotalus scutulatus*
4. Tamaulipan Province:	
Amphibia:	
Salientia:	1. *Bufo speciosus*
5. Sonoran Province:	
Reptilia:	
Lacertilia:	1. *Crotaphytus wislizeni*
	2. *Sceloporus magister*
	3. *Uta stansburiana*
	4. *Cnemidophorus tigris*
	5. *Xantusia vigilis*
Serpentes:	1. *Leptotyphlops humilis*
	2. *Salvadora hexalepis*
	3. *Sonora semiannulata*
	4. *Trimorphodon biscutatus* (somewhat arbitrary — alternatively western Mexican province

* endemics.

In summary, five lines of quantitative evidence support the validity of the Chihuahuan faunal province as depicted by the base map. These evidences are as follows: (1) IFC values determined for 80 kilometer squares along the longest axis of the Mexican Plateau indicate that regions of maximal and minimal faunal change occur as predicted by the base map. (2) The primary area affirmed by the IFC values is also distinctive from all contiguous provinces at the 0.60-0.65 level for C.C. values. This qualifies the Chihuahuan Desert as a faunal province by the quantitative standards established by HAGMEIER, (1966). (3) A quantified comparison of all North American deserts, including Chihuahuan subdivisions, employed both S.C. and C.C. values. It affirmed the provincial level distinctiveness of the primary area relative to other deserts. (4) The three Chihuahuan Desert subdivisions cluster together at C.C. values of about 0.60. They are far less differentiated from one another than is the Chihuahuan Desert differentiated from any of the western American deserts. Furthermore, the most distinctive subdivision by C.C. value, the Saladan Desert, is revealed by S.C. value to harbor a herpetofauna almost totally shared with the adjacent Mapimian Division. It differs only in the depauperate state of its fauna. It is concluded that the Chihuahuan Desert has great internal homogeneity, and that its subdivisions differ from one another only at the subprovince level. (5) The defined Chihuahuan Desert was also analyzed in terms of the general distributions of its herpetofauna. Species with Chihuahuan centered distributions made the largest single contribution with 33%, of which 79% are endemics.

Table 17. Recent Distributional Affinities of the Chihuahuan Desert Scrub Association Herpetofauna.

Category	Chihuahuan Province (N:58)	Trans-Pecos Sub-Province (N:47)	Mapimian Sub-province (N:48)	Saladan Sub-province (N:29)
Nearctic Region				
A. Ubiquitous	9%	11%	8%	14%
B. American Highland Sub-Region				
I. Sierra Madre Super-province				
1. Ubiquitous	5%	4%	4%	10%
2. Arizonian Province	2%	0%	4%	0%
II. Desert and Plains Super-province				
1. Ubiquitous	26%	32%	31%	41%
2. Kansan (Great Plains) Province	8%	9%	7%	7%
3. Chihuahuan	33%	25%	31%	28%
(a. Endemics)	26%	6%	12%	3%
4. Tamaulipan Province	2%	2%	0%	0%
5. Sonoran Province	15%	17%	17%	0%

Subdivisions of the Chihuahuan Desert

I have recognized the same subdivisions for herpetofaunas for reasons stated in a separate analysis (MORAFKA, 1977). Table 18 makes a brief review of the characteristics of each subdivision. The Trans-Pecos is the most poorly defined unit being transitional with prairie and populated by parthenogenetic clonal *Cnemidophorus* of hybrid origin. The Mapimian subprovince appears to be the center of endemism -- the stable core of the desert. The Saladan Subdivision is virtually devoid of endemics. It is little more than a depauperate (and recent) extension of the Mapimian Desert fauna.

All three units cluster appropriately at the subprovincial level (of C.C.) as defined by HAGMEIER & STULTS (1964) as seen in Table 15.

The Affinities of the Chihuahuan Desert Herpetofauna with those of Adjacent Non-Desert Provinces

These relationships have already been alluded to piecemeal in resolving the primary area and in the previous resolution of subprovinces. The C.C. and S.C. values as indicators of faunal affinity are listed in Table 14, and the percentage breakdown of the Chihuahuan Desert herpetofauna in terms of regional derivation is provided in Table 17. This review will start with the Kansan Province and

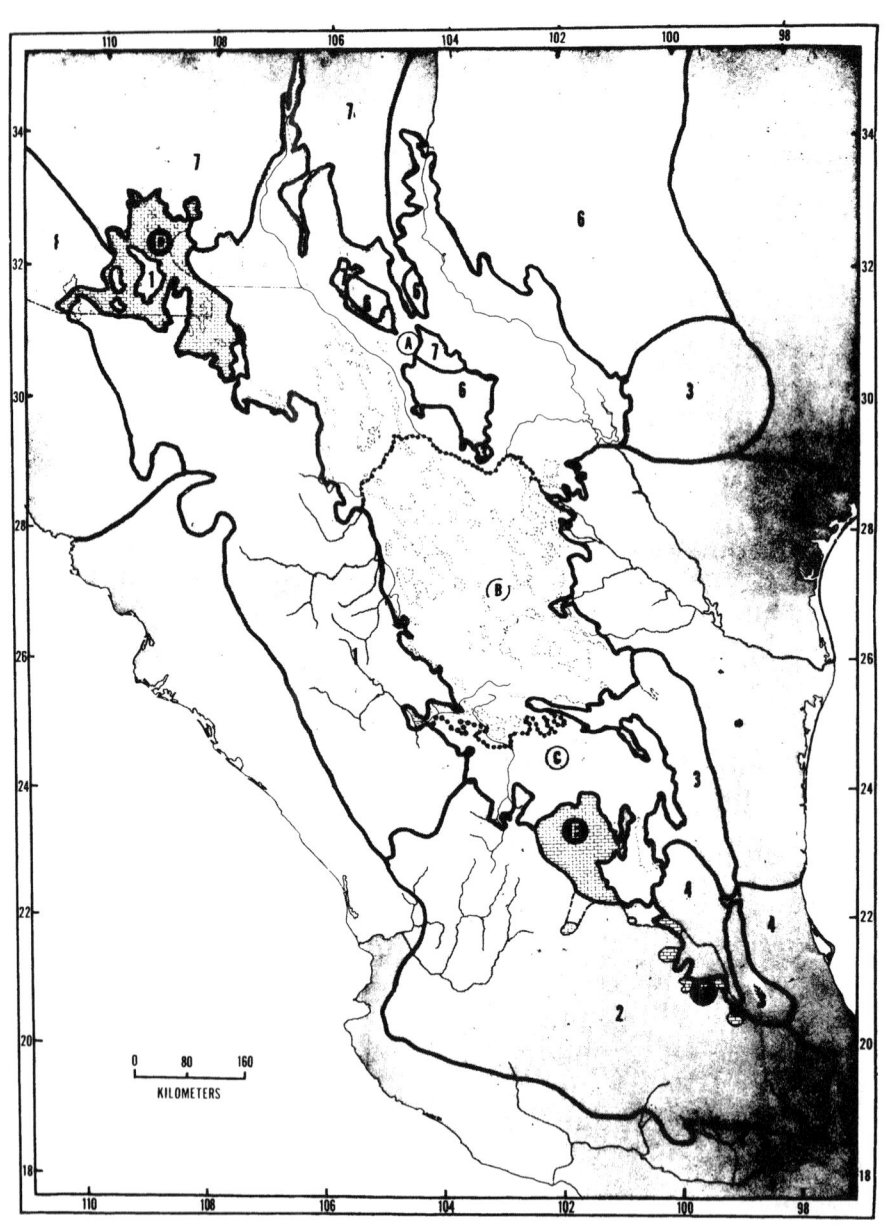

Map 6 . Biotic Provinces and Subdivisions of South Central North America. Key: Clear area = Chihuahuan Desert Province; A. = Trans-Pecos Subprovince; B. = Mapimian Subprovince; C. = Saladan Subprovince; D. = Cochise Filter Barrier; E. = Saladan Filter Barrier; F. = Rio Panuco Relict Desert; 1. = Sierra Madre Occidental Province; 2. = Trans-Volcanic Province; 3-A. = Sierra Madre Province; 3-B. = Balconian Province; 4. = Vera Cruzian Province; 5. = Tamaulipan Province; 6. = Kansan Province; 7. = Navahonian Province.

122

Table 18. A Comparison of the Participant Herpetofaunas of the Three Chihuahuan Desert Subprovinces.

Endemics	Trans-Pecos	Mapimian	Saladan
	1. *Coleonyx reticulatus*	1. *Gopherus flavomarginatus*	1. *Sceloporus cautus*
	2. *"Cnemidophorus tesselatus"**	2. *Sceloporus maculosus*	
	3. *"Cnemidophorus neomexicanus"**	3. *Sceloporus merriami***	
1. Number of Species	47	48	29
2. Area (kms^2)	240,000	160,000	60,000 + 10,000 in the filter barrier
3. Percent of Endemics within Sub-province Herpeto-fauna	6	10	3

* Parthenogenetic clones (geographically *"C. exanguis"* and *"C. uniparens"* could also be included)

** Extends slightly into Trans-Pecos Texas

continue clockwise in its treatment of adjacent provinces. The provinces dealt with here are defined on Map 6 and drawn largely from DICE (1943), HAGMEIER & STULTS (1964), SAVAGE (1960), and STUART (1964).

The Kansan province by virtue of its transitional physiography, geography and vegetation is particularly difficult to distinguish from the Chihuahuan Desert. The physiographical transition from one to the other is determined largely by subtle climatic changes. Like the transition itself, the local basins and ranges and river systems run north-south in parallel, none constituting a discrete perpendicular border between the two. The region of contact between the two provinces is broad, approximately 600 kilometers across eastern New Mexico and west Texas. The ecological discrimination between true Chihuahuan Desert grasslands and Kansan prairie relicts contributes further difficulties, as does the mesquite-grassland ecotone that forms a virtual corona around the desert proper. Disjunct prairie relicts are depicted in the Big Bend of Trans-Pecos Texas on the base map. To these might be added other grasslands along the Sierra del Hueso in northeastern Chihuahua and the Sierra Madre foothills between El Sueco and Nueva Casas Grandes in the northwestern part of the same state. The C.C. for a Kansan-Chihuahuan herpetofauna is 0.32, clearly maintaining their distinction on the province level. The S.C. value, emphasizing common derivation more than ecological identity, indicates a closer relationship, at 0.55 level. In Table 17 the Great Plains (Kansan) centered component of the Chihuahuan herpetofauna is considered. Only 8% of the Chihuahuan fauna has more than half its range in the Kansan plains. For the most proximal Trans-Pecos subprovince the percentage is still only 9%. Thus the two provinces are clearly distinct, though they share a major component of common derivation.

The Balconian province has been subject to several previous analyses of its herpetofauna, the most comprehensive being that of SMITH & BUECHNER (1947) and RAUN (1959). The province is essentially a pinyon-juniper woodland set on a limestone plateau. Ecologically and biogeographically it is essentially an extension of the Sierra Madre Oriental foothills. This was noted for its woodland vegetation by AXELROD (1958, p. 486). One might view the province as foothill vegetation and fauna in search of a mountain range. Its herpetofauna differs from that of the Sierra Madre proper only in the greater importance of Chihuahuan Desert saxicolous species (e.g., *Sceloporus poinsetti*) and in the presence of an Austro-riparian woodland component. The latter is represented by mesic relict stocks such as *Plethodon, Eurycea, Typhlomolge, Graptemys, Virginia, Elaphe, Natrix,* and *Agkistrodon*. Tongues of prairie may contribute a few Kansan forms such as *Troidopclonion* as well. Viewing the Edwards Plateau as a limestone foothill it is not surprising that the Balconian province has a greater shared herpetofauna with the Chihuahuan Desert than any other contiguous unit. This occurs despite rather narrow suture of contacts – about 120 kms. long. The C.C. value of the two compared is 0.42, removing the possibility of treating the two as a single province. However, the S.C. value of 0.59 affirms the importance of the shared fauna. Table 17 does not list a specific Balconian component to the Chihuahuan herpetofauna, but includes it in the Sierra Madres, a maximum component of 5%. While the Balconian herpetofauna shows significant overlap with Chihuahuan, no predominantly Balconian species participate in the Chihuahuan fauna. The general lack of endemic species on the Edwards Plateau, except mesic relicts, makes its treatment as a faunal province somewhat dubious, especially when coupled with general similarities to the Sierra Madrean foothills to its south.

To the south of the Balconian province lies the Tamaulipan Plain. This province is essentially a mesquite-grassland, at some points approaching a tropical savanna. It is not a desert despite the somewhat sclerophytic and succulent aspects of its vegetation. It can be rejected as desert by both P/T ratios and incident solar radiation (see Climate). The Tamaulipan province can also be rejected as desert by assessing its faunal relationships to the Chihuahuan Desert with which it is contiguous. The two maintain contact along the 300 kilometers lowland gap in the northern Coahuila Folded Belt (from Amistad Reservoir to south of Monclova). Tamaulipan biota extends west along riparian peninsulas, generally mixed with an Austro-riparian vegetation and fauna along the Rio Sabinas (Musquiz) and Rio Salado (Cuatro Cienegas) both in Coahuila. The shared herpetofauna as indicated by C.C. values indicate less faunal affinity than that seen between Chihuahuan and Sonoran Deserts (see Table 14). The relatively limited relationship between the two provinces is further indicated in Table 17 by the low percentage, 2%, of Chihuahuan species which have Tamaulipan centered distribution, again contrasting to the 15% derived from the only adjacent true desert, the Sonoran.

The Sierra Madre Oriental is a difficult and complex province to define both ecologically and geographically. As employed here, it includes chaparral, pinyon-juniper woodland, coniferous forest, upland tropical deciduous and evergreen forests, oak pine woodlands, and piedmont (see *Vegetation*, and MARTIN, 1958). Mesquite grassland is treated as an ecotone and geographical transition zone, not as part of either province, for reasons discussed in the last chapter. Geographically, problems result from the fragmentation of montane ecosystem along the disjunct ranges of the Coahuila Folded Belt. Ranges above 1500 meters elevation and along an axis running northwest from the Cumbres of Monterey to the Chisos Mountains, of Texas (Big Bend) have all been considered as part of this province, though some isolated units (Chisos and Carmens) have been evaluated separately as well. Total contact is about 500 kilometers. The fauna of the Anticlinorium of Arteaga has also been included here and the southern limits of the province have been set along the Rio Verde in San Luis Potosi. This montane fauna and the Chihuahuan participant fauna are as alien to one another as their respective environments. By C.C. values, only 0.09 of the faunas are shared, and even S.C. values indicate a minor common component at the 0.18 level. Only 5% of the Chihuahuan participant fauna is Sierra Madrean in distribution, and this increases to 10% only in the most mesic and elevated Saladan subprovince. *Sceloporus grammicus* is the lone conspicuous Chihuahuan Desert form which is ultimately centered in the Sierra Madres in its general distribution.

The Vera Cruz province is again poorly defined. It is set along the Gulf Coast plain from the Rio Tamesi (near Tampico, Vera Cruz) south approximately to the Isthmus of Tehuantepec. The Vera Cruz province includes vegetation from below 1000 meters elevation of both mesic tropical woodlands and semi-arid columnar cactus and mesquite woodland. This essentially tropical mosaic does not conform well with temperate concepts of simple discrete communities with dominant species, especially in its more mesic aspects. Its contact with the Chihuahuan Desert is confined to a southern gap in the Sierra Madre Oriental of San Luis Potosi in the upper drainage of the Rio Verde. The gap is about 40 kilometers in width (north to south) and mixed with Sierra Madrean vegetation and faunas. The region is centered between Arista and Cerritos, San Louis Potosi Faunal relationships are minimal between this province and the desert. C.C. and S.C. values are 0.04 and 0.11, respectively. There are no Chihuahuan Desert participant species which are predominantly Vera Cruzian in distribution.

The Trans-Volcanic province is again somewhat heterogenous and poorly defined. It has a fauna that shares surprisingly few species with the adjacent Chihuahuan Desert, into which it grades gently along a 250 kilometer wide southeast-northwest running border. The details of this border and its constituent filter barrier and adjacent desert relict sites will be discussed in the subsequent section of this chapter. The Trans-Volcanic province is 2000 meters high and tectonically active with vulcanism characterizing the region since the late Pliocene Epoch. Basaltic flows and derived soils dominate the landscape. Arid woodlands of acacia, arborescent yucca (*Yucca filifera*), and platyopuntias (nopals) are typical vegetation in the lower basins. Further elevation (2200 meters and up) sustain encinal-oak woodlands and ultimately coniferous forest (2500 meters and above). Complex mosaics of vegetation occur here and in the Vera Cruz province, the situation is essentially tropical and community classification is inadequate to the complexity of organization. It is probably most accurately viewed as an arid high tropical plateau involving flora and fauna with many alliances with temperate stocks at least on the generic level. Its shared herpetofauna with the desert is only 0.09 for C.C. and 0.18 for S.C. The contact between the two is made over a considerable length. It forms a transition across the southern peneplain of the Mexican Plateau which is devoid of sudden disruptive features in its physiography. Fundamental climatic changes in both temperature and P/T ratios have probably done much to reduce the common elements in herpetofaunas of the two provinces. No single species whose range is centered in the Trans-Volcanic province is a characteristic participant in the Chihuahuan herpetofauna. However, the 5% ubiquitous Sierra Madrean component does include forms occurring in the Trans-Volcanic province.

The Sierra Madre Occidental is another artificial province. Much that has just been discussed in relation to the Oriental applies here as well. All non-desert associations and their faunas above 1500 meters within the mountain system have been treated as portions of the province. Again the mesquite-grassland has not been included, but consigned to ecotone status. The vegetation of the region has been reviewed in the appropriate chapter (also see MARSHALL, 1957), as have its geographical limits. Contact with the Desert is maintained across an elevational gradient along an 800 kilometer western wall of the Mexican Plateau. The mesquite-grassland forms a transitional belt -- sometimes 100 kilometers in breadth (east-west) -- across the entire length. The shared herpetofauna constitutes a C.C. of 0.07 and a S.C. of 0.13. No species characteristic of the Occidental in particular participates in the Chihuahuan Desert herpetofauna. The same 5% ubiquitous Sierra Madrean stock can be applied to this situation just as it was to the Oriental and Trans-Volcanic Province.

The Sonoran Desert will be discussed as part of a general review of relationship and affinity within North American deserts later in this chapter. It may be noted here, that along with the Kansan and Balconian provinces, this province has a particularly high affinity with the Chihuahuan fauna as indicated by both coefficients and percentage of Chihuahuan fauna of Sonoran centered distributions.

The Navahonian (or Colorado Plateau) province is a montane unit with vegetation and fauna of very distinct derivation and ecological structure from that of the Chihuahuan Desert. Its biota appear to be of mixed affinities both from the Sierra Madre Occidental and the western Rocky Mountains. The province is centered on the cold arid to semi-arid Colorado Plateau. It includes sagebrush steppe, coniferous forest, pinyon-juniper woodland and narrow local stretches of chaparral. Again, generally elevation is above 1500 meters. Its contact with Chihuahuan Desert follows a 1000 kilometer tortuous zigzag up and down the river valleys of the upper Rio Grande and Pecos and west along the Cochise filter barrier. Most areas of contact involve abrupt vertical transitions in biota, excepting the upper Rio Grande between Elephant Butte Dam and Albuquerque, New Mexico (see next section). Shared faunas with the Chihuahuan Desert consists on only 0.12 by C.C. and 0.26 by S.C. This is slightly higher than for other montane provinces and is due to the presence of ubiquitous arid southwestern species in dry cold plateau habitats (e.g., *Crotaphytus collaris* and *Arizona elegans* in sagebrush flats). No Navahonian centered species participate in the Chihuahuan herpetofauna.

Chihuahuan Desert Filter Barriers and Extralimital Desert Relict Herpetofaunas

The crucifix analysis of the desert and borders using IFC values indicate that borders of the base map, while rational, are pretentious in their exactness. Complex transitional areas, here treated as filter barriers, exist both as indicated on the base map and in a number of other positions as well (see Maps 6 and 7). In addition, the definition of primary area and an endemic herpetofauna allows the analyst to search beyond the continuous desert borders for external relict desert faunas. These have been discovered. In this section a more detailed, though largely descriptive review of the major border complexes will be undertaken.

Filter Barriers

The most complex and best known of these barriers is the Cochise filter barrier of Cochise and Santa Cruz Counties, Arizona, Hidalgo County, New Mexico, the northeastern tip of Sonora and extreme northwestern Chihuahua (Municipio Janos; see Map 7). Physiographically, the Cochise filter barrier is a 100 to 200 kilometer wide gap between the Sierra Madre Occidental and the Colorado Plateau. It is about 200 kilometers east-west as well. It is a broad, 1500 meter high plain disrupted by steep but discontinuous north-south ranges (Chiricahua, Huachucas, Peloncillos), up to 3000 meters at their peaks. It is the only physiographic portal along the western wall of the desert. Its elevation and cool arid climate is such that the region would be transitional between desert grassland, and pinyon-juniper woodland even if it were not accessible to other biotas. The Cochise barrier is continuous with the mesquite-grassland ecotone that surrounds the entire Chihuahuan Desert. Only occasional pockets of Chihuahuan vegetation, such as *Agave lechiguilla, Flourensia cernua,* or *Parthenium* appear in the lower valleys and outcroppings. Other true desert species, such as *Larrea tridentata* and *Fouquieria splendens* are ambiguous in their significance -- they indicate only warm desert affinities, not being peculiar to either Chihuahuan or Sonoran Deserts. For a more definitive picture, the faunal affinities must be sorted, defined, and evaluated. High

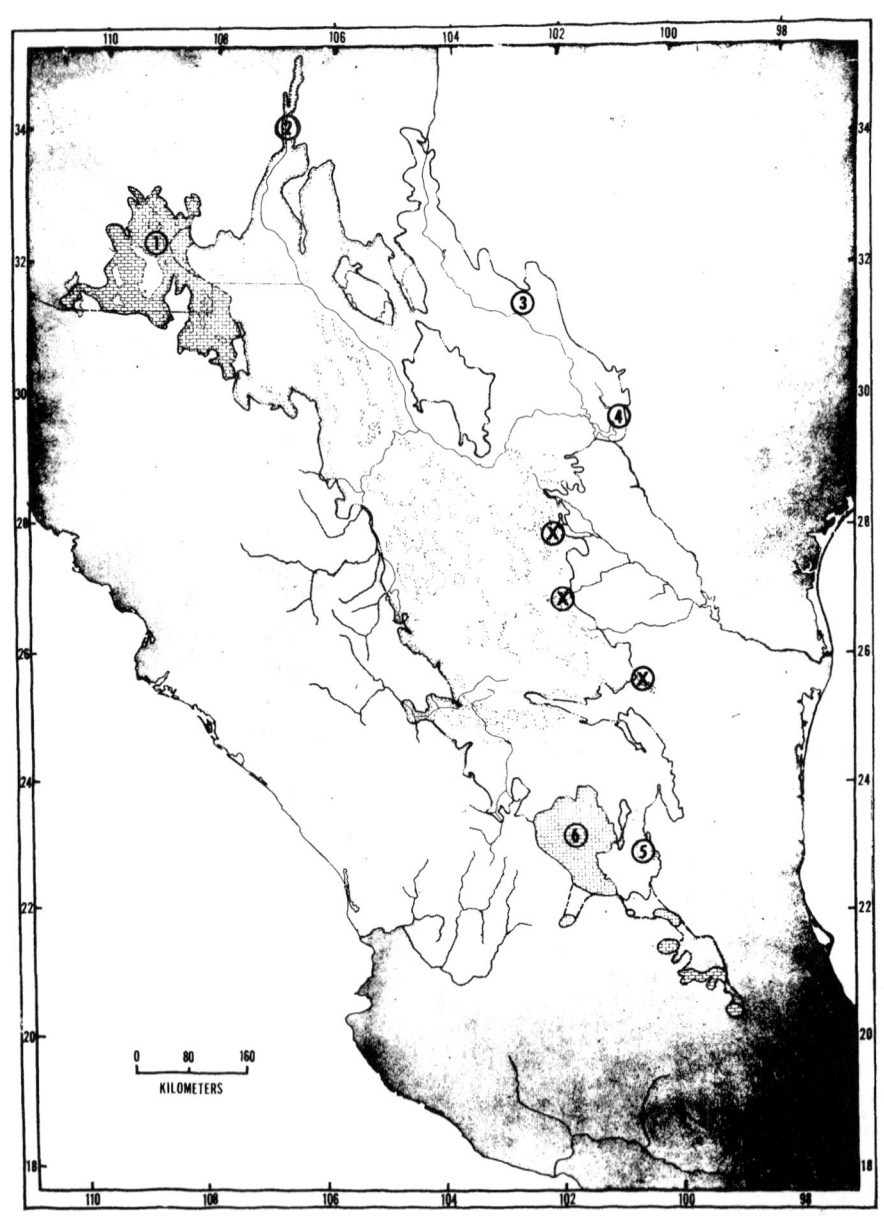

Map 7. Filter Barriers and Major Ecotonal Transitions Between Chihuahuan Herpetofaunas and Those of Adjacent Provinces. Key: 1. = The Cochise Filter Barrier; 2. = Upper Rio Grande Filter Barrier (Elephant Butte to Albuquerque); 3. = Trans-Pecos Filter Barrier; 4. = Devils River-Val Verde Filter Barrier; 5. =Cerritos-Arista Filter Barrier (Chihuahuan-Vera Cruzian); 6. = Saladan Filter Barrier; x = Chihuahuan-Tamaulipan Ecotonal Transitions (upper to lower: Musquiz, Cuatro Cienegas, Cumbres de Monterrey).

128

species diversity is due both to the ecotonal nature of the region and its geographical/physiographical accessibility.

Several very productive biogeographical accounts of the Cochise filter barrier or its political subdivisions already stand in the literature. DARROW (1944) and LOWE (1964) contributed detailed ecological background in terms of vegetation. MARSHALL (1957) provided an excellent discussion on montane forests and their birds. FINDLEY (1969) gave a lucid presentation of local mammal distributional patterns. Herpetological analysis included LOWE (1955) and ZWEIFEL (1962), POUGH (1966), and BOGERT & DEGENHARDT (1961). FOWLIE (1965) offered a speculative model for Crotalid snake dispersal through the region. My purpose here is not to recapitulate the conclusions of these sources, but to place them in slightly different context, particularly in relation to the Chihuahuan Province. Early treatments of the herpetofaunas of this region are understandably simplistic. LOWE (1955) noted that two typically Sonoran species, *Heloderma suspectum* and *Micruroides euryxanthus* reach the upper Gila River drainage in southwestern New Mexico, and on this feeble basis set the eastern limits of the Sonoran Desert. He failed to recognize any more basic characteristics of the Sonoran Desert, and he further ignored the presence or even existence of a distinctive Chihuahuan herpetofauna. Were this logic applied to setting the western limits of the Chihuahuan Desert, using *Cnemidophorus inornatus* and *Phrynosoma modestum* as indicator species, these limits could extend to the Grand Canyon of Arizona south to Nogales, Sonora. LOWE (1964) later demonstrated far more profound understanding of the complexity of the situation. BOGERT & DEGENHARDT (1961) considered a much greater spectrum of the herpetofauna, including both Sonoran and Chihuahuan stocks and suggest an approximate demarcation between the two along the San Simon Valley-Peloncillo Mountains (about 32°N, 109°W.) along the New Mexico-Arizona boundary. DESSAUER, FOX & POUGH (1962) and ZWEIFEL (1962), extended this same line northward (see broken lines on base map). Their criteria centered on meristic and electrophoretic data regarding the secondary contact between Sonoran and Chihuahuan stocks of *Cnemidophorus tigris*. However, the true complexity of the situation can only be described when all faunas, including northern and southern, are considered.

The Cochise filter barrier contains four distinctive herpetofaunas by regional affinity. These include a Chihuahuan Desert herpetofauna, a Sonoran Desert herpetofauna, a Sierra Madrean herpetofauna, and a Kansan (Great Plains) herpetofauna. The edge effect of these sympatric components (at least geographically sympatric) produces a total herpetofauna higher in diversity than that of the adjacent part of the Chihuahuan province proper. Square 37 on the IFC crucifix has 59 species and is centered in the middle of the Cochise filter barrier. In the adjacent square within the desert proper, the number of species drops to 40. A breakdown of these 59 species into the herpetofaunas is listed below (square 37 has been chosen because it has not only a high species diversity but the highest IFC, 53%, within the filter barrier):

Herpetofauna	Percentage	Examples
I. Chihuahuan	17%	1. *Cnemidophorus inornatus*
		2. *Sceloporus poinsetti*
		3. *Phrynosoma modestum*
		4. *Ficimia cana*

129

Herpetofauna		Percentage	Examples
II.	Sonoran	10%	1. *Bufo alvarius*
			2. *Heloderma suspectum*
			3. *Callisaurus draconoides*
			4. *Micruroides euryxthanus*
III.	Sierra Madrean	27%	1. *Phrynosoma douglassi*
	(Occidental and		2. *Sceloporus jarrovi*
	Navahonian)		3. *Lampropeltis pyromelana*
			4. *Elaphe triaspis*
			5. *Crotalus willardi*
IV.	Kansan	14%	1. *Scaphiopus bombifrons*
			2. *Terrepene ornata*
			3. *Crotalus viridis*
			4. *Heterodon nasicus*
			5. *Sistrurus catenatus*
V.	Ubiquitous	32%	1. *Bufo punctatus*
	(two or more provinces)		2. *Cnemidophorus tigris*
			3. *Sceloporus undulatus*
			4. *Pituophis melanoleucus*

The percentage breakdown indicates a fairly even four-way distribution of the fauna. The Sonoran is least significant in numerical representation. A full third of the herpetofauna, like much of the vegetation, is so widely distributed as to resist provincial categorization. If the considerable montane Sierra Madrean fauna is removed from consideration, a more decisive re-evaluation of lowland fauna is possible. First, the IFC value for square 37 (Map 4) is lowered from 53% to 26%. This is still higher than either contiguous square along the filter barrier, but it does indicate that the intervening Chiracahua and Pelloncillo montane faunas exaggerate the degree of faunal change. Furthermore, of the remaining fauna, characteristic of the mesquite-grassland basins of the filter barrier, the following percentage analysis can be made: Eastern (Chihuahuan & Kansas): 42%; Western (Sonoran) exclusively: 14%; Ubiquitus: 44% In general then, at this central position the filter barrier is still apparently dominated by eastern grassland, desert, and ubiquitos desert-plains faunas much as the rest of the grassland ecotone which surrounds the Chihuahuan Desert. A relatively minor and possibly relict Sonoran fauna is the only exceptional aspect of this ecotone. The squares are too large to discriminate the exact sites of major shifts, but inspection of ranges indicates that all but two of the Chihuahuan forms drop out between the Pelloncillo and Huachuca Mountains. Similarly, *Heloderma* and *Micruroides* appear to be the only purely Sonoran stocks to penetrate across (or around) the Pelloncillos and even then only to the very next valley east (BOGERT & DEGENHARDT, 1961). Very little further can be said in terms of spatial biogeography with the distributional information presently available. The IFC values affirm the Huachura-Pelloncillo region as the area of maximal fauna change within the filter barrier, and a breakdown of non-montane faunas indicates that they are typical of the mesquite-grassland ecotone of the Chihuahuan Desert except for the 14% component of clearly Sonoran derivation. Previously suggested break points such as the Pelloncillos and the line west of Lordsburg (ZWEIFEL, 1962) fall well within square 37 and are considered reasonable, if somewhat arbitrary. A paleoecological interpretation of the conditions described here will be postponed to the subsequent chapter.

The river valley of the upper Rio Grande, specifically the 300 kilometer long strip between Truth or Consequences and Albuquerque, New Mexico, is another zone of extended transition. I have restricted this zone to Socorro, Bernalillo, and the southeastern tip of Sandoval Counties, New Mexico. Generally this narrow strip is no more than 50-75 kilometers across. A considerable and generally endemic complement of Chihuahuan reptiles occur in this valley. They include *Cnemidophorus inornatus, Phrynosoma modestum, Coleonyx brevis,* and *Ficimia cana.* However, several other distinctive Chihuahua forms *Cophosaurus texana, Trimorphodon biscutatus,* and *Elaphe subocularis* fail to extend north into the barrier. Furthermore, Chihuahuan scrub (*Larrea, Flourensia, Fouquieria*) drops out dramatically in this filter barrier, replaced by flood plain scrub, prairie grasses, and cold steppe vegetation including *Artemesia tridentata.* With this change comes a cold northern montane (Navahonian) herpetofauna along the river, including *Bufo woodhousei, Chrysemys picta, Phrynosoma douglassi,* and *Thamnophis sirtalis.* The Socorro filter barrier appears to be more of a riparian flood plain than an extension of the mesquite-grassland corona. The mixture of faunas indicate a transition between dry, warm Chihuahuan Desert forms and dry, cold Navahonian sagebrush scrub and montane riparian forms.

The border between Chihuahuan Desert and Kansan prairie along the valley of the Pecos River may constitute another transition. For 50 to 100 kilometers on either side of the river, especially in Texas, the landscape alternates between desert patched with islands of grassland and a prairie moth-eaten with pockets of desert sclerophylls.

Pockets of *"Cnemidophorus tesselatus", Crotaphytus wislizeni, Phrynosoma modestum,* and *Ficimia cana* are indicators of a disjunct desert periphery. *Ambystoma tigrinum, Scaphiopus bombifrons, Kinosternon flavescens, Sonora episcopa,* and *Tantilla nigriceps* may be their prairie counterparts. To some degree the entire Trans-Pecos subprovince could be included in the filter barrier. Its most extreme form, where grassland and desert maintain an almost 50/50 mosaic, is confined to Culberson, Reeves, Loving, Ward, Ector, Winkler, Midland, Upton, Reagan, Crockett, Terrell, Brewster, and Jeff Davis Counties, Texas, and Eddy and Chaves Counties, New Mexico. I frankly believe detailed local vegetation and faunal distribution maps are inadequate to resolve the details of this mosaic at the present time.

Continuous with this Pecos grassland-desert exchange and in some respects part of it is the Devil's River region of south-central Texas. It differs from the previous filter barrier in the presence of two added faunal components, the Balconian and the Tamaulipan. The area centers around Val Verde County but includes also Terrell, Edwards, Kinney, and Uvalde, all in Texas. The Balconian contribution (or Sierra Madrean) is confined to two saxicolous anurans, *Syrrhophus marnocki* and *Eleutherodactylus augusti.* Tamaulipan forms include *Gopherus berlandieri, Sceloporus olivaceus, Sceloporus cyanogenys* (or *S. cyanogenys* x *S. poinsetti*), *Drymarchon corias,* and *Micrurus fulvius.* There is a reciprocal presence of typical Chihuahuan, especially saxicolous forms along the western edge of the Balconian province in Edwards and Kinney Counties. Representatives of this latter group include: *Phrynosoma modestum, Sceloporus poinsetti* (throughout the Edwards Plateau), *Elaphe subocularis, Ficimia cana,* and *Tantilla atriceps.* Square 29 has the highest IFC value, 32%, of the units imposed on this area and indicates that the greatest faunal change within the transition is not at Devil's River or the mouth of the Pecos River, but along the western rim of the Edwards Plateau. A detailed account, though incomplete, of Val Verde County has been undertaken by PROVINE (1969). The longest axis of the transitional area is probably no more than 125 kilometers from west to east. The rim of the Edwards Plateau provides a strong explicit break within the transition (apparent in the vegetation shift from desert scrub to juniper woodland). The scrub transition from Chihuahuan Desert to Tamaulipan mesquite-grassland south of the rim is relatively abrupt at the same longitude.

Further south, also along the eastern perimeter of the desert, numerous portals in the discontinuous Sierra Madre Oriental bring Tamaulipan and Chihuahuan scrub vegetation and faunas into direct contact. However, transitions between the two are narrow, generally within 25 kilometers in breadth. Documented sites of overlap between Chihuahuan and Tamaulipan herpetofaunas include: Melchor Musquiz, between the Puerto Salado and Puerto Sacramento (along the Rio Salado, 20 kilometers east of Cuatro Cienegas), the valleys approximately 30 kilometers southeast of Monclova on Mexico 53 (to Monterrey, all in Coahuila). Further south (of Saltillo) the Oriental is more abrupt and continuous until about the 23rd parallel north where a portal opens between Charcas and Cerritos in San Luis Potosi. This exposes a short flank segment of the desert to contact with a Vera Cruz fauna. No uniquely Vera Cruzian forms occur here. However, the presence of *Scaphiopus couchi*, *Leiolopisma* (both at Charcas) and *Sceloporus olivaceus* at Arista indicate some lowland influence on the herpetofauna indigenous to this portion of the desert.

The Chihuahuan Desert terminates abruptly in both vegetation and fauna at its southern end around the city of San Luis Potosi. Further west, however, between San Luis Potosi and Zacatecas and the latter city and Majoma (in Zacatecas) stretches a dilation of the mesquite-grassland ecotone running 150 kilometers north-south and approximately 100 kilometers east-west. The transitional decline in desert biota across a very gently tilted (up toward 2000 meters in the south) peneplain is quite visible in the vegetation along highway Mexico 57 between Majoma and Zacatecas. *Larrea*, *Fouquieria*, *Flourensia* and *Parthenium* all become discontinuous and increasingly restricted to drier lower slopes. Such typical Chihuahuan reptiles as *Cnemidophorus inornatus, Cophosaurus texanus, Phryno-soma modestum*[1] and *Ficimia cana* have not been taken more than 20 kilometers southwest of Majoma at about 1900 meters elevation. Like the Cochise filter barrier, this extended ecotone between the Saladan Desert and the Trans-Volcanic province vegetation is predominantly mesquite-grassland with patches of desert scrub and associated herpetofauna. The southward transition drops out one desert species (plant and animal) after another. The *Prosopis* grassland development begins as edaphic arroyo and swale phenomenon gradually expanding across the higher, more southern valleys to replace the desert scrub. Eventually it covers even the better drained bajadas and fans. While lack of local distributions make a detailed account impossible, several species of reptiles show excellent correlation with the transitions in vegetation. *Cnemidophorus inornatus* and *Ficimia cana* appear to be limited to the Chihuahuan Desert scrub association. *Sceloporus poinsetti* and *Cophosaurus texanus* show a similar correlation, but have the added indicator value of being replaced by *Sceloporus torquatus* and *Holbrookia maculata* respectively, wherever the transition to grassland takes place (these

1. One record from Aguascalientes.

statements apply only to the region in question).

The Sierra Madre Occidental forms a continuous western perimeter to the Chihuahuan Desert herpetofauna until about the 30th parallel north where the portals of the Cochise filter barrier develop.

All of the transitions and filter barriers of the Chihuahuan Desert involve a mesquite-grassland ecotone (the Balconian exchange excepted). The concept of a filter barrier is applied to those portions of the ecotone which have expanded as a result of climate and topography and involve a transitional gradient with some overlap between species characteristic to adjacent provinces. In many respects the grassland corona of the desert is like a silhouette margin of chromatograph paper into which adjacent provinces diffuse their faunas. The differential penetrance by these alien species results in the complex gradients that have been described. Characteristically Chihuahuan forms illustrate the same phenomenon in the reverse direction. While this assessment of the filter barrier problem is admittedly simplistic, it does offer a fundamental model from which more discriminate analyses can be built.

Extralimital Desert Relict Herpetofaunas

Adjacent but discontinuous with the Chihuahuan Desert as defined in this analysis are five major regions harboring scattered and depauperate complements of Chihuahuan herpetofauna (Map 8). These are the panhandle region of northwest Texas (and Cimarron County, Oklahoma), the flood plains of the lower Rio Grande (Webb County, Texas), the sclerophytic vegetation patches (especially along river gorge walls) of Guanajuato, Queretero, and Hidalgo, and lastly, the sclerophyll of northeastern Aguascalientes and the edges of Mogollon Rim of Arizona (Navahonian Province). In all of these regions some typically desert herpetofaunas are represented by disjunct populations. In some, but not all cases, they are associated with isolated sclerophyll vegetation. The emphasis here will be on establishing the spatial limits and content of these districts. The temporal (paleoecological) significance of these situations is assigned to the subsequent chapter.

The desert relict herpetofauna of the Kansan biotic province is summarized in Table 19. The table is undoubtedly incomplete and relict or disjunct occurrence could conceivably be the result of inadequate sampling. Only six species are involved, all reptiles, and these are not so tightly clustered as to form discrete desert islands within the grassland sea. Rather, the pattern is of differential penetrance or survival of individual desert scrub species, and the model of the chromatograph paper would seem more applicable.

Other ubiquitous arid southwestern species could be included in the analysis, including the lizards *Crotaphytus collaris* and *Phrynosoma cornutum*, and the snakes *Arizona elegans, Rhinocheilus lecontei* and *Crotalus atrox*. But these forms are not so confined either to deserts ecologically nor to the Chihuahuan Desert biogeographically.

Relict species from the Texas panhandle and adjacent Oklahoma do indicate a strong association with the open flood plains of the Canadian and Red River drainages. These situations afford greater solar radiation, scarce and sometimes sclerophytic vegetation, and substrates of sandy or gravelly alluvium. They may provide local edaphic "desert" habitats.

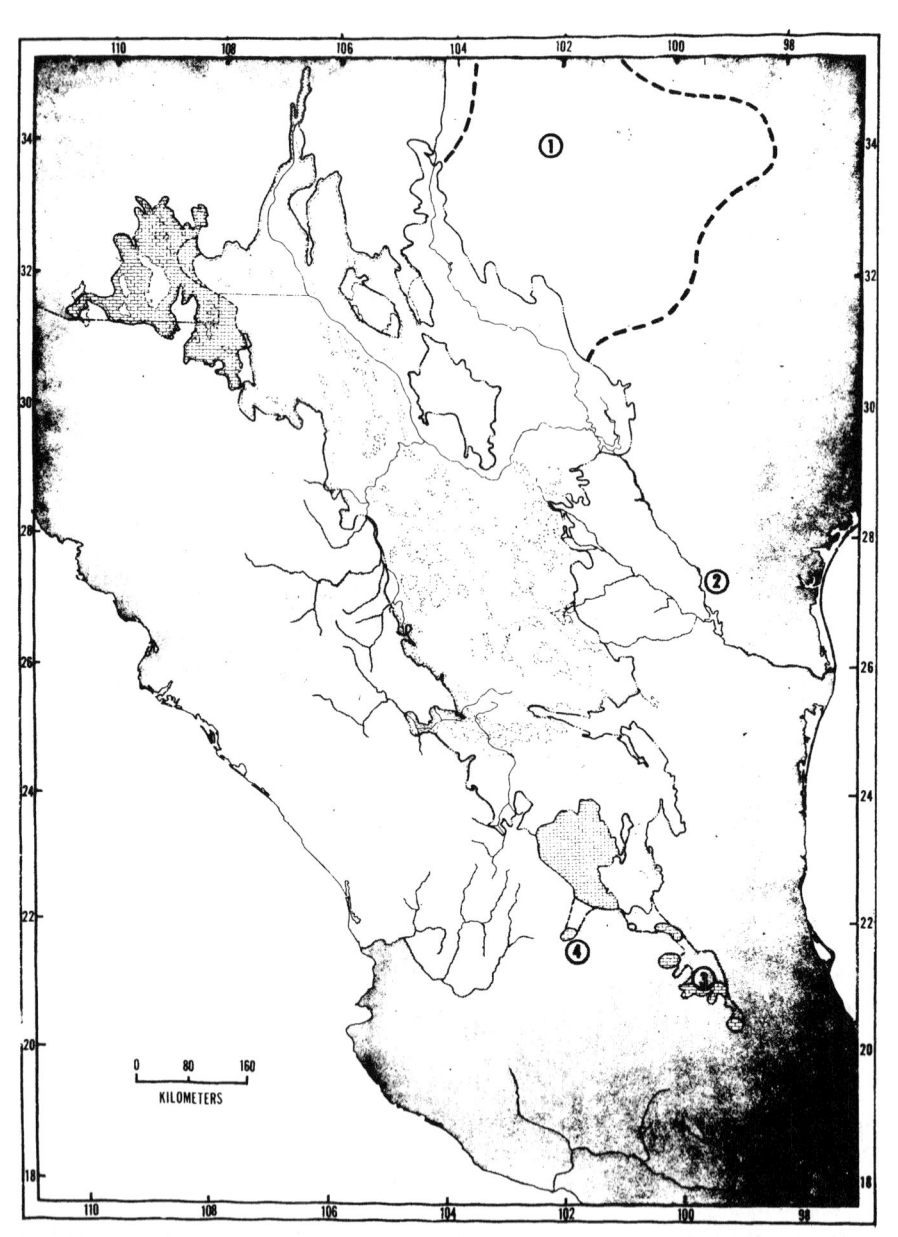

Map 8. Extralimital Chihuahuan Desert Herpetofaunas. Key: 1 . = Desert relicts in Kansan Province Prairie; 2. = Desert relicts along Rio Grande riparian plains in the Tamaulipan province; 3. = Desert relicts in steep slope edaphic sclerophyll, in the Sierra Madre Oriental province; 4. = Desert relicts in sclerophyll of the acacia-grassland of the Trans-Volcanic province.

134

Table 19. Chihuahuan Desert Relict Herpetofauna in the Kansan Biotic Province.

Counties (from north to south)	Species	Cophosaurus texanus	Phrynosoma modestum	Uta stansburiana	Cnemidophorus tigris	Cnemidophorus tesselatus	Ficimia cana
Oklahoma:							
1. Cimmeron			X			X	
Texas:							
1. Hemphill			X				
2. Potter			X			X	
3. Randall			X	X		X	
4. Armstrong		X		X		X	
5. Donley		X	X				
6. Swischer						X	
7. Brisco		X		X		X	
8. Hardeman		X		X			
9. Motley			X	X			
10. Dickens							X
11. King							X
12. Garza		X		X	X		
13. Scurry		X					
14. Fisher		X					
15. Jones			X				
16. Andres		X					
17. Howard			X	X			
18. Tom Green			X				

Further information on the herpetofauna of this region can be found in FOQUETTE & LINDSAY (1955), TINKLE (1959), and in the maps of RAUN & GEHLBACH (1972).

Along the flood plains of the lower Rio Grande may be another refugium for desert species within the more mesic Tamaulipan thornscrub and grassland. The representation of a Chihuahuan Desert herpetofauna is apparently confined to *Cnemidophorus tigris*, and *Phrynosoma modestum*, both occurring in distinct pockets in Webb County, Texas. Distinct patches of *Larrea* and *Agave lechiguilla* in northern Tamaulipas may indicate a related eastern extension of desert vegetation (in the last Xerothermic Period).

While continuous Chihuahuan Desert terminates at the city of San Luis Potosi, scattered and isolated populations of desert scrub and herpetofauna occur further south and east along the edge of the plateau. In the upper Rio Verde drainage of San Luis Potosi are isolated patches of *Larrea* (see Vegetation) and the Saladan lizard *Sceloporus cautus* have been recorded there. Other *Larrea* patches occur in extreme southern San Luis Potosi, Guanajuato, Hidalgo, and Queretero (see bricked pockets on base Map 6). Most of these are unsampled for herpetofauna. Fortunately, a recent and extensive sampling of the Queretero herpetofauna has been presented by DIXON, KETCHERSID & LIEB (1972a). These authors reported the slopes of the Rio Extoras (a tributary of the Rio Panuco) support a desert sclerophyll of *Prosopis*, *Acacia*, and *Jatropha*. To their account should be added the presence of the dense stands of *Larrea*, the saxicolous bromiliad *Hectia*, and Chihuahuan cactus, genus *Lophophora*. The

[1] and the parthenoform *"Cnemidophorus laredoensis"*

ocotillo *Fouquieria splendens* is conspecific with that of the Sonoran-Chihuahuan Deserts. The dramatic affinity of this vegetation to Chihuahuan scrub is reflected to a much lesser degree in its herpetofauna. The only certain representative is *Ficimia cana*. *Sceloporus exsul*, clearly derivative of the *undulatus* group of the genus, may be a relict form of *Sceloporus cautus* as indicated by the meristic characters compared by DIXON, KETCHERSID & LIEB(1972b). The 275 kilometers gap between the two forms suggested by the authors does not consider a Rio Verde locality for *S. cautus* which effectively halves the separating distance. Several other arid land associates, (i.e. *Crotalus atrox*) are ambiguous indicators of faunal origins, since they could have been derived from either Chihuahuan or Tamaulipan populations. Ecologically, these southeastern relict districts are edaphic faciations, reflecting shallow soils, and rapid water runoff. They are not, however, situated on the alluvium of flood plains, but rather represent steep slope vegetation of river canyons. The Pena Blanca locality supported this vertically tilted sclerophyll from 1350-2000 meters elevation (DIXON, *et al.* 1972a). The Pena Blanca herpetofauna shares 54% of its 13 terrestrial species with the Saladan subprovince, 150 kilometers northwest. However, except for *Ficimia* and *Sceloporus exsul*, this shared fauna could as easily be derived from ubiquitous Trans-Volcanic Plateau, Tamaulipan, or Chihuahuan stocks. Certainly, present knowledge of the Queretero herpetofauna is far from complete and most information comes from less than a half dozen localities. Further quantification at this point would be of little value. Nonetheless, it is clear from both vegetation and herpetofauna that distinctly Chihuahuan species do occur as relict populations along the walls of the Rio Extoras. By extrapolation similar situations probably exist at the other *Larrea* relict stands previously noted.

Below the southwestern edge of the Saladan filter barrier is another pocket of sclerophyll vegetation and a depauperate representation of an otherwise Chihuahuan herpetofauna. This region, centered in northeastern Aguascalientes, might be more simply viewed as an extension of the Saladan filter barrier. Recent herpetological accounts of the region include: WILLIAMS, CHRAPLIWY & SMITH (1963), BANTA (1962a), and ANDERSON & LIDICKER (1963). The last provided a summary of previous knowledge and brief ecological characterization of localities. They noted the following localities within Aguascalientes (between 1700-23 kilometers elevation) as mesquite-grassland (or "desert" scrub): Cuidad de Las Ninos, Rincon de Romos, Tepezala, and Venadero. To these localities might be added Santa Maria Gallardo (GARCIA, SOTO & MIRANDA, 1960), 35 kilometers southwest of Loreto, (San Louis Potosi), Juan Batista, and Canada Honda.

From these localities a composite terrestrial herpetofauna can be constructed. Of the thirty species of amphibians and reptiles occurring in Aguascalientes, only 18 have been reported from mesquite-grassland localities. Of these, 65% (i.e., 100 X the Similarity Coefficient) are shared with the Chihuahuan Desert. All but three of these species however, *Bufo cognatus*, *Phrynosoma modestum*, and *Arizona elegans*, are ubiquitous to both the Chihuahuan and Trans-Volcanic provinces. Even these three occur in the mesquite-grassland elsewhere along the desert border and filter barriers. No genuine Chihuahuan endemic has been reported for Aguascalientes. This applies for vegetation as well. Even the ubiquitous desert indicator, *Larrea* is absent. The most striking disjunct record is for *Arizona elegans* near Loreto. It is 120 kilometers from the nearest locality at Venado in San Luis

Potosi. Considering the uniformity of the intervening landscape, climate and vegetation, and the known occurence of this species in grassland elsewhere, the disjunction is more probably an artifact of inadequate collection than a biological reality. As depicted on the base map, only 40 kilometers separate the Aguascalientes relict area from the filter barrier at the south edge of the Saladan subprovince. The circumstantial and incomplete evidence before me forces the conclusion that this region is but a southern extension of that filter barrier which lies between Chihuahuan Desert and Trans-Volcanic provinces.

The Mogollon Rim of the Colorado Plateau in Arizona forms the south and southwestern border of the montane Navahonian Province. According to LOWE (1964) the Rim supports a chaparral vegetation between 1200 and 1800 meters elevation. It is sclerophyll dominated by scrub oak (*Quercus turbinella*) with other typical North American chaparal genera present (*Arctostaphylos, Rhus, Cercocarpus, Ceanothus,* and *Rhamnus*). Pinyon-juniper woodlands also occur in this region, often interspersed with local blue grama (*Bouteloua gracilis*) grasslands. This vegetational setting is very similar (the grass being conspecific) to the western borderlands of the Chihuahuan Desert. In this context peripheral extensions of Chihuahuan herpetofauna are expected. Four distinctive species, all reptiles, establish the presence of a Chihuahuan component. These are *Cophosaurus texanus, Cnemidophorus inornatus, Tantilla atriceps* and *Xantusia arizonae. Cophosaurus* appears to have continuous distribution from the grassland and scrub of the Cochise filter barrier, skirting the northeastern edge of the Sonoran Desert in local sympatry with its western sibling *Callisaurus draconoides* (SAVAGE, MS). *Cophosaurus* penetrates as far northwest as Williams River, Arizona (STEBBINS, 1966) on the periphery of the Mojave Desert. *Cnemidophorus inornatus* is represented by an apparently disjunct population in the floor of the Grand Canyon of the Colorado River (JOHN R. WRIGHT, pers. comm.). The scattered Mogollon populations of *Tantilla atriceps* and *Xantusia arizonae* are apparently distinct also from conspecific Chihuahuan stocks. In Arizona this is a chaparral species dwelling in granite. In Durango, the race *bolsoni* utilizes andesite as the saxicolous substrate. The latter form occurs with a saxicolous sclerophyll that appears to be transitional between an edaphic Chihuahuan Desert scrub and a Sierra Madrean chaparral. These species represent a northwestern extension of Chihuahuan fauna spatially, but historically may be *in situ* relicts.

The Affinities Among North American Desert Herpetofaunas

In Table 14 the Chihuahuan Desert province had no greater similarity to the Sonoran Desert province than it did to the Kansan Plains province; in fact, its very closest faunal affinities were with the herpetofauna of the Balconian province, a juniper woodland. Does the Chihuahuan province have a desert herpetofauna, and if so, what are its affinities with other North American deserts? These questions will now be revolved, and in addition, the relationship between spatial and psysiographical position to faunal affinity will be assessed for these deserts.

As a preliminary, the Tamaulipan, Navahonian (Colorado Plateau) and Great Basin Artemsian provinces can be eliminated from the warm deserts on the basis of climate alone (see Climate Chapter) and the Tamaulipan has also been rejected as a desert in terms of herpetofaunal affinities. These discriminations confine North

American deserts to the warm creosote (*Larrea tridentata*) scrub land. The extent of this region has been documented by the classic study of GARCIA, SOTO & MIRANDA (1960). Coupling the definitions of JAEGER (1957) and SHREVE (1951) with the faunal provinces recognized by HAGMEIER (1966) and SAVAGE (1960), I have derived the following synthetic solution: I recognized three distinctive warm deserts west of the Continental Divide on the basis of climate, vegetation, and fauna (see Climate Chapter). These are the Mojavian, the Peninsular, and the Sonoran. This approach is unorthodox only in its treatment of the Peninsular Deserts (including the Baja Gulf, Yuman, Magdalenan, and Vizcaino of JAEGER) as a single unit. I believe climatic and biotic considerations, especially the absence of a strong summer rain and its dependent biota, justify this approach, at least provisionally. The same treatment of a Peninsular Desert ("Central Desert Phytogeographic Area") is made by WIEGMAN (1960, p. 150) on the basis of vegetation.

Table 15 and the accompanying dendrogram, Figure 9, illustrate the same basic affinities between North American deserts, using both C.C. and S.C. values. Generally, the deserts east of the Continental Divide cluster closely together, at about the 0.6 C.C. level and are best treated as subprovinces of a single Chihuahuan Desert. The western deserts are not so homogeneous and cluster only at the 0.4 to 0.5 C.C. level. Each of the three can be considered a distinct province in its own right. They can be categorized together at the superprovince level or above (criteria of HAGMEIER, 1966). The Mojave Desert and Sonoran Deserts have a shared herpetofauna with a C.C. of 0.52 and S.C. of 0.87 which does make the recognition of a Mojavian province questionable. The Mojave herpetofauna as indicated by the S.C. value, could logically be treated as a depauperate northern and elevated derivative of the Sonoran. This is essentially the treatment I have offered the Saladan herpetofauna in relation to the Mapimian in recognizing them as mutual subprovinces. However, both quantitative and qualitative evidence compels a different decision in this case. Quantitatively the Mojave herpetofauna has a very high S.C. value when compared with the Chihuahuan Desert, 0.61. The Saladan herpetofauna has no such dual affinities. Qualitatively, the Mojave shares several highly characteristic generic stocks, e.g. *Xantusia* and *Uma*, with the Chihuahuan Desert herpetofauna which are entirely absent fromt the intervening Sonoran (excluding the Peninsular-Colorado Desert). The Sonoran in turn also has several distinctive genera, *Heloderma* and *Micruroides,* which are absent from both Mojave and Chihuahuan faunas. Further consideration of climate, vegetation, physiography and mammalian provinces (HAGMEIER, 1966) all serve to reaffirm the validity of Sonoran and Mojave Desert as separate biotic provinces.

In affirming the distinction between the Sonoran and Mojave herpetofaunas, a corollary inquiry has been made. If the Mojave fauna is not merely a depauperate extension of the Sonoran Desert, where then do its true affinities lie?

As a generality, the deserts east of the Continental Divide not only cluster together in their herpetofaunas, but only resemble the western cluster at an average C.C. of about 0.20 and S.C. of 0.40.

It was just noted that a much more extensive shared component exists between Chihuahuan and Mojavian herpetofaunas. It is further intriguing that the greatest affinity is not between adjacent desert provinces or subprovinces, not between the Trans-Pecos and Mojavian faunas. Rather, the greatest shared element, especially as measured by S.C. values (see Table 15), is between the Mojavian and Mapimian faunas, some 500 airline kilometers apart. This peculiar situation is furthermore reflected in climate, physiography, and vegetation.

As noted in the earlier comparisons of desert climates (see Climate) the Mojave shares six of ten climatic characteristics with both Sonoran and Chihuahuan (specifically the Mapimian) Deserts. However, the shared characteristics are not the same in each case. Two characters not shared with the Chihuahuan Desert are in fact reversed in the Mojave, namely the absence of summer precipitation and the direction of storm winds. Mojave storm fronts come from the northwest and Chihuahuan fronts (especially in summer) from the southeast.

Physiographically the northern edge of the Peninsular (Colorado) and Mojave Deserts form a reversely symmetrical pattern in relation to the Mapimian and Saladan subdivisions of the Chihuahuan Desert. Both systems are characterized by a desert basin range topography with fragmented ranges in parallel and tilted to the northwest. Both systems are interrupted by an abrupt transverse ridge, marking a boundary between low desert and high plateau. In the Mojave-Peninsular system the ridge is the Transverse Ranges (e.g.,Little San Bernardino Mountains) north of which lies the high desert plateau of the Mojave (about 700 meters above the Peninsular desert). The Mapimian-Saladan system has a reversely symmetrical Anticlinorium of Arteage south of which lies the high desert plateau of the Saladan Desert, about 700 meters above the Mapimian. The two systems differ from one another radically in elevation and latitude, but here again the differences are largely reciprocal, especially in their effects on temperature. The Mojave-north Peninsular system averages about 650 meters lower in elevation than its Mapimian-Saladan counterpart. If the constant of $-6°C$ (average annual temperature) per 1000 meter increase in elevation is applied, elevation, considered alone, should make the Mojave system $3.9°C$ warmer than the Mapimian. If latitude alone is considered, the Mojavian system is centered about 7 degrees north of the Mapimian. Using the constant of $0.5°C$ increase per degree toward the equator, the predicted difference would make the Mapimian $3.5°C$ warmer than the Mojavian system. Thus elevation virtually cancels the effect of latitude in altering the average annual temperatures between the two systems; the net effect is predicted at about $0.4°C$ warmer in the Mojavian system. In fact, as noted in Table 2, the average temperature range for the Mojave-Peninsular (Colorado) systems is $19.2 - 22°C$ and for the Mapimian-Saladan it is $17.7 - 20.9°C$. The net difference based on these averages is $1.2°C$ warmer in the Mojave-Peninsular system. Considering the inherently crude and uneven sources implicit in the calculation of average temperatures for large regions, the difference between the predicted $0.4°C$ and the averaged $1.2°C$ is probably not significant.[1]

The symmetry of the land and climate are intimately interdependent, the discussion above illustrating but one aspect of their relationships. The vegetation, especially through its dominates and associations, shows a profound obedience to the interactions of land and climate (MULLER, 1939). Figure 10 illustrates the fundamental isomeric symmetry between Mojavian and Mapimian systems. It has been extended to include the Great Basin steppe to the north, peripheral xeric associations, and the Saladan subprovince to the south. The diagram reveals coarse reversely symmetrical patterns between the distributions of creosote scrub, Joshua tree (arborescent *Yucca*), edaphic dune associations, *Artemisia* scrub, and even palm oases. The fit is by no means exact as reverse images, but does illustrate a

[1] Other reversely symmetrical factors are presented in Chapter on Climate.

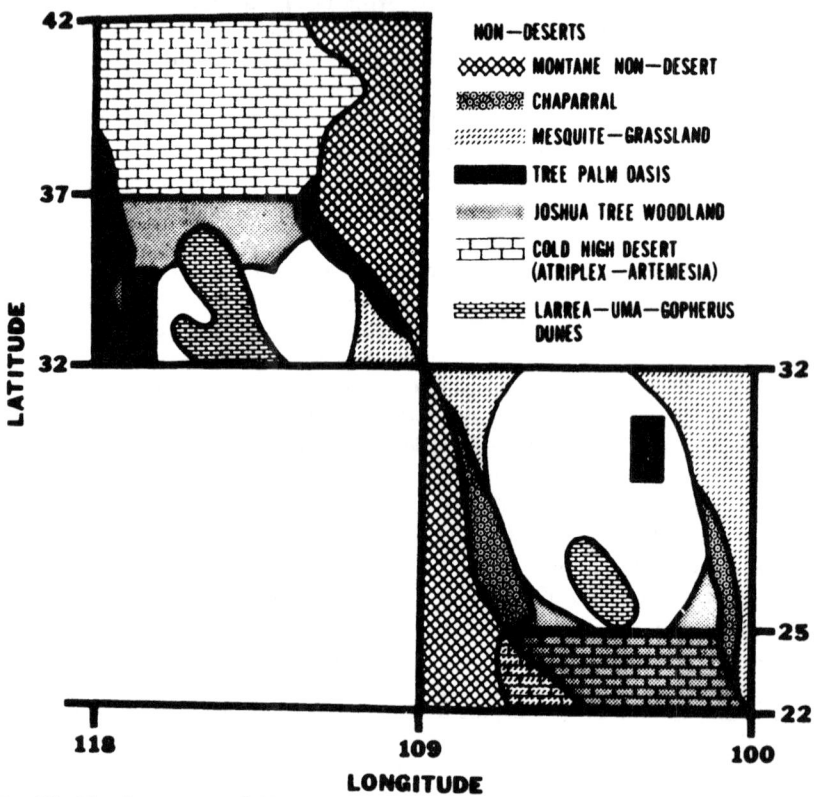

THE SYMMETRY OF NORTH AMERICAN DESERT PLANT ASSOCIATIONS

Legend:
NON—DESERTS
- MONTANE NON—DESERT
- CHAPARRAL
- MESQUITE—GRASSLAND
- TREE PALM OASIS
- JOSHUA TREE WOODLAND
- COLD HIGH DESERT (ATRIPLEX —ARTEMESIA)
- LARREA—UMA—GOPHERUS DUNES

Fig. 10. The Symmetry of North American Desert Plant Associations (Including Selected Edaphic Associations and Faciations).

basic reversal of low and high desert vegetation with certain edaphic features in approximately inverted and reverse positions. Major discrepancies include the extension of creosote scrub (*Larrea*) four degrees north of the 32nd parallel but a full ten degrees south. Also, the Saladan Desert sustains a superimposition of artemisian "cold desert" and "high desert" yucca woodland. This compression of vegetation toward the tropics may reflect a general tendency toward narrower elevational zonation in tropical systems.

Map 9 and Table 20 make an explicit comparison between associations in the Mojavian and Mapimian systems. It is based on points on a 1000 kilometer transect with the 500 kilometer midpoint set at the Continental Divide crossing the 30th parallel north. The points are compared in pairs (with the same designating numbers) equidistant from the Continental Divide, one northwest in the Mojavian system and the other southeast in the Mapimian. The points, consisting of five sets, were chosen as showing the best possible reverse symmetry between the two systems. Equidistant positions were fixed for mesquite-grassland (or oak savanna), creosote scrub, creosote-mesquite dunes, palm tree oases, arborescent yucca

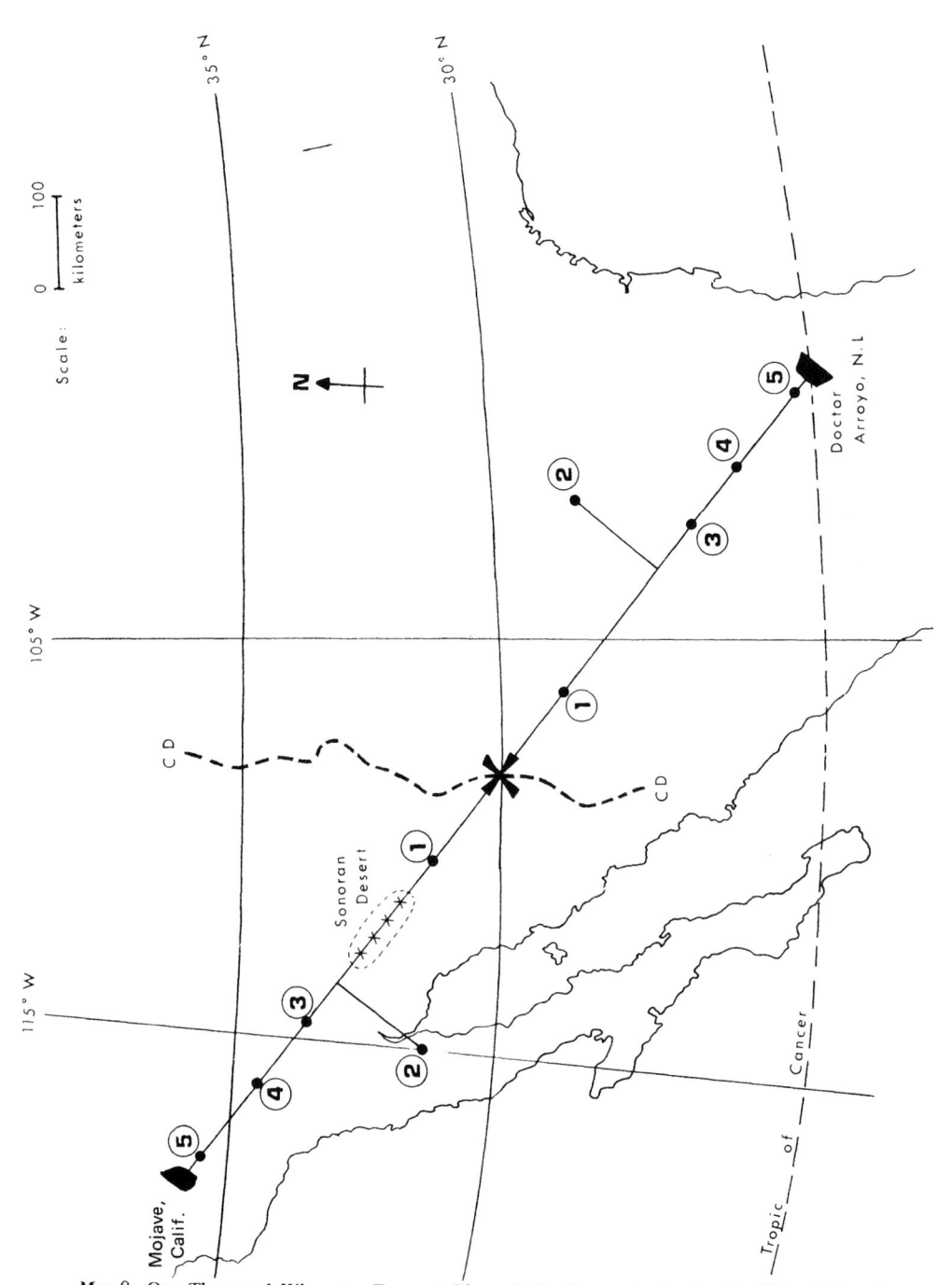

Map 9. One Thousand Kilometer Transect Through the Reversely Symmetrical Vegetation of North American Deserts. Key: 1. = Grassland (Oak Savanna); 2. = Relict Palm Tree Stands; 3. = Mesquite/Creosote Dunes; 4. = Joshua Tree Woodland; 5. = Chaparral/Pinyon-Juniper Woodland; CD = Continental Divide.

Table 20. The Reverse Symmetry of North American Desert Biotas East and West of the Continental Divide. Based on a 1000 kilometer transect from the vicinity of Mojave, California to Dr. Arroyo, Neulo Leon and with its midpoint at the intercept between the Continental Divide and the 30th parallel north.

Station No.	Association (see plates)	Herpetofauna (examples)	Transect Distance from Cont. Divide (NW & SE)	Lat./Long.	Elevation
1.	Grassland (Mesquite-Oak Savanna)	1. Holbrookia maculata	110 kms	31N/110W 27N/108W	1700 m 1900 m
2.	Palm Oasis (West: Washingtonia East: Brahea)	1. West a. Elaphe rosaliae[1] East b. Elaphe subocularis[1]	290 kms	32N/115W 29N/102 1/2W	100 m 100 m
3.	Creosote-Mesquite Silicate Dunes (Larrea-Prosopis)	1. West a. Uma scoparia[1] East b. Uma exsul[1]	310 kms	34N/115W 26N/106N	100 m 1100 m
4.	Arborescent Yucca Woodland (West: Y. brevifolia East: Y. filifera)	1. Hypsiglena torquata	400 kms	34N/116W 25/N/101W	1100 m 1750 m
5.	Pinyon-Juniper (& Oak) (West: Pinus monophylla East: Pinus cembroides)	1. West a. Masticophis lateralis[1] East b. Masticophis taeniatus[1]	500 kms	36N/117W 23N/100W	1100 m 1700 m

1. Sibling species.

Table 20. The Reverse Symmetry of North American Desert Biotas East and West of the Continental Divide (continued).

Station No.	Association (see plates)	Herpetofauna (examples)	Transect Distance from Cont. Divide (NW & SE)	Lat./Long.	Elevation
Non-Symmetrical					
6.	Sahuaro Cactus Woodland (*Cereus*)	1. *Heloderma suspectum*	150-250 kms NW only	33N/113N	100-2000 m
	A. Average Difference in Latitude between 5 symmetrical stations on the transect: 7 degrees		Average Expected Temperature[2] increase to the south 7 x .5 C = + 3.5 C		
	B. Average Difference in Elevation between the 5 symmetrical stations on the transect: 650 m		Average Expected Temperature Decrease with greater elevation (in south) 6.5 x 6 C = −3.9 C		
			Average Net Temperature Difference: −.4 C in the more southern station		

2. Average annual temperature

woodland, and chaparral/pinyon woodland associations. A number of other equidistant points that support the same dominant vegetation can be plotted along the transect. The only major discrepancy is the segment from 150 kilometers to 250 kilometers northwest of the Continental Divide which is dominated by a Sonoran tubular cactus (*Cereus*) association. It alone is without parallel on the other side of the divide (at least within the range of the transect line).

Climate, physiography and vegetation all affirm and partially explain the affinities already noted between Mojavian and Mapimian herpetofaunas. These affinities may now be examined in more detail and with further quantification, using the transect just discussed. The herpetofaunas of paired points no. 3 (dunes) have been compared in Table 21. The results, quantified into Coefficients of Community, indicate a value of 0.43, overall with 0.29 for lizards and 0.66 for snakes. These indicate a stronger affinity than characterizes the fauna of the Mojavian and Mapimian units taken as a whole. Furthermore, the shared component in the snake fauna is so great that they could be treated as samples from the same biotic province using the 0.60 level of HAGMEIER (1966) for mammals. Snakes are very different ecological entities from lizards. They are larger, operate over larger areas, draw food on a higher trophic level, and are quite possibly less restricted by thermal limits (i.e., don't require the high body temperatures of heliothermic lizards for sustained activity). It is not surprising then that snakes appear to have been more conservative in differentiation between eastern and western deserts and that a higher percentage is conspecific at both points. If the list of compared lizards is corrected (by synonomy) for sibling species in the sense of MAYR (1942) and/or allopatric semispecies, the C.C. value is raised from 0.29 to 0.80. Thus, the low shared lizard fauna is almost entirely accounted for by closely related allopatric forms which may or may not be valid biological species. If the value corrected for sibling lizards is incorporated into a raised total C.C., the new value is 0.73. Therefore, only the distinction of sibling lizard stocks lower the shared fauna values below the minimum for consideration of both samples as part of a single faunal province.

A second comparison in Table 22 between the herpetofaunas at points no. 4 (arborescent yucca) on the transect reveals these general patterns of relationship, but to a lesser degree. The C.C. value for the two points of no. 4 are lower than for the two major provincial (or subprovincial) units being compared, but only slightly (0.16 versus 0.18). Corrections for sibling species dramatically raise the C.C. values, totals from 0.16 to 0.31, but these are still below the minimum (0.60) for single province status. Here too, most of the siblings are lizards (two out of three). Coefficients both before and after correction do indicate much closer affinities between snake faunas than lizard faunas, 0.42 and 0.21 respectively -- after correction for siblings.

Appendix I provides illustration of the actual localities at paired points 4.

If sibling species are treated as conspecific, the whole set of quantified comparisons, C.C. and S.C. for all North American deserts, may be recalculated to determine the influence of this one factor in the differentiation of desert herpetofaunas. A list of sibling pairs (one west and one east of the Divide) have been provided in Table 23. The C.C. and S.C. values are generally raised by an additive factor of 0.20 between eastern and western provinces and subprovinces (see Table 24).

Table 21. A Comparison of the Herpetofaunas of Chihuahuan and Colorado Desert Dune Edaphic Associations.

Localities:	Mojave Desert Colorado River Valley Riverside, California	Chihuahuan Desert Bolson of Laguna Mayran (vicinity of San Pedro) Coahuila
Latitude:	34° 00'	25° 45'
Longitude:	11° 30'	103° 00'

Herpetofauna
  Reptilia:

Lacertilia:

1. *Coleonyx variegatus*	1. *Coleonyx brevis*
2. *Crotaphytus wislizeni*	2. *Crotaphytus wislizeni*
3. *Phrynosoma m'calli*	3. *Phrynosoma modestum*
4. *Phrynosoma platyrhinos*	4. *Phrynosoma cornutum*
5. *Sceloporus magister*	5. *Sceloporus magister* (?)
6. *Uma inornata*	6. *Uma exsul*
7. *Cnemidophorus tigris*	7. *Cnemidorphorus tigris*
8. *Uta stansburiana*	8. *Uta stansburiana*
9. *Dipsosaurus dorsalis*	
10. *Urosaurus graciosus*	

Serpentes:

1. *Leptotyphlops humilis*	1. *Leptotyphlops humilis*
2. *Arizona elegans*	2. *Arizona elegans*
3. *Masticophis flagellum*	3. *Masticophis flagellum*
4. *Hypsiglena torquata*	4. *Hypsiglena torquata*
5. *Rhinocheilus lecontei*	5. *Rhinocheilus lecontei*
6. *Chionactis occipitalis*	6. *Crotalus atrox*
7. *Phyllorhynchus decurtatus*	
8. *Crotalus atrox*	
9. *Crotalus cerastes*	

Coefficient of Community:		Coefficient of Community Corrected for Sibling and Allopatric Semi-Species:
Lacertilia:	.29	.80
Serpentes:	.66	.66
Total:	.43	.73

While the fraction of shared faunas is raised significantly in all of these comparisons, the sibling corrected Coefficient of Community of the Mapimian-Peninsular comparison is a striking 0.55 in contrast to the original value, 0.22.´ While other shifts in corrected values for both S.C. and C.C. are uniform and proportional, the corrected Peninsular-Mapimian relationship radically changes the rank affinities between the Peninsular Desert and the other units. The shared Peninsular-Mapimian siblings account for so much of the difference between them, that the corrected C.C. values indicate that they are virtually units of the same faunal province. Furthermore the corrected Peninsular fauna has a higher affinity for the Mapimian than for any of the other North-American desert units, including the more adjacent Mojavian and Sonoran. S.C. values rank Sonoran and Mapimian herpetofaunas at approximately equal status in relation to the Peninsular fauna. The depauperate Mojavian fauna is revealed as almost totally shared with the Peninsular (S.C. = 0.97). The specific sibling sets involved include the following list (modified from Table 23):

Peninsular	Mapimian

Reptilia:
  Lacertilia:

1. *Coleonyx variegatus*	1. *Coleonyx brevis*
2. *Callisaurus draconoides*	2. *Cophosaurus texanus*
3. *Phrynosoma platyrhinos*	3. *Phrynosoma cornutum*
4. *Phrynosoma m'cally*	4. *Phrynosoma modestum*
5. *Cnemidophorus labialis*	5. *Cnemidophorus inornatus*
6. *Xantusia henshawi*	6. *Xantusia arizonae (bolsoni)*

Serpentes:

1. *Elaphe rosaliae*	1. *Elaphe subocularis*
2. *Masticophis lateralis*	2. *Masticophis taeniatus*
3. *Tantilla planiceps*	3. *Tantilla atriceps*

Table 22. A comparison of the Herpetofaunas of Chihuahuan and Mojave Desert Creosote and Arborescent Yucca Associations.

Localities	Mojave Desert[1] (Within 40 kms. radius of center) Joshua Tree National Monument Riverside County, California	Chihuahuan Desert (within 40 kms.) Rocamonte, 75 kms. SW of Saltillo on Mex. 54, Coahuila
Latitude:	34 00	24 45
Longitude:	116 00	101 15
Herpetofauna:		
Amphibia:		
Salientia:	1. *Bufo punctatus*	1. *Bufo cognatus*
		2. *Bufo punctatus*
		3. *Scaphiopus hammondi*
Reptilia:		
Testudinata:	1. *Gopherus agassizi*	
Squamata:		
Lacertilia:	1. *Coleonyx variegatus*	1. *Cophosaurus texanus*
	2. *Callisaurus draconoides*	2. *Phrynosoma cornutum* (?)
	3. *Crotaphytus wislizeni*	3. *Phrynosoma modestrum*
	4. *Dipsosaurus dorsalis*	4. *Sceloporus grammicus*
	5. *Phrynosoma platyrhinos*	5. *Sceloporus cautus*
	6. *Masticophis flagellum*	6. *Ficimia cana*
	7. *Phyllorhynchus decuratus*	7. *Rhinocheilus lecontei*
	8. *Pituophis melanoleucus*	8. *Salvadora grahamiae*
	9. *Rhinocheilus leconti*	9. *Tantilla atriceps* (?)
	10. *Salvadora hexalepis*	10. *Crotalus atrox*
	11. *Crotalus cerastes* (?)	11. *Crotalus scutulatus*
	12. *Crotalus scutulatus*	12. *Lampropeltis getulus*
	13. *Tantilla planiceps* (?)	13. *Pituophis melanoleucus*
	14. *Arizona elegans*	14. *Masticophis taeniatus*
	15. *Chionactis occipitalis*	15. *Arizona elegans*
	16. *Lampropeitis getulus*	
Coefficient of Community:		Coefficient of Community Corrected for Sibling and Allopatric Semi-Species:
Lacertilia:	.0	.20
Serpentes:	.24	.31
Total:	.16	.28

[1] *Uta stansburiana and Xantusia vigilis* should be included on this list.

In addition, several non-desert species in Peninsular California have siblings on the Central Plateau of Mexico. These include: *Hyla cadaverina-Hyla*

Table 23. Pairs of Sibling Species Allopatric and Centered East or West of the Continental Divide.

	WEST	EAST
Amphibia		
Salientia:	1. *Bufo retiformis*	1. *Bufo debilis*
Reptilia		
Squamata:	1. *Coleonyx variegatus*	1. *Coleonyx brevis*
	2. *Callisaurus draconoides*	2. *Cophosaurus texanus*
	3. *Phrynosoma platyrhinos*	3. *Phrynosoma cornutum*
	4. *Phrynosoma m'calli*	4. *Phrynosoma modestum*
	5. *Sceloporus occidentalis*	5. *Sceloporus undulatus (S. cautus)*
	6. *Cnemidophorus labialis*	6. *Cnemidophorus inornatus*
Serpentes:	1. *Elaphe rosaliae*	1. *Elaphe subocularis*
	2. *Ficimia quadrangularis*	2. *Facimia cana*
	3. *Masticophis bilineatus*	3. *Masticophis taeniatus*
	(Masticophis lateralis)	

arenicolor, Hyla regilla-Hyla eximia, Phrynosoma cornutum-Phrynosoma orbiculare, Eumeces skiltonianus-Eumeces brevirostris, and possibly *Gerrhonotus multicarinatus* (plus *paucicarinatus)-Gerrhonotus liocephalus.* Furthermore, the mesic turtle genus *Chrysemys (scripta)* occurs in the rivers of both regions with very restricted distribution in the Sonoran province. On the racial level further evidences of a special Peninsular-Mapimian affinity in herpetofaunas may be found. WEBB (1965) indicated that *Xantusia vigilis gilberti,* a Peninsular race, most closely resembled the Mapimian race, *extorris.*

Geographically the Peninsular and Mapimian Deserts are about 600 airline kilometers apart. However, they do fall at parallel latitudes, basically between 25 and 30 degrees north. The Colorado Desert herpetofauna represents a depauperate northern extension of the Peninsular, terminating about the 33rd parallel. The herpetofauna of the Trans-Pecos Desert could also be viewed as a depauperate version of the Mapimian (note that the former is virtually devoid of sexual endemics) which also extends north to the 33rd parallel along the Pecos and Rio Grande Rivers (farther north extensions are very much depauperate and often devoid of true desert vegetation, e.g., *Larrea*). JAEGER (1957) has noted the striking climatic and vegetative similarities between central Peninsular (Magdelenan) and Chihuahuan Deserts. He particularly noted similar precipitation and average temperature.

The similar climates of Peninsular and Chihuahuan Desert seems hardly justified by their relative geographical and topographical positions. While they both occur along the same latitudes, the Mapimian is a full 1000 meters above the Peninsular. Applying the constants used previously, the Peninsular Desert should be about 6°C higher in average annual temperature. This is not the case. As shown in Table 2, the two deserts average temperatures within 1.1°C of one another. Greater exposure to cooler moist gulf winds and coastal fog banks probably contribute greatly to lowering Peninsular temperature.

The two regions also share diversified *Agave* floras (which contrasts with the poor Sonoran representation). Arborescent yucca woodlands, *Y. brevifolia* in the Mojave and *Y. filifera* in the Chihuahuan Desert, are joined by *Y. valida* in central Baja California. No parallel or sibling yucca associations characterize the intervening Sonoran Desert. JOHNSTON (1971) indicated that the scrub *Colubrina viridis* is Peninsular centered in its distribution, but has a few scattered populations in the Sonoran Desert and three along the Nazas River in the Mapimian Desert. The species is a low disperal form, and populations were quite possibly connected by overland dispersal, not bird or wind carried (JOHNSTON, pers. comm.). The Mapimian localities correspond almost perfectly with the range of *Xantusia vigilis extorris* which was just noted as a form possibly linked to Peninsular populations (possibly in the Miocene Epoch).

Table 24. Matrix of Resemblance Coefficients Corrected for Sibling Species for the Compared Herpetofaunas of North American Deserts. Coefficients of Community appear in upper right triangle; Coefficients of Similarity in lower left; circled figures indicate number of samples (N).

	Trans-Pecos	Mapimian	Saladan	Sonoran	Peninsular	Mojavian
Trans-Pecos	㊼	.61	.55	.50	.30	.34
Mapimian	.71	㊶	.56	.59	.55	.46
Saladan	1.0	1.0	㉙	.34	.32	.21
Sonoran	.69	.60	.67	㊽	.41	.55
Peninsular	.48	.56	.55	.60	㊼	.52
Mojavian	.68	.81	.39	.90	.97	㉝

There are also obvious affinities between the Chihuahuan and Sonoran Desert herpetofaunas. Despite climatic and vegetative differences, the unmodified C.C. value for the shared Trans-Pecos-Sonoran fauna is 0.32, higher than between any of the other pairs of eastern and western deserts. Many of these shared species also penetrate into the Mapimian Desert. Even the high, cool and distant Saladan subprovince shares 52% of its herpetofauna with the Sonoran province (S.C.). When corrected for sibling species (Table 24) all of these herpetofaunas produce higher shared values, but not in equal proportions. The most dramatic change is the increased affinity between Sonoran and Mapimian at 0.59 C.C. (even higher than Sonoran-Trans-Pecos). The corrected S.C. values maintain the same general pattern and rank of relationships as uncorrected. The physical proximity of Chihuahuan and Sonoran Deserts and their historical mutual accessibility via the Cochise filter barrier does much to offset ecological differences.

The relationships between North American desert biotas have been poorly explored in the past, largely because of a near total ignorance of the existence, position, and biotic composition of the Chihuahuan Desert. Zoologists in particular have demonstrated a rather "ethnocentric" prejudice in favor of the Sonoran Desert. Peninsular and Mojave Desert have been viewed solely in terms of being an extended or depauperate version of the Sonoran motherland. Since many taxa, *Xantusia* and *Uma* being good examples among the reptiles, are virtually absent from the core of the Sonoran Desert, this attitude can be no more than a half-truth. The foregoing discussions have purposely placed a counter-emphasis on the relationships between the Mojave and Peninsular Desert biotas to those of the Chihuahuan Desert and especially its Mapimian core. A few desert biologists have been aware of these relationships, at least in a general intuitive way. SHREVE (1951) offered an early insight into the potential significance of these relationships with the following statement:

The Chihuahuan and the Sonoran Deserts constitute parallel paths along which plant races have moved or receded under dissimilar sets of conditions. Their histories have undoubtedly been linked by a common participation in the effects of climatic change. The results of investigation of the two regions would reveal many relationships and contrasts of value in the study of each of the regions, and also be of use in deciphering as much as possible of the history of the entire arid region of North America.

SHREVE used the term "Sonoran Desert" to include all deserts west of the Continental Divide. Still his meaning is clear and the predictive insight of his statement is reflected in the previous discussions of desert symmetry and differentiation. The topography and paleoecological history of arid western North America have combined, sometimes by incredible coincidences, to provide the biologist with exceptional opportunities to examine related but isolated biotas. These assemblages of common evolutionary origin have been fragmented in parallel but reversely symmetrical environmental settings. The conformity and predictability with which they reassemble into symmetrical communities tends to affirm the validity of the community concept as a biological tool. The symmetry observed in this analysis, especially along the 1000 kilometer transect of Map 9, supplies a strong basis for the use of communities in characterizing the biota of arid temperate North America. Not only do dominant or indicator plant species assemble in predictable patterns relative to the climate and topography, but so do whole faunas, such as the dune herpetofaunas of points no. 3.

In summary, three models of relationship between the Chihuahuan and three western deserts can be discriminated. One, between the Mapimian and Mojavian Deserts reveals a pattern of reverse and inverse symmetry both in vegetation associations and herpetofaunas. The model involves peripheral portions of the Saladan and Peninsular Deserts as well. It further indicates a generally reciprocal compensation in climate due to lower elevation in the Mojavian system and lower latitude in the Mapimian.

A second model is that of the Peninsular versus Mapimian Deserts. Here symmetrical distributions of biotic communities are directed along parallel axes between 25 and 33 degrees north latitude. Both axes tilt northwest and run about 600 kilometers apart (east-west).

The third model demonstrates the relationship between Sonoran and Chihuahuan Deserts. Here, the main theme is not ecological, climatic or physiographical symmetry, but rather geographical proximity and contact specifically through the Cochise filter barrier. Much of the differentiation between the herpetofaunas of these deserts can be accounted for by the presence of sibling species east and west of the Continental Divide. Correcting for these forms radically effects statistical comparisons of shared faunas, in all east-west desert comparisons. Especially between Mapimian-Peninsular faunas is this the case. Between these two units, the Coefficient of Community was raised from 0.22 to 0.55 by correcting for sibling forms.

Montane Relict Herpetofaunas

The herpetofaunas here under consideration are ecological island inhabitants, populations occurring in upland mesic non-desert associations (chaparral, pinyon-juniper, encinal, and coniferous forests). Physiographically these islands are situated on isolated blocks above 1500 meters in the Trans-Pecos and Mapimian subprovinces and 1800 meters in the Saladan. Several of the seven ranges being considered might better be viewed as peninsular extensions of major mountain systems rather than completely distinct islands. Map 10 shows the general arrangement of the montane systems under review. As it indicates, the Chirachuas, Huachuca, and Animas ranges, here combined into a single unit, are virtually northern extensions of the Sierra Madre Occidental. The Sacramento Mountains, similarly, could be viewed as a topographical southern extension of the Rocky Mountains. The Anticlinorium of Arteaga is an unquestionable montane peninsula of the Sierra Madre Oriental. Other peripheral montane regions could have been included, but have been deleted since they have already been reviewed as contiguous faunal provinces: The Sierra Madres, the Navahonian (consisting of the Colorado Plateau and Southern Rockies), the Balconian (Edwards Plateau), and Trans-Volcanic provinces in particular. The emphasis in this analysis is placed on montane units of comparable areas that are immediately peripheral or to within the Chihuahuan Desert.

Several major and numerous minor montane relict localities within the desert are not considered here simply for lack of adequate sampling. Many are virtually without a known herpetofauna despite their imposing area and elevation (the Sierra Mojada of Coahuila being a good example). Some of these more major isolated uplands are outlined by broken contour lines, 1500 meter and 1800 meter levels. One fairly well sampled range, the Davis Mountains of Jeff Davis County, Texas, was omitted for simplification and because its herpetofauna is nearly identical to that of the Guadalupe Mountains, Culberson County, Texas.[1]

Within the seven selected montane systems, adequate herpetofaunal sampling has been undertaken for the Sierra Madres of Arizona-New Mexico (see BOGERT & DEGENHART, 1961; FOWLIE, 1965), and for the Chisos (and to a lesser extent the Carmens) (see MINTON, 1959; SCHMIDT & OWENS, 1944; SCHMIDT & SMITH, 1944). A moderate sampling effort has been undertaken for the Sierra del Nido (ANDERSON, 1962 and in preparation) and for the Guadalupe Mountains (GEHLBACH, 1964). The Sacramento Mountains and Anticlinorium have moderately and irregularly been sampled by several workers. The plateaus near Charcas and Sierra Catorce are very poorly collected, the few available records coming from scatterd museums, TAYLOR (1949, 1950, 1952, and 1953), and the author's field work. The values presented here represent only a preliminary analysis, obviously based on uneven sampling, and should be construed to depict only very generalized patterns of faunal affinity.

Both C.C. and S.C. have been presented for the compared relict faunas in Table 25. Traditionally (SIMPSON, 1960) S.C. is preferred for comparisons such as these where faunal assemblages are of small and uneven sizes (13 to 31 species). I have decided to use both coefficients since the C.C. emphasizes existing ecological difference in terms of total compared diversity. Therefore it can differentiate a particularly depauperate system from a similar but more diverse system where the S.C. value might fail to do so. In contrast, S.C. values are useful in their emphasis of common (paleoecological) derivation. Thus both coefficients may provide useful patterns of relationship, the former for spatial and structural patterns, the latter for paleoecological and phyletic relationships.

The dendrograms generated from these coefficients are in Figure 11. They

[1] *Hyla arenicolor, Eumeces brevilineatus,* and *E. multivirgatus* excepted.

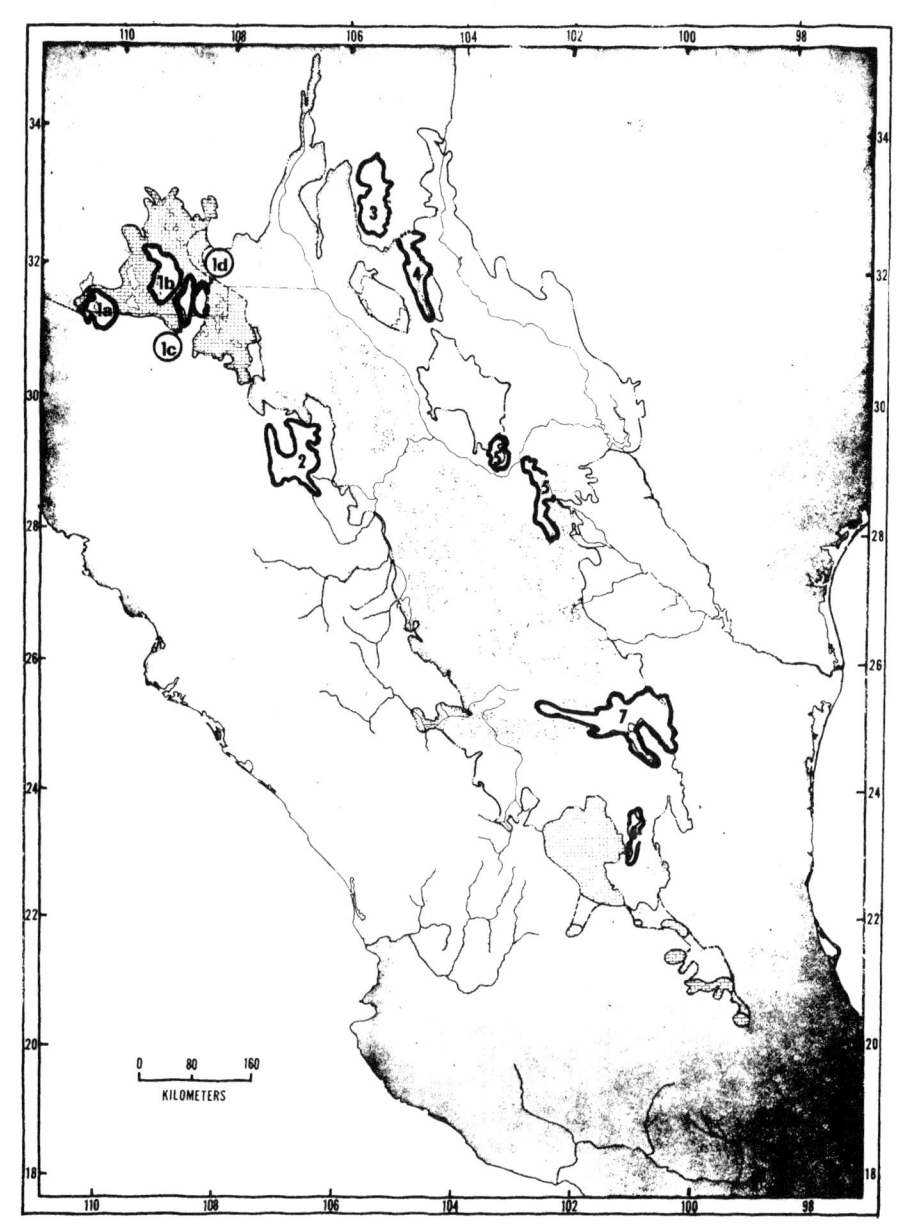

Map 10. Major Mountain Ranges in the Vicinity of Chihuahuan Desert. Key: 1. =Chirachaua-Huachuca; 2. = Sierra del Nido; 3. = Sacramento Mountains; 4. = **Guadalupe** Mountains; 5. = Chisos – Sierra del Carmen; 6. =Sierra de Catorce; 7. =Anticlinorium de Arteaga.

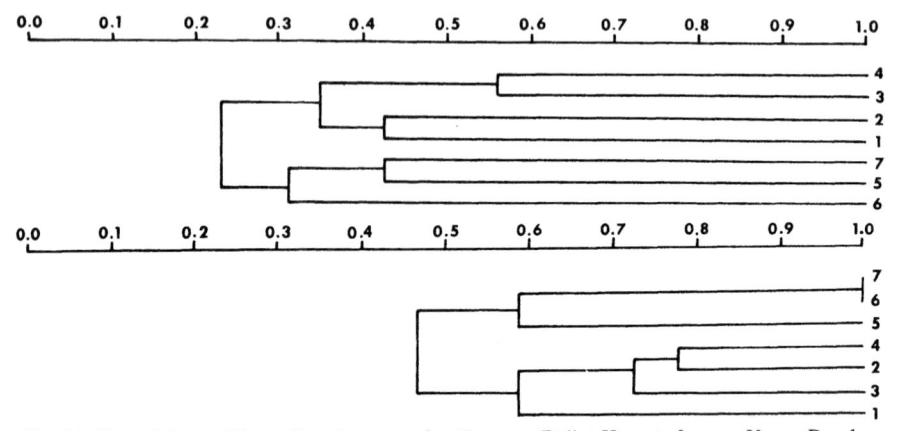

Fig. 11. Faunal Resemblance Dendrograms for Montane Relict Herpetofaunas; Upper Dendrogram: Coefficients of Community; Lower Dendrogram: Coefficients of Similarity; Key: 1. = Sierra Madre Occidental, Arizona Herpetofauna; 2. = Sierra del Nido Herpetofauna; 3. = Sacramento Mountains, New Mexico Herpetofauna; 4. = Guadalupe Mountains, Texas-New Mexico Herpetofauna; 5. = Chisos-Carment Mountains Herpetofauna; 6. = Sierra de Catorce (-Charcas) Herpetofauna; 7. = Anticlinorium de Arteaga Herpetofauna.

Table 25. Matrix of Resemblance Coefficients for Montane Relict Herpetofaunas of the Chihuahuan Desert. Coefficients of Community appear in upper right triangle; Coefficients of Similarity in lower left; circled figures indicate number of samples (N).

	Sierra Madre Occ. Ariz.	Sierra del Nido	Sacramento Mts. N.M.	Guadalupe Mts. Texas	Chisos Mts. (and Carmen)	Catorce-Charcas S.L.P.	Anticlinorium of Arteaga Coahuila
Sierra Madre Occ. Ariz.	③①	.42	.27	.32	.18	.13	.26
Sierra del Nido	.60	③⓪	.42	.41	.33	.25	.27
Sacramento Mountains New Mexico	.52	.52	②①	.56	.30	.17	.23
Guadalupe Mountains Texas	.67	.78	.78	⑱	.48	.19	.28
Chisos and Carmen Mountains Texas-Coahuila	.36	.59	.48	.72	㉒	.21	.42
Catorce Charcas S.L.P.	.38	.70	.34	.38	.46	⑬	.42
Anti-clinorium of Arteaga Coahuila	.42	.30	.48	.61	.72	1.0	㉜

indicate the presence of an eastern and western cluster of montane relict faunas, regardless of the coefficient applied. Most of the compared units would be treated as highly distinct units on the basis of C.C. values within each cluster. Only the Sacramento and Guadalupe Mountains, at 0.56, approach faunal resemblance at the province level. The two clusters differ from one another at the C.C. level of 0.23 and S.C. level of 0.47.

The western cluster consists of the Sierra Madre Occidental of Arizona-New Mexico, the Sierra Del Nido, the Sacramento Mountains and the Guadalupes. Among the species that are typically (but not universally) present are the lizards *Phrynosoma douglassi*, *Urosaurus ornatus* (also in the Chisos) and *Eumeces multivirigatus*; the snake *Thamnophis elegans (errans)* and the northwestern race of the rock rattlesnake, *Crotalus lepidus klauberi*. The toad, *Bufo woodhousei*, also characterizes this western montane fauna. In general, they are forms that traverse the northern Sierra Madre Occidental into the southern Rockies and Colorado Plateau.

The eastern cluster includes the Chisos-Carmen Mountains, the Anticlinorium de Arteaga, and the plateau north of Charcas through the Sierra Catorce. Typical herpetofauna include the lizards: *Sceloporus grammicus*, *Sceloporus goldmani*, *Sceloporus parvus*, *Eumeces brevilineatus* (Chisos-Carmens) or *Eumeces brevirostris* (Anticlinorium), *Leiolopisma silvicolum*, *Gerrhonotus liocephalus* and *Phrynosoma orbiculare*. Within the Saladan Desert are numerous local montane ranges of pinyon-oak-juniper woodland. Populations of *Sceloporus jarrovi* and *Sceloporus torquatus* stud these mountains along with the smaller *Sceloporus parvus* and the snakes *Salvadora grahamiae* and *Crotalus molossus*. *Tantilla wilcoxi* is also represented by montane populations within the Saladan Desert. The absolute resemblance (S.C. = 100%) of the Charcas-Catorce fauna, to the large Anticlinorium assemblages may be in part an artifact of inadequate sampling in the former unit.

The relationships between the two clusters indicate a gradient of change between the two along the northern rim of the Trans-Pecos subprovince. Table 25 indicates a very gradual reduction in affinity between montane faunas, generally along the northern east-west axis. The northwest-southeast transition between Guadalupes (and Davis Mountains) and Chisos mountains is reflective of this very gradual shift. Between these two ranges the herpetofaunas produce a C.C. of 0.48 and S.C. of 0.78. Both are very high values of resemblance, at least relative to other comparisons. The physiographical gap between the two is the Marathon Basin, a plateau still generally above 1000 meters elevation and probably easily traversed by pinyon-juniper community during glacial maxima. Thus the Marathon Basin is better viewed as midpoint in a northern arc of transitional montane faunas than as a major biogeographical barrier.

In general, the western montane faunas become most expansive to the north while those of the east are of broadest distribution in the south, especially in the Saladan Desert. (The Balconian Province might be interpreted as another northern expansion to the east). The result is that montane systems appear to be separated by a diagonal running northwest across the desert. The lowland barriers running in the same direction as the separating diagonal from south to north are as follows: the valley of the Rio Aguavanal, the Bolson de Mapimi, and the Marathon Basin. Actually the separation runs close to the 103rd longitude west and is nearly linear except for the northern tip. It is the Chihuahuan Desert that is skewed to the northwest. This separation is incomplete and bridges or portals

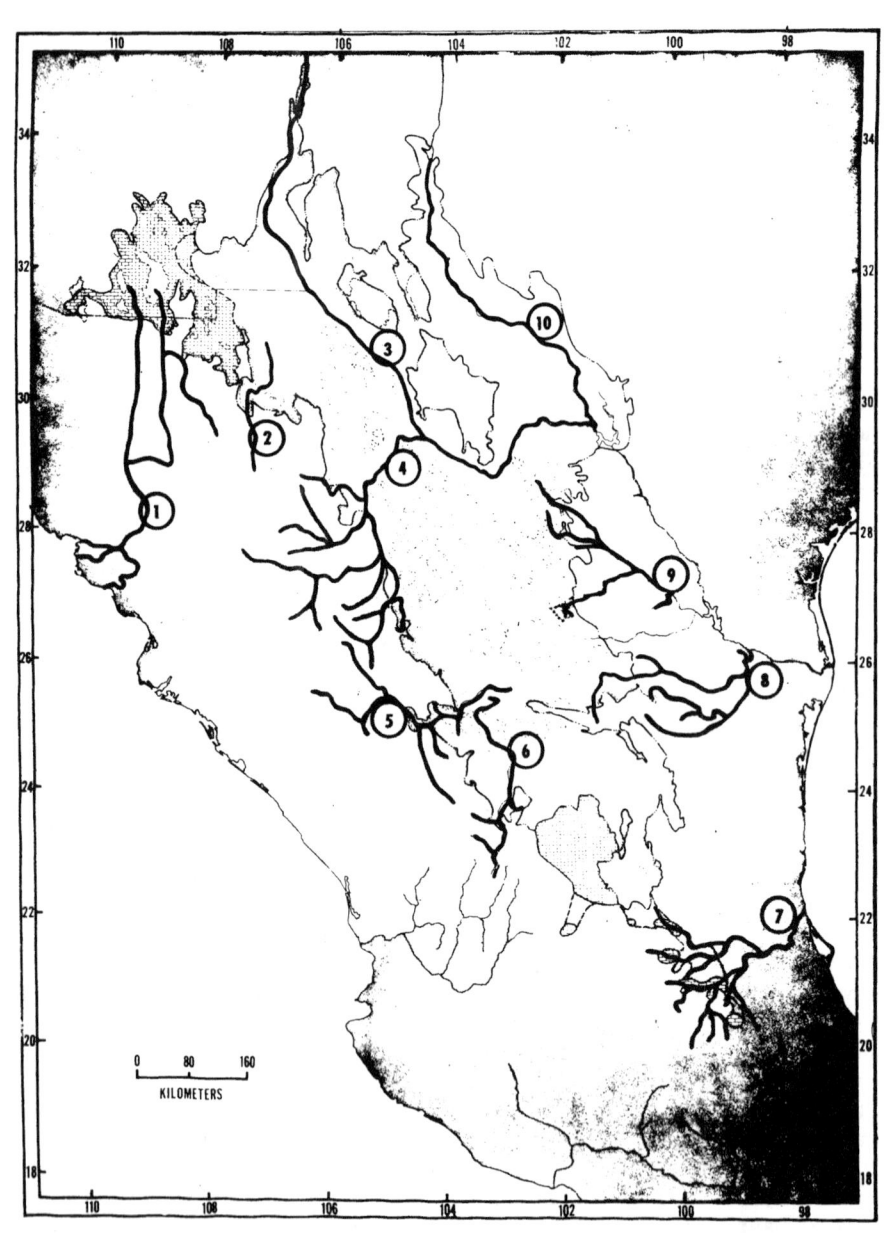

Map 11. Present Major Drainage Systems of the Mexican Plateau. Key: 1 = Rio Yaqui; 2 = Rio Santa Maria; 3 = Rio Grande; 4 = Rio Conchos; 5 = Rio Nazas; 6 = Rio Aguavanal; 7 = Rio Panuco; 8 = Rio Pesquieria; 9 = Rio Sabinas (include R. Salado and Cuatro Cienegas).

154

across the lowland desert barriers have been hypothesized. These will be noted and evaluated in the paleoecological discussion of the subsequent chapter.

Riparian Herpetofaunas within the Chihuahuan Desert

This spatial ecological analysis is confined to the six major permanent drainage systems penetrating the geographical Chihuahuan Desert. The species considered were either physical aquatic (frequent and voluntary swimmers) or repeated and abundant participants in riparian vegetation. These forms were treated as the participant fauna of aquatic and riparian ecosystems. This herpetofauna was not confined to obligatory participants, such as *Chrysemys scripta*, but included abundant but facultative riparian associates such as *"Cnemidophorus tesselatus"*. The contemporary drainage of the region is depicted on Map 11.

Permanent water systems that were omitted include numerous "Ojos de Agua" or natural springs that dot the Chihuahuan Desert (especially in a limestone landscape) and the Bolson. The former are both too numerous and too poorly known for an adequate analysis even on a descriptive level. The Parras Bolson has a well developed riparian woodland of *Fraxinus, Salix* and *Populus*, but its streams and their vegetation have been highly modified by agriculture, irrigation for the development of corn, beans, and vineyards being particularly important. Collecting in the Parras Bolson has been very limited, though the following riparian forms can be listed: *Rana pipiens, Scaphiopus hammondi* (not particularly associated with permanent water), *Kinosternon flavescens, Sceloporus undulatus, Thamnophis marcianus*, and *Thamnophis proximus*. The list is far too incomplete to be included in a quantitative analysis. The Rio Aguavanal is another poorly sampled riparian system. It has been assigned 10 species but much of its length is totally unknown biologically as is its terminal basin, Laguna Viesca (itself an ephemeral body of water). While it is included in the quantitative analysis, it has been deleted from the dendrograms constructed from this analysis.

The Cuatro Cienegas Bolson is another major system of permanent water that is not listed in either the analysis matrix of Table 26 or the accompanying dendrogram, Figure 12. The Cuatro Cienegas Bolson has not been ignored for lack of herpetofaunal sampling nor because of small size. MINCKLEY (1969) and MCCOY (1970 and checklist in preparation)[1] have intensively investigated the fauna and ecology of this bolson for more than a decade. However, the Cuatro Cienegas Bolson is denied discrete recognition here simply because it lacks a distinctive herpetofauna and has had obvious and repeated connections in the recent past with the Rio Salado. Thus the Bolson has been treated here as the upper terminal discharge of the Salado drainage (not to be confused with the Saladan Desert). While it is true that only a man-made aqueduct presently connects the two systems, the separating distance involved, less than 20 kilometers, and intervening topography is such that a natural connection must have occurred through the Puerto Salado in recent glacial and/or pluvial times. I accept the validity of a single riparian species, *Terrepene coahuilae*, for the Bolson. As noted by MILSTEAD (1969) this form is morphologically no more than a subspecies of *Terrepene carolina* and only its ecology, behavior and isolation justify its continued treatment as a distinct species. Other endemics such as *"Trionyx ater"* and

1. and MCCOY and MINCKLEY (1969).

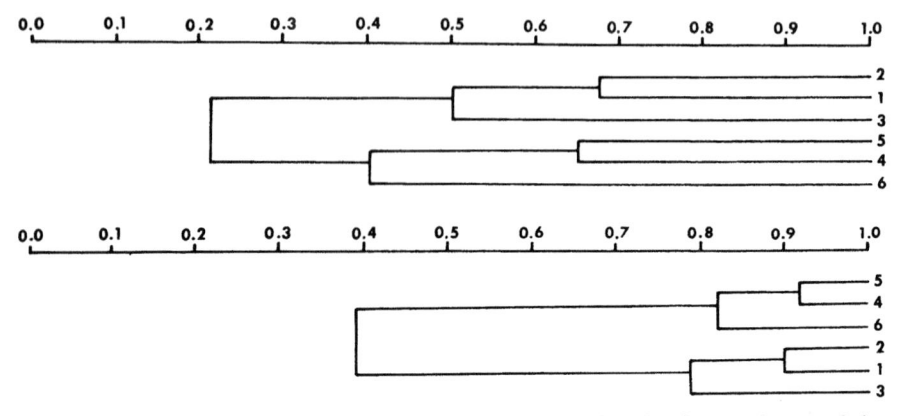

Fig. 12. Faunal Resemblance Dendrogram for Aquatic and Riparian Herpetofaunas of the Chihuahuan Desert; Upper Dendrogram: Coefficients of Community; Lower Dendrogram: Coefficients of Similarity; Key: 1. = Rio Grande (proper west of Pecos); 2. = Rio Conchos; 3. = Rio Nazas; 4. = Rio Salado; 5. = Rio Pesqueiria; 6. = Pecos River.

Table 26. Matrix of Resemblance Coefficient for the Aquatic and Riparian Herpetofaunas of the Chihuahuan Desert.
Coefficients of Community appear in upper right triangle; Coefficients of Similarity in lower triangle; circled figures indicate number of samples (N).

	Rio Grande (above Pecos)	Rio Conchos	Rio Nazas	Rio Aqua-naval	Rio Salado (Cuatro Cienegas)	Rio Pesqueria	Pecos River
Rio Grande	㉖	.68	.48	.30	.24	.12	.30
Rio Conchos	.90	㉑	.52	.30	.28	.10	.38
Rio Nazas	.82	.76	⑰	.44	.14	.16	.26
Rio Aquanaval	.90	.88	.88	⑩	.19	.24	.35
Rio Salado	.50	.31	.25	.44	⑯	.65	.44
Rio Pesquieria	.33	.25	.33	.44	.92	⑫	.36
Pecos River	.46	.63	.53	1.0	.81	.83	㉖

Chrysemys scripta taylori are just melanistic stocks of widespread species of the Rio Grande drainage. Therefore, in this analysis, Cuatro Cienegas is consigned to the Rio Salado drainage system.

Figure 12 provides dendrograms based on both C.C. and S.C. coefficients in Table 26 for the same reasons discussed in justifying their application to montane relicts. The results largely parallel those for montane herpetofaunas. Again the dendrogram produces an eastern and western cluster of drainage systems. However, the clusters are more completely differentiated from one another and more homogeneous within their internal subdivisions. The between cluster difference is at a C.C. level of about 0.20 and S.C. level of 0.39. The three internal units of the western cluster all share faunas at the C.C. level of 0.48 or above and S.C. level of 0.78 or above. The eastern three cluster at a C.C. of 0.36 and an S.C. of 0.81.

The distribution of these units is depicted on Map 6. The 103rd longitude west adequately separates western and eastern systems just as it did for the montane herpetofaunas. The Bolson de Mapimi also probably serves as a barrier to riparian fauna communication or dispersal just as it has to montane species. Other barriers along this axis include the porous limestone topography of the Saladan Desert, a region virtually without major surface drainages, and the Parras Bolson. The latter Bolson has appreciable internal drainage and permanent surface water, but its drainage patterns may have remained internal long enough to create a formidable barrier to eastward communication of riparian faunas. Well above the northern end of the Bolson de Mapimi the two systems do in fact connect through the Big Bend of the Rio Grande (Big Bend National Park, Brewster County, Texas). But the connection is carved through narrow limestone walled steep canyons (i.e., Santa Elena at Big Bend) and probably is relatively recent (mid to late Pleistocene). The paleogeography and ecology of this union will be evaluated in the subsequent chapter.

The western drainage systems consist of the upper (west of Big Bend) Rio Grande, the Rio Conchos, and the Rio Nazas (almost certainly including the Rio Aguavanal). The Rio Grande and Rio Conchos are very closely linked systems with a contemporary connection at Ojinga, Chihuahua. The resemblance coefficients for these two are a C.C. of 0.68 and an S.C. of 0.90. They are virtually arms of the same herpetofaunal province, continuous ecological peninsulas of mesic biota extending northwest and southwest respectively across the desert. The Nazas (and Aguavanal) systems are presently closed basin drainages. CONANT (1963) hypothesized three possible former connections to account for the present Nazas fauna (natricine snakes in particular). He suggested a possible northern connection with the Conchos, a northeastern connection with the Rio Salado via Cuatro Cienegas, and an eastern route of union to the Rio Pesqueria via the Parras Bolson and the pass between Saltillo and Monterrey. The quantitative results presented here strongly favor the first alternative. The Nazas is clearly a distinctive member of the Rio Grande-Conchos system (Conchos-Nazas C.C.: 0.52; S.C.: 0.76), none of its comparisons to eastern riparian assemblages approach these levels of resemblance. Species such as the toad, *Bufo woodhousei*, the turtle, *Kinosternon flavescens*, and the garter snake, *Thamnophis cyrtopsis*, distinguish this western riparian herpetofauna, occurring in three river components. In addition, several northern prairie stocks, the spadefoot *Scaphiopus bombifrons* and the snake *Tantilla nigriceps* penetrate the Nazas region, apparently from the north. The only disharmonious fauna is the presence of the snake, *Natrix erythrogaster*, in the Nazas and Aguavanal systems. This stock is exclusively aquatic and otherwise confined to the eastern drainage. Its occurrence in this portion of the western system will be reviewed in light of paleoecological factors in the following chapter.

The eastern river systems all drain into the lower (east of Big Bend) Rio Grande. Its three units are the Pecos River, the Rio Salado (including Cuatro Cienegas), and the Rio Pesqueria. The Pecos is the most diverse in its herpetofauna and the most distinctive, differing from the other two with C.C. values 0.46 or below and S.C. values 0.83 or below. Distinctive elements include the turtles, *Cheyldra sepentina* and *Chrysemys concinna*, both probably of Austro-riparian origin. The Salado and Pesqueria resemble one another at a C.C. of 0.65 and an S.C. of 0.92 indicating affinity at subprovincial level. The trio of eastern drainages should be slightly extended to include the Rio Grande proper east of Big Bend and the contributing Devil's River and Independence Creek. The entire eastern system is characterized by the presence of the snakes *Natrix erythrogaster, N. rhombifera, Thamnophis proximus,* and the toad *Bufo speciosus.* Some of the herpetofaunas restricted to eastern riparian systems within the Chihuahuan Desert are otherwise ecologically widespread in the more mesic Tamaulipan province or Austro-riparian plain, *Sceloporus olivaceus*, *Acris crepitans*, *Leiolopisma laterale*, *Agkistrodon contortrix* and *Bufo valliceps* being particular examples.

MILSTEAD's (1960) treatment of mesic relicts in the Chihuahuan Desert provided a partial résumé of localized mesic relict herpetofaunas. Several localities, however, are better viewed as peninsular extensions of continuous rivers intruding into the desert rather than as isolated relicts. This is the case at Musquiz and Cuatro Cienegas, both part of the Rio Salado drainage. Furthermore, several other localities, specifically in the vicinity of the cities of San Luis Potosi and Aguascalientes, are ecologically part of the cooler and more mesic Trans-Volcanic province.

X. A HISTORICAL BIOGEOGRAPHY OF THE CHIHUAHUAN HERPETOFAUNA

Introduction

Ultimately, life is a process, not a static object. Biological phenomena must be analyzed through the parameter of time if they are to be understood. Only by understanding process, is it possible for the biologist to derive some predictive value from his science. This chapter, then, represents the most comprehensive overview of the Chihuahuan Desert, a synthesis of objects, assemblages and temporal events in order to discriminate the desert as an on-going dynamic process through time and space.

Since this investigator is confined far more narrowly in time than he has been in space, investigations in a temporal parameter require the use of fragmentary and circumstantial lines of evidence. The conclusions drawn should be judged as the most probable explanation based on current evidence, not as statements of observable fact. The several lines of evidence tapped here include the following: (1) paleogeographical and paleophysiographical evidence as indicated by stratigraphy: locating and qualifying tectonic events and determining the former positions of coastlines and pluvial and lacustrine conditions (for a résumé of tectonic history of the region, see The Physiography of the Chihuahuan Desert); (2) paleoclimatic conditions, as extrapolated from fossil floras (including living species with known physiological-climatic limits), from foraminiferan-radiolarian marine deposits, and from chemical compositions of associated sedimentary matrices; (3) paleobotanical evidence indicating paleoecological conditions: vegetation directly; climate, soil structure, and physiography indirectly; and (4) a direct fossil fauna, both general faunal assemblages (especially mammalian and gastropod) as well as the paleontological record for North American herpetofaunas.

Obviously, the best basis for evolutionary and paleoecological reconstruction lies with the direct fossil record. However, the record is incomplete in the extreme. In addition, many named forms are based on feeble microfossil evidence (e.g., a single vertebra) requiring very speculative systematic diagnoses and often the invoking of subjective chronospecies (qualitative divisions of a single phyletic stock through time). Much more extensive and objective information is provided by the paleobotanical and paleogeographical record of western and central North America. The synthetic use of all available sources of relevant information, with special regard for the direct record, is the most responsible basis for analysis. SAVAGE (1960) has made a particularly lucid explanation and justification for this synthetic approach as applied to the North American herpetofauna.

The ensuing analysis will be focused on the participant herpetofauna of the Chihuahuan Desert proper (as defined in the spatial account). Its purposes will be the determination of biogeographical origins of the component genera and species and their dispersals. Peripheral montane and riparian assemblages will also be discussed, but more briefly and largely as units of contrast. A biogeographical analysis of the proportions here at hand requires the manipulation and synthesis of extremely diverse data over 70 millions years of time. Lines of evidence will be presented in the one to four sequence previously listed. Critical historical events will be presented in temporal order toward the Recent, starting with the Cretaceous Period of the Mesozoic Era and continuing through all the epochs of the Cenozoic Era.

The Cretaceous Period

Before the very Late Cretaceous the paleogeography of North America provided the biogeographer with a "clean slate" in several different and crucial parameters. Earlier Cretaceous strata indicate shallow seas covering virtually all of the middle longitudinal aspect of the continent. Furthermore, a mountainous basin and range topography with associated rain shadows and arid valleys was unknown from this period for what remained of western North America (STIRTON, 1964; KING, 1958). The leveled peneplains that composed the terrestrial Cretaceous West were exposed to a tropical to subtropical climate (heavy precipitation, average annual temperature probably $24°C$ plus, narrow seasonal perturbations)[1]. They supported a subtropical to warm temperate vegetation at the present latitude of Alberta to Wyoming, including some modern familial and generic stocks, but many more that are now extinct or absent from temperate North America. The Cretaceous vegetation was a complex gradient from Alberta to Big Bend, Texas, with Arcto-Tertiary stocks dominating in the north and gradually giving way to a Neotropical Geoflora in the south. The shifts in the continuum were extremely graded. Both Geofloras were probably equally important at 45 degrees north latitude. The same may be said of the fauna in general, and herpetofauna in particular. Therefore, the Cretaceous slate was generally devoid of modern conditions in terms of geography, physiography, and the composition and distributions of its biota.

ESTES (1970) provided an excellent review of the direct herpetofaunal record for the Cretaceous of North America as it relates to existing phyletic lines.

The Lance Formation of Wyoming is again the best single source of evidence (ESTES, 1964). Excepting several archosaurian stocks (eosuchians, ornithischian and saurischian dinosaurs, and pterosaurs) and baenid turtles, the herpetofauna is unquestionably modern on the familial level. Modern amphibian stocks include the families: Ascaphidae, Discoglossidae, Pelobatidae, near Hylids(?), Leptodactyloids, Plethodontidae, and Sirenidae. The turtle fauna consists of Dermatemydidae, Emydidae(?), and Trionychidae. The families of the squamate order are Scincidae, Teiidae, Xenosauridae, Iguanidae, Anguidae (including the Diploglossinae), Parasaniwidae (an extinct helodermatoid family), Varanidae, and the snake family Aniliidae. Crocodyline and alligatorine crocodilians were also represented.

This essentially modern assemblage in terms of extant families differs from a modern fauna in two important features. First, only two modern genera can be identified, *Gerrhonotus* among the anguid lizards, and debatably, a form very close to *Trionyx* (= *Aspiderectes*) among the turtles. Secondly, the assemblage demonstrates the geographical and ecological sympatry of stocks which never occur together in present conditions. Geographically, modern Palearctic stocks, such as the discoglossids then co-existed with Neotropical leptodactyloids.

Generally summarizing, the Cretaceous herpetofauna of middle North America

[1] See ESTES, (1964).

was dominated by most of the modern families presently occurring on the continent. However, modern genera were generally absent. In addition, the herpetofauna, just as the vegetation, reflected a complex mixture of Old Northern and Middle American Elements with additional stocks now completely restricted to the Old World (Varanidae). ESTES (1964) suggested that tropical herpetofaunas, Middle American, or even South American Elements (e.g., the macroteiids) were predominant, more so than would have been indicated by the relative importance of the Neotropical Geoflora alone. No indications of a Young Northern Element were reported. Otherwise the faunal and floral systems are largely co-ordinated in differentiation, organizations and biogeographical affinities, affirming the correlations of SAVAGE (1960).

Paleogene Period

The approach of dividing the Cenozoic Era into Paleogene and Neogene Periods has long held the support of many European stratigraphers. This system essentially clusters the three older epochs, Paleocene, Eocene, and Oligocene into a single historical unit, and the Miocene to Recent into a second such unit. This organization is highly compatible with the paleoecological history of North American biota and places a proper emphasis on the crucial Oligocene-Miocene transition. For this reason, I have employed these terms here.

Paleogeography

The geographical and physiographical history of Paleogene middle North America is essentially a gradual continuum of processes and trends established by the initiation of the Laramide Orogeny in the Cretaceous. The remaining extensions of the Rio Grande Embayment that covered future Mexican Plateau landscape were permenantly obliterated when their underlying geosynclines were folded, faulted, and subjected to volcanism in the Paleocene to Eocene (IMLAY, 1936; MALDONADO-KOERDELL, 1964). Other than this process, the tectonics of the Paleogene may have left the landscape level and generally stable (KING, 1958). Sediments and paleobotanical evidence indicate low aggregational plains and the absence of major ranges or their extended rain shadows. The Continental Divide, as a discrete division, may have been absent. Local ranges farther northwest probably did not rise above 1500 m (KING, 1958). Orogenic forces (especially volcanism) (DREWES, 1968) may have been more active in the western part of the present Chihuahuan Desert, especially in the Oligocene, but evidence is inadequate for either certainty or detail (MALDONADO-KOERDELL, 1964; SINGEWALD, 1936; MAXWELL *et al.*, 1967). To the north, Bering (Eurasian) landbridges maintained terrestrial contact.

Paleoclimate

Paleoclimate trends suggest a temporary warming and precipitation increase between late Cretaceous and Eocene time as indicated by the paleobotanical record and possibly marine deposits as well (see COHN, 1965, p. 129). This reversal in the general post-Laramide climatic cooling-drying trend was overcome by Oligocene time.

Geofloral representatives for the Paleogene of southwestern to central North America are predominantly Neotropical, extending as far north as Oregon and Colorado (40 to 45 degrees north latitude) even into Oligocene time. Major assemblages vary from broadleaf tropical woodlands to tropical savannas, with the general climatic trend (toward aridity) favoring the latter. The inland situations developed savannas and in more mesic (often riparian) sites a warm temperate woodland of predominantly Arcto-Tertiary stocks was sustained. AXELROD (1966a) indicated an elevational zonation of temperate Arcto-Tertiary forests, but displaced about 1000 meters below present levels.

The Middle Eocene Green River assemblage provides the first explicit indication of Madro-Tertiary stocks (AXELROD, 1950). Madro-Tertiary Geoflora appears in the Oligocene of Colorado for the first time as recognizable modern assemblages: woodland, savanna, chaparral, and thorn scrub (MACGINITE, 1953). AXELROD emphasized that these associations were at least geographically (and presumably ecologically) ecotonal between adjacent Arcto-Tertiary and Neotropical Geofloras. Secondly, he observed that many of the stocks present were highly adapted xeric sclerophylls, such as *Ceanothus, Cercocarpus, Quercus* (scrub oak habitats) and *Mahonia*. AXELROD concluded that these specialized stocks must have evolved well before the Oligocene, hypothetically in edaphic xeric islands (rain shadows, lithosol exposures, steep slopes) of the Late Cretaceous-Paleocene. He went on to suggest that these edaphic islands, scattered across southwestern North America between the 25th and 40th parallels north, gradually coalesced into a reticulum, as cooling and aridity favored their expansion through the surrounding Neotropical-Arcto-Tertiary transition. Various localized stocks dispersed between interconnecting islands until a more complex and ubiquitous Madro-Tertiary Geoflora was established.

The implications of Late Paleogene Geofloras could have great importance regarding the origins of a modern xeric herpetofauna. In particular the following conclusions should be considered: (1) essentially modern temperate climates, though less severe, are indicated in the middle latitudes of western North America, even back to the Eocene Epoch; (2) low, possibly localized mountains did occur in regions during the Paleogene; (3) the earliest xeric temperate biota, the Madro-Tertiary Geoflora, included modern and highly adapted sclerophylls; (4) this vegetation apparently first appeared in local edaphic pockets as a result of immediate xeric climates (or microclimates), later interconnecting and exchanging components; and finally (5) that Madro-Tertiary Geofloras first appeared in localities geographically ecotonal between Arcto-Tertiary and Neotropical vegetation, and in fact supported mixed assemblages with representatives of all three geofloras present.

Fauna

The herpetofaunal evidence is unquestionably feeble, but does elucidate trends indicated by the paleoclimatic evidence, geofloras, and previous Late Cretaceous faunas. Just as the Arcto-Tertiary Geoflora continued to differentiate in Paleogene, so did its associated herpetofauna, the Old Northern Element. Emydid turtles and ambystomatid salamanders joined the stocks previously reported for the Cretaceous. Modern genera such as *Bufo, Hyla* and *Alligator* all made their appearance in North America before the end of the Oligocene (ESTES, 1970). Some of these stocks may have dispersed across the Bering land bridge from Eurasian centers and entered the central continental region via Canada. *Ophisaurus*, present in North America by the Miocene, may be one of these forms.

ESTES (1970) documented a progressive regression southward of tropical herpetofaunas, with increasing extinctions of major phyletic lines toward the end of the period. This tropical fauna includes four discernable units, now wholly extinct Mesozoic stocks, stocks now restricted to the Old World, a South American Element, and a Middle American Element. The extinct Mesozoic stocks include eosuchians and baenid turtles, both gone by the end of Eocene. The varanid lizards, now restricted to the Old World warm temperate and tropics, retreated south from Alberta in the Cretaceous to New Mexico in the Oligocene, when their North American stock ultimately disappeared. The diverse macroteiid lizard fauna described by ESTES (1964) from the Cretaceous was apparently eliminated in the Paleogene, along with the pelomedusid (side-necked) turtles. Both are South American Elements. The Middle American Element includes the frog family Rhinophrynidae, and the lizard families Xenosauridae and Helodermatidae. The latter is represented by the modern genus *Heloderma* of Oligocene Colorado (GILMORE, 1928). All three families are now associated with the more xeric thorn scrub aspects of Neotropical derivative vegetation.

The previous discussion of Paleogene climate, topography, and geofloras all indicate the first appearance of a very distinctly semi-arid land herpetofauna, the Young Northern Element. The fossils admirably affirm the implications of these more circumstantial evidences. The above mentioned Middle American Elements suggest the evolution of a xeric adapted herpetofauna. More convincing documentation is provided by the earliest presence of the frog genus *Scaphiopus*, the tortoises *Gopherus* (and ancestor, *Stylemys*), and the lizard genera *Paleoxantusia*, and *Paradisposaurus*. The former two genera are fundamentally Young Northern stocks, being associated almost exclusively with Madro-Tertiary vegetation both past and present. Even modern eastern members of these general stocks occur in xeric microhabitats, such as sandy soil pine barrens, or open grassland. The latter two genera are both extinct but appear to represent phyletic transitions between older Middle American tropical xantusiids and iguanids and temperate semi-arid Young Northern forms. Of the four, *Paleoxantusia* is the oldest (Eocene: HECHT, 1956). The genera *Scaphiopus* and *Gopherus* are Oligocene in their first occurrences. Geographically, all four genera are from middle longitudes of North America, and all but *Paradipsosaurus* (Mexican) are from 40 to 50 degrees north latitude in the Great Plains. *Scaphiopus* might be derived from a Holarctic Old Northern pelobatid stock, but the three reptilian genera are clearly transitional forms of tropical derivation.

Thus the history of the Paleogene Young Northern herpetofauna parallels that of its botanical analogue, the Madro-Tertiary Geoflora very closely in time, space, and evolutionary derivation. Both appear to be evolved as specialized arid edaphic derivatives of tropical, and to a much lesser extent, temperate holarctic stocks (perhaps in ecotonal combinations). The definitions and descendent genera of the major faunal elements are listed on Table 27 and Table 28.

Table 27. Historical Elements of North American Herpetofaunas[1]

Unit	Geofloral Association	Climate	Origin	Examples
A. South American Element (complexes not listed)	Neotropical Tertiary	Tropical	From generalized Tropical American unit; ultimately from Paleotropics	*Atelopus* *Atractus* *Caiman*
B. Old Northern Element[2]	Arcto-Tertiary		Palearctic	
1. Eastern American Complex		Cool temperate		*Cryptobranchus* *Gyrinophilus*
2. Western American Complex[2]		Cool temperate		*Dicamptodon* *Ascaphus*
3. Central American Complex		Warm temperate		*Pseudoeurycea* *Staurotypus*
4. Southeastern American Complex[2]		Warm temperate		*Siren* *Graptemys*
C. Middle American Element (complexes not listed)	Neotropical Tertiary	Tropical, subtropical	From generalized Tropical American	*Basciliscus* *Xenosaurus*
D. Young Northern Element	Madro-Tertiary		From generalized Tropical American unit	
1. Madrean Complex		Subhumid		*Gerrhonotus*
2. Desert and Plains Complex		Semiarid, arid		*Crotaphytus* *Phyllorhynchus*
E. Holarctic Element	Arcto-Tertiary	Cool temperate	Recent invaders from Eurasia	*Bufo boreas* *Rana sylvatica*

Arranged in approximate order from most ancient to most recent in America.

1. Modified from SAVAGE (1960).
2. The arid southwestern colubroid snake genera (e.g., *Pituophis*) may represent a fifth and unnamed complex of Old Northern derivation.

Table 29 provides a percentage breakdown by genus and species of present affinities of Chihuahuan herpetofaunas. It further affirms the intermediate condition of modern xeric land assemblages. The (ecological) participant genera are 64% Young Northern with the remainder being equally divided between Old Northern and Middle American Elements. The same generally holds valid for the derived species. The breakdown indicates a contemporary system with a predominately distinctive fauna, but influenced by stocks from both north and south. Such a system could easily have been initiated in an area of ecotonal contact between its older alternatives. Geographically and ecologically, the Mexican Plateau is in a position very conducive to the differentation and isolation of assemblages originally edaphic and ecotonal in nature. The table indicates a similar but less balanced derivation for relict montane genera, the Old Northern Element being favored here over the Middle American. The Chihuahuan riparian fauna appears to have strong Old Northern affinities (50%), indicating its early establishment and differentiation. This, too, is consistent with the paleobotanical record. Other parallels between vegetation and herpetofauna include the Paleogene presence of

Table 28. Component Genera of Principal Historical Assemblages of the Chihuahuan Desert Herpetofauna.

Unit	Participant Desert Genera	Relict Montane Genera	Riparian Genera
A. South American Element	none	none	none
B. Old Northern Element	1. *Ambystoma* 2. *Bufo* (part) 3. *Terrepene* 4. *Eumeces* 5. *Gerrhonotus* 6. *Elaphe*	1. *Ambystoma* 2. *Aneides* 3. *Bufo* 4. *Gerrhonotus* 5. *Eumeces* 6. *Opheodrys* 7. *Storeria* 8. *Leiolopisma* 9. *Agkistrodon* 10. *Tropidoclonium*	1. *Acris* 2. *Bufo* (part) 3. *Rana* 4. *Chelydra* 5. *Chrysemys* 6. *Terrepene* 7. *Eumeces* 8. *Elaphe* 9. *Natrix* 10. *Agkistrodon* 11. *Leiolopisma* 12. *Lampropeltis*
C. Middle American Element	1. *Gastrophryne* 2. *Cnemidophorus* 3. *Coleonyx* 4. *Leptotyphlops* 5. *Tantilla* 6. *Trimorphodon*	1. *Hyla* 2. *Eleutherodactylus* 3. *Kinosternon* 4. *Cnemidophorus*[1] 5. *Rhadinaea* 6. *Tantilla*	1. *Eleutherodactylus* 2. *Gastrophryne* 3. *Kinosternon* 4. *Cnemidophorus*[1] 5. *Drymarchon* 6. *Micrurus*
D. Young Northern Element	1. *Gopherus* 2. *Scaphiopus* 3. *Cophosaurus* 4. *Crotaphytus* 5. *Holbrookia* 6. *Phrynosoma* 7. *Sceloporus* 8. *Uma* 9. *Uta* 10. *Xantusia* 11. *Arizona* 12. *Crotalus* 13. *Ficimia* 14. *Heterodon* 15. *Hypsiglena* 16. *Lampropeltis* 17. *Masticophis* 18. *Pituophis* 19. *Rhinocheilus* 20. *Salvadora* 21. *Sonora* 22. *Thamnophis*	1. *Chiropterotriton* 2. *Pseudoeurycea* 3. *Phrynosoma* 4. *Sceloporus* 5. *Urosaurus* 6. *Crotalus* 7. *Lampropeltis* 8. *Masticophis* 9. *Pituophis* 10. *Thamnophis* 11. *Diadophis*	1. *Scaphiopus* 2. *Sceloporus* 3. *Urosaurus* 4. *Lampropeltis* 5. *Thamnophis* 6. *Sistrurus*
E. Holarctic Elements	none	none	none

1. Could also be considered Young Northern Element.

Table 29. Percentage Derivation of Chihuahuan Herpetofaunas by Principal Historical Element

Historical Element: Contemporary Herpetofauna	Old Northern in % of Contemporary Herpetofauna	Middle American in %	Young Northern in %
Participant Chihuahuan Desert			
genera (N:34)	18	18	64
genera (N:38)	17	21	62
Montane Relict			
genera (N:26)	35	23	42
genera (N:68)	26	18	56
Riparian			
genera (N:24)	50	25	25
genera (N:49)	51	24	25

[1]genera

virtually all modern families (except colubrid snakes) and several modern genera, especially riparian forms. Both systems appear to have undergone a gradual diversification, differentation, and possibly modest dispersal in the Oligocene Great Plains region about 40 to 45 degrees north latitude. In both systems the importance of tropical components deteriorated steadily. In conclusion, the herpetofaunal fossil record and the derivation of contemporary assemblages are again consistent with the model of origin for a Madro-Tertiary Geoflora constructed by AXELROD (1958).

Early Neogene Period
(Miocene through Middle Pliocene Epochs)

The Early Neogene of North America was a time of geographical and ecological isolation and differentiation in which modern species, modern assemblages, and modern dominants first assumed the positions they now hold. Events after this time merely rearranged spatially the stocks that were already established on the Mexican Plateau.

Paleogeography

North America was stabilized, isolated, and newly obtained topographical features took on a modern aspect, at least in their positions, if not in their dimensions. Coastal perturbations occurred, but embayments were generally less extensive along the Gulf Plain than in the Paleogene (SCHUCHERT, 1955). To the west narrow and localized precursors of the Modern Gulf of California began to wedge between Baja California and the contiguous mainland (ATWATER, 1970; ANDERSON, 1971). The Isthmus of Tehuantepec, then a marine portal, formed the southern limit of the continent, separating it from both South America and nuclear Central America. To the north, the Bering land bridge continued as a geographical feature, but polar climatic deterioration probably made the bridge ecologically inaccessible to warm temperate and xeric adapted biotas. The continent as whole maintained a northward drift — possibly contributing to the progressively temperate climatic trends of these times (LE PICHON, 1968).

Continental tectonic events in the Miocene Epoch erected the modern Sierra Madre Occidental. By Middle Miocene the whole Mexican Plateau may have been uplifted, predominantly by faulting, but supplemented by volcanism (see Geological History under The Physiography of the Chihuahuan Desert). COHN (1965, pp. 138-9) provided an excellent series of maps documenting the sequence of Cenozoic changes in the physiography of the Mexican Plateau (based largely on KING, 1958). I take a single exception to the maps in that they make no recognition of alternations in paleogeographical coast lines, the most relevant being the Rio Grande Embayment on the east and the formation of the Gulf of California on the west.

The Mio-Pliocene formation of a modern plateau topography, was completed by the orogeny of the Coahuila Folded Belt (northern Sierra Madre Oriental) (SEGERSTROM, 1962). However, this region was apparently modern topographically by the late Pliocene Epoch (SCHUCHERT, 1935). The dimensions and details of early Neogene mountains in western North America are unknown. However, AXELROD (1964) extrapolated from geofloral evidence an estimatic of 1400 meters elevation (minimum) for the Colombia Plateau of Miocene Idaho. KING (1958) indicated that Miocene Cordilleran uplifts resulted in the aggregation of eroded sands and gravels in the high plains.

Paleoclimate

The onset of the Neogene was apparently marked by rapidly cooling and drying climates. Two major factors contributed to the radical alteration of Neogene climates. One was the previously noted orogeny of major montane systems. Faulted blocks coupled with volcanic extrusives extended in both local dimensions and geographical distribution. As a result of elevation, land surface climates cooled, and, at least on the leeward side of these uplands, extensive arid rain shadows were cast across the lowlands. Secondly, a brief middle Miocene rise in paleotemperature was followed by a very sharp drop, as documented by a vast array of foraminiferal and geofloral evidence (AXELROD, 1964; BERGGREN, 1969; ADDICOTT, 1969; BANDY, CASEY & WRIGHT, 1970). The general Eocene to Pleistocene trend toward deteriorating temperatures was temporarily reversed about 15,000,000 years B.P. (Barstovian stage). But after several million years of the reversal, temperatures had dropped far lower than had ever been indicated previous to that time. The sudden temperature decline following the reversal reflected both a basic drop in temperate North American temperatures and had associated with it, a parallel decrease in the moisture bearing capacity of the cooler oceanic wind currents. This second condition meant that the absolute humidity, or the water bearing capacity of marine wind was sharply reduced. Thus, the impinging precipitation on terrestrial ecosystems was also reduced sharply. The cooling, and drying trends re-established in Middle Miocene continued until Late Pliocene time. The biological impact of these changes is best evaluated in terms of P/T ratios. Whenever cooling accelerated more rapidly than drying, the P/T ratios were raised and cooler mesic climates resulted. When the rates were reversed (but not the directions of the trends!) warmer and drier climatic conditions prevailed. Coastal and elevational factors could have had strong local effects on this crucial ratio. The early Neogene may be classified into two climatic stages. At currently warm temperate latitudes, 20 to 40 degrees north, early Neogene lowland climates were apparently mesic warm temperate to subtropical with mild seasonal fluctuations in both temperature and precipitation. Precipitation was probably both stable in monthly averages and relatively abundant in an absolute sense, generally two to five times the current annual values. This early Miocene continuation of Oligocene circumstances terminated abruptly following the Middle Miocene temperature inversion. After the shortlived inversion, cooling probably outstripped drying and thus raised the P/T ratio, judging from the expansion of Late Pliocene Arcto-Tertiary Geofloras (AXELROD, 1956, 1958) along coastal and montane areas[1], while the reverse was true for the interior low plains and incipient Mexican Plateau. This second stage, coupled with the increasingly modern topography, obviously did much to forge the contemporary arid continental biota of North America.

Throughout this latter stage Neogene climatic trends continued as follows: (1) paleotemperatures deteriorated. By way of example, the $25°C$ marine surface isotherm was displaced to the south, perhaps 10 degrees latitude between Mid-Miocene and Late Pliocene (DURHAM, 1950; SAVAGE, 1960). By Late Pliocene the $20°C$ marine isotherm was approximately 5 degrees north of its present position along the Pacific Coast. (2) Even where essentially modern cool temperate conditions prevailed, seasonal extremes were less than at present. AXELROD (1956) surmises from west central Nevada (approximately 40 degrees north latitude) that summers were about $7°C$ cooler and winters $5°C$ warmer than at present. (3) Precipitation declined not only in absolute annual amount but also became increasingly seasonal with a significant deterioration of summer rains (accompanying the cooling of adjacent marine waters). In the early Neogene drying outstripped cooling to lower P/T ratios only in the lowlands. (i.e., Mojavia of AXELROD, 1958). But by the end of the Miocene it had probably reversed early raises in montane P/T ratios as well, restricting and fragmenting the early Miocene expansions southward of Arcto-Tertiary forest. (4) Glacial episodes already existed (DENTON & ARMSTRONG 1969).

Geoflora

Geofloral evidence strongly affirms the trends toward modern conditions indicated by terrestrial stratigraphy and marine deposits. AXELROD (1956, 1958, 1962, 1968) has reconstructed the major geofloral trends for Neogene western North America. DORF (1960) and CHANEY (1938, 1940, 1947), and JENKINS (1951) and MACGINITE (1962) also reconstructed Neogene floras. In addition, pa-leobotanical distribution maps for the region and time under consideration are presented by AXELROD (1958), SAVAGE (1960), MCDIARMID (1964), and COHN (1965). Another detailed recapitulation of the information presented in these sources is unnecessary. However, I will present a brief summary of their conclusions for the sake of coherence, and as basis for reinterpretation.

I have presented a Mio-Pliocene map (Map 12) of xeric Madro-Tertiary vegeta-tion for southwestern North America. It differs from previous maps in two features. It indicates that Baja California was not a well formed peninsula, but

1. DENTON & ARMSTRONG (1969) reported Alaskan glaciation during early Neogene times.

rather an abutting ridge along the now Sonoran-Sinaloan coast (based on ANDERSON, 1971). Secondly, some maps (SAVAGE, 1960, p. 205) did not indicate a continuum of desert vegetation to the east of the modern desert shown here.

The major features of the early Neogene Geofloras are as follows: (1) Madro-Tertiary Geoflora, in response to an uplifted topography throughout Cordilleran North America, and a steady deterioration of lowland P/T ratios, continued to expand from Middle Miocene onward. It rapidly completed the transition from an edaphic and ecotonal faciation to a major system of geographical plant formations. This system included arid woodland (piñon-juniper), grasslands, oak savannas, grassland-thornscrubs, and chaparral. Earlier records indicate these communities may have occurred as complex edaphic alternatives to one another, not in broad regional stretches. FRYE & LEONARD (1957) indicated a transition from Arcto-Tertiary deciduous woodland to a modern grassland in the Miocene (Valentine) of Kansas and Nebraska based upon floral and molluskan evidence. Probable, at least in Mojavia, a mesquite grassland precursor was dominant (AXELROD, 1958). By Middle Pliocene, a specifically modern chaparral assemblage was present at the same latitude (and probable elevation) at which it occurs today in California (CHANEY in JENKINS, 1951). AXELROD (1958) speculated that true desert vegetation first appeared in especially dry edaphic pockets within the mesquite grasslands of Mohavia in the Miocene. These edaphic pockets may have included the rain shadows of mountain blocks, and xeric lithosols such as dunes and alluvial fans. MACGINITE (1962) indicated that Miocene Nebraska had warm temperate vegetation – a more mesic Madro-Tertiary Geoflora than Axelrod's Mojavia. It included a near modern riparian woodland and a dominant vegetation of oak-pine savanna. Geographically, the Cordilleran and plains regions, from 25-40 degrees north latitude may have been involved. See Map 12 for details.

Apparently, not until Late Pliocene did a true desert scrub differentiate into a major geographical formation. The origins and expansions of desert scrub from a Madro-Tertiary grassland-thornscrub precursor during the Mio-Pliocene was probably analogous to the expansions and coalescence of bald spots (the desert) in a progressively moth-eaten carpet (the mesquite-grassland). This model is a repetition of the Oligocene-Miocene origins and development of the Madro-Tertiary Geoflora itself. (2) Neotropical Geoflora generally receded throughout the Neogene, both in elevational and latitudinal distribution. Uplift, cooling, drying (especially the deterioration of summer precipitation), and increasing seasonality of climate, all worked against the survival of this flora in the Cordilleran region. Madro-Tertiary Geofloras expanded at its expense in the valleys, while Arcto-Tertiary forests, now well differentiated (AXELROD, 1956, 1962), did likewise in cool mesic montane situations. (3) Arcto-Tertiary forests expanded in cool mesic montane situations, probably forming a virtually continuous reticulum of deep forests (i.e., *Pseudotsuga, Abies,* and tall *Pinus*). Their movements were governed by a Middle to Late Miocene trend for more rapid cooling than drying in upland situations (thus higher P/T ratios). The reticulum was probably never reconstructed during the Pleistocene or post glacial Pluvial that followed. MARTIN & HARRELL (1957) made a particularly strong circumstantial case for a Late Neogene connection of Arcto-Tertiary forests of the Sierra Madres and eastern North America (probably via the Edwards Plateau). WELLS (1966) documents, at least for the Chisos Mountains (Big Bend National Park), that cold coniferous forest (*Pseudotsuga taxifolia* and

Pinus ponderosa) never descended below 1200 meters elevation during the Wisconsin Pluvial, and thus failed to interconnect with adjacent ranges in Trans-Pecos Texas. Miocene, rather than Quaternary, dispersal of montane biotas has had a profound influence on the affinities between montane assemblages and the degrees to which they have become differentiated. The specific significance of this probability for the herpetofauna will be discussed later in this section. (4) A modern riparian woodland (of Arcto-Tertiary derivation) was further differentiated. Neotropical components, previously present, were absent from the Neogene flora. Assemblages of essentially modern species have been described from the Neogene record, differing from their contemporary descendants much more in distribution than in composition. Warm temperate deciduous woodland, characterized by cypress (*Taxodium*) still survived in Miocene central coastal California (CHANEY in JENKINS, 1951), but a more contemporary (and arid adapted) willow, cottonwood, and sycamore (*Salix, Populus*, and *Platanus*) association had replaced it by Middle Pliocene times. In general, even with deteriorating precipitation, conditions in the Cordilleran region were more mesic than at present. Drainage systems and their associated vegetation were therefore more continuous and extensive than at present.

General Paleoecology

FRYE & LEONARD (1957) and HIBBARD (1960) synthesized data from stratigraphy, substrate, and fossil biotas (especially vegetation, gastropods and tortoises) into an explicit reconstruction of ecological transitions in the Pliocene Great Plains immediately north of the modern Chihuahuan Desert. Their works may well offer the best available insight to Middle Neogene conditions on the northern Mexican Plateau. According to these authors, the Texas to Kansan Plains were stable tectonic features throughout the Neogene. Maximum relief, at least in Miocene, may not have exceeded 80 meters, and apparently underwent progressive leveling through erosional spread of alluviums on to valley slopes. Hydrographical patterns indicated extensive flood plains producing only moderate incisions on the landscape. Drainage patterns ran west to east as they do now. Both sources emphasized the presence of a more moderate climate (at both extremes) with greater rainfall than at present. Both also indicated a gradual deterioration which had produced, for both, an essentially modern Great Plains climate by very late Pliocene times. During this climatic deterioration, declines in precipitation and temperature alternately outstripped one another in their effect on P/T ratios. During Middle and Late Pliocene, precipitation reduction effectively lowered the P/T levels sufficiently to produce a caliche soil and presumably desert conditions. These temporary xeric conditions (the earliest is Hemphillan-Middle Pliocene of Kansas to Texas) indicate the first pulses of desert conditions in the southern Great Plains, and possibly heralded the beginning of a continuous Chihuahuan Desert to their south. Fossil assemblages further support these arid episodes in the Pliocene plains. Neogene vegetation apparently began as pine-oak savanna and gradually took on an increasingly modern prairie grassland aspect by Middle Pliocene. A riparian woodland was present and generally held an Austro-riparian biota, including palms, alligators, and a mammalian megafauna of forest horses, rhinos and mastodons. Characterizations of this environment as subtropical or warm-temperate are largely a matter of definition.

The Neogene ushered the differentiation and domination of a modern North American herpetofauna. This period was well documented and evaluated by TIHEN (1964). Virtually all Neogene assemblages were modern on the generic level and excluded tropical and Old World stocks previously present. Altered conditions of topography, climate, and vegetation all conspired to hasten the elimination of the older faunas and accelerate the expansion in diversity and distribution of modern genera, including the first colubrid and viperid snakes.

I will present several hypothetical patterns based on inference from paleoclimatic, topographical and geofloral evidence. Geographically most of the Neogene fossil record is from the Cordilleran western United States and from the southern (Kansan) Great Plains. Evidence from the Mexican Plateau proper is virtually confined to stratigraphic dating of tectonic events (e.g., uplift of Sierra Madre Occidental or Durango andesite flows). An additional line of circumstantial evidence is the fossil herpetofauna. The Miocene record includes the following terrestrial genera from the southern Great Plains: *Bufo, Scaphiopus, Gerrhonotus, Gopherus, Leiocephalus*(?), *Phrynosoma, Sceloporus,* and *Cnemidophorus* (TIHEN & ESTES, 1964; CHANTELL, 1964; HOLMAN, 1966; GEHLBACH, 1965; ROBINSON & VAN DEVENDER, 1973). AUFFENBERG (1963) and HOLMAN (1969) reported the modern colubrid snake genera *Elaphe* and *Lampropeltis* from Florida. Early Pliocene Kansas added *Gerrhonotus, Coluber, Heterodon, Pituophis* and *Thamnophis* (WILSON, 1968) and a *Sceloporus* of the *undulatus* group (TWENTE, 1952). *Terrepene ornata* may be added to the Middle Pliocene fauna as well (MILSTEAD, 1969).

In addition, large tortoises, *Geochelone*, were present throughout the Neogene, at least as far north as the 40th parallel (Nebraska). HIBBARD (1960) interpreted this latter evidence as strong documentation of a subtropical savanna climate. I accept this interpretation only with qualification. Virtually all vegetation and fossil fishes (MACGINITE, 1962; TIHEN & ESTES 1964, respectively), were unquestionably temperate stocks. The Neogene plains were habitats of pine and oak savannas, modern riparian woodlands, sunfish, bowfins and gar pike. The herpetofauna was generally small in size and temperate in character. The rodent fauna and snails were also predominantly temperate. No evidence of tropical lateritic soils was presented by HIBBARD. In contrast, arid temperate caliches are recorded (FRYE & LEONARD, 1957). Again, HIBBARD's evidence for a tropical climate consisted of the giant tortoises, the alligator (really a warm temperate Old Northern genus), and several spectacular and now tropical mammal stocks of dubious significance. (To these could be added *Leiocephalus* reported by ROBINSON & VAN DEVENDER, 1973). His suggestion that early Neogene plains climates, and by extrapolation the Mexican Plateau climates as well, were more mesic and less subject to climate fluctuations than it now experiences is doubtless true. The presence of large and presumably non-fossorial tortoises suggested the absence of annual extremes, especially winter frosts. These conclusions largely parallel those of AXELROD (1956, 1962). But even here the presumption is feeble. The modern *Gopherus flavomarginatus* doubtless reaches a shell length of two-thirds of a meter (fossils possibly conspecific are over a meter long, AUFFENBERG, 1969) and does in fact live on high temperate plains (1400 meters elevation). This extant form excavates a burrow two meters deep and three meters in length. In conclusion, the Mio-Pliocene climate of the Great Plains and contiguous northern Mexican Plateau was more mesic and moderate than at present, but it was probably no subtropical Serengeti Plain in terms of average annual temperature.

Chelonians contribute modestly to the Chihuahuan participant herpetofauna, but they have good fossil record. The chelonian genus *Terrepene* is fundamentally Old Northern (MILSTEAD, 1969), with a single species, *T. ornata* associated with the Madro-Tertiary descendant grasslands of the Kansan Plain. These populations are little more than glacial or pluvial prairie relicts. Its Chihuahuan populations are best derived from Quaternary events to be described subsequently. The genus *Gopherus* is an old (Oligocene) Young Northern stock. It is fundamentally a warm dry grassland associate which persists in true deserts as a fossorial form — still with some affinity for arid grasslands, or as a dweller in arid microhabitats in the sandy pine barrens of the forests of the Austro-riparian plain.

The amphibians, *Scaphiopus*, *Bufo cognatus*, and *Ambystoma tigrinum* show somewhat similar ecological and biogeographical patterns of distribution. In the case of *Scaphiopus* the direct fossil record largely parallels *Gopherus* (back into the Oligocene Period). All of these stocks may ultimately be of Old Northern ancestry.

The participant lizard fauna of the Chihuahuan Desert was probably also expanded and differentiated, at least to modern species groups, at this time as well. Since the vast majority are from the fundamentally Neotropical family, Iguanidae (all known Old World fossils having been rejected by ESTES, 1970), I presume the Young Northern genera to be descendants of the Middle American Element. This would include the genera, *Crotaphytus*, *Cophosaurus*, *Holbrookia*, *Callisaurus* — *Uma* stock (probably represented by modern species groups by Middle Pliocene), *Phrynosoma* (known from Miocene Nebraska), *Sceloporus*, and *Uta*. Macroteiids are known from the Late Cretaceous of Wyoming and Montana (see Cretaceous account) but are eclipsed in the subsequent fossil record of North America until Miocene (TIHEN & ESTES, 1964). The modern Chihuahuan representative, *Cnemidophorus*, may represent a Neogene invasion from a Middle American Element precursor, either in the Neotropical Geoflora of the Mexican lowlands or from stocks south of the Isthmus of Tehuantepec in nuclear Central America (the genus itself is Young Northern). If the latter, *Cnemidophorus* antecedent stock would have had to make a fortuitous (rafting?) crossing of the Tehuantepec portal in order to be present in Miocene North America (ESTES, 1970). *Xantusia* most surely evolved from *Paleoxantusia* in association with expanding Madro-Tertiary scrub, both pre-desert and chaparral. The fundamental distribution of *Xantusia arizonae* may have been established in the early Neogene. Climatic conditions, topographical features including the andesite outcroppings which Chihuahuan populations now utilize, and chaparral vegetation (with which they are now associated) were all present before the end of the Miocene. Late events may have done little more than disrupt an arched distribution from the Sierra Madre foothills of eastern Durango northwest along the Mogollon Rim of Arizona. *Coleonyx* is probably an old Middle American Element which gradually differentiated, *in situ*, to the desert adapted species within the primitive genus as tropical environments progressively deteriorated on the Mexican Plateau. The first occurrence of North American colubrid snakes in the Miocene indicates a coupling of previous evolutionary events with the paleogeographical accessibility of the continent. The invasion of apparently modern genera suggests previous Old World evolution. Considering the time and route of entrance, virtually all of these early arrivals must have been Old Northern stocks moving through Arcto-Tertiary vegetation (SAVAGE & SCOTT, personal communication). They could have come via Beringia or Europe[1]. The presence of *Elaphe* and *Lampropeltis* in Miocene is already documented. The general pattern of dispersal from Old World centers is illustrated by the colubrids *Coluber, Elaphe, Natrix*, and the crotaline *Agkistrodon*. These stocks may have made a northern entrance onto the newly erected Mexican Plateau (and other portions of the Cordilleran region) by way of the Arcto-Tertiary (especially riparian) corridors fingering southward in early Neogene. As a Madro-Tertiary vegetation took form and expanded on the Plateau, these stocks gave rise, *in situ*, to the Young Northern (Mexican Plateau centered and Madro-Tertiary vegetation associated) genera. Explicitly, *Elaphe* gave rise to *Pituophis, Natrix* to *Thamnophis*, and *Agkistrodon* to *Crotalus* and *Sistrurus*. Paleoecological conditions throughout the plains and southern Cordilleran North America could have been suitable sites for these developments.

I make no particular designation of the Mexican Plateau as "the" center of differentiation on the generic level. However, the plateau does make an attractive hypothetical model for subsequent radiations of xeric derivatives of these stocks because of its size, extensive rain shadowed valleys, and proximity to both Arcto-Tertiary and Neotropical Geofloras. The derivation of a Young Northern *Pituophis* on the southern plateau from the Old Northern *Elaphe* has already been discussed from a systematic perspective as point 26 of the systematic problems chapter. The genus *Elaphe* offers a living transition of stocks, namely from Old World Arcto-Tertiary stocks, to an arid Madro-Tertiary species group *rosaliae*, with a Young Northern derivative genus *Pituophis*. This may be a fundamental pattern in the early Neogene origins of Young Northern colubrid, viperid, and even elapid genera, involving such genera as *Arizona, Crotalus, Masticophis, Rhinocheilus, Salvadora, Sistrurus*, and *Thamnophis*. Stocks such as *Hypsiglena, Ficimia*(?), *Trimorphodon*, and *Tantilla* appear to have fundamental affinities with the Neotropical derivative Middle American Element. Their derivation, at least on the level of modern genera, may be from the south (SAVAGE, 1967). *Heterodon* has been present, (represented by a pre-*nasicus* stock) since the later Pliocene Epoch of the southern Great Plains (Kansas) (PETERS, 1955). Its ultimate derivation is uncertain. I tentatively consider the majority of Young Northern Chihuahuan snake genera Old Northern derivatives from Oligocene-Miocene invaders from Eurasia. *Leptotyphlops* represents a much older Cretaceous Neotropical Element. Another minority of genera, *Trimorphodon et al.*, appear to be much more recent invaders from the south, probably in early to middle Neogene times.

The Mexican Plateau as an Evolutionary Center

Since the fossil record positively documents the geographical expansion and increased diversity of the Madro-Tertiary Geoflora and its associated Young Northern Element herpetofauna, it is now appropriate to ask, "does this documentation indicate the Mexican Plateau (including the Sierra Madres) to be the center of origin for arid North American biotas? "Certainly the implication seems intrinsic to the name "Madro-Tertiary". The answer must be tentative because of the absence of an extensive early Neogene fossil biota from the Mexican Plateau. The

[1] as part of the Madro-Tethyan biota described by Axelrod [1975. Ann. Missouri Bot. Garden 62(2): 280-334].

only report of significance was an early Miocene vertebrate fauna from Big Bend National Park by STEVENS, STEVENS & DAWSON (1969). Considerations of stratigraphy, structure of substrate, and the fauna indicate essentially a continuation of its contemporary Great Plains biota, but perhaps slightly more xeric. The authors suggest that the absence of horses implies a more xeric state than further north, since Miocene horses were still largely confined to more mesic woodlands. In general, this fauna affirms the validity of extrapolating from the southern Great Plains to the Northern Mexican Plateau.

I will now return to the question, "Was the Mexican Plateau the center of origin for xeric biota of Neogene North America?" The answer is that the Mexican Plateau unquestionably sustained a coalescing reticulum of expanding xeric biotas that stretched west to the Mojavia of eastern California. The reticulum took on the form of a major regional belt across Cordilleran North America between 25 and 35 degrees north and (generally) west of the 100th meridian. Modern generic stocks probably differentiated throughout the belt (see Map 12), not exclusively on the Mexican Plateau. As noted in the previous paragraph, mammalian fossils indicate a trend toward increasing arid climates (lowered P/T) from south to north. This would favor some earlier development of xeric biota in the south and their subsequent dispersal north following progressive shifts in Neogene climate. Thus, the Mexican Plateau would be favored as *one* center of origin for Madro-Tertiary vegetation and Young Northern faunas, but certainly not as the center. The lizard genus *Uma* is a good example where the primitive member is Chihuahuan. Massive area, the presence of mountains and rain shadows, arid caliche and alluvial soils, and its southern inland position, all make the plateau a likely site for the Neogene differentation of desert organisms. Other important centers may have been Mohavia of California, the lowland of Sonora, and local xeric pockets formed by rain shadows of the southern Rockies and adjacent water poor azonal soils of alluvium, sand, and volcanic extrusives.

I believe the role of the Mexican Plateau (and the Chihuahuan Desert in particular) as a center of origin should be re-evaluated in terms of its present climate and biota. The Chihuahuan Desert cradles at its center a huge, dry, hot lowland system referred to as the Bolson de Mapimi (actually several contiguous bolsons). The climatic analysis of this investigation has already emphasized the importance of this region as the hottest portion of the desert. The present situation is largely the result of Neogene tectonic events. Climatic conditions in the Mapimian region may have been highly insulated and continuous since middle Neogene time. Vegetation in the region is strikingly similar in both organization, and, to a lesser extent, content, to the Mojavia Madro-Tertiary Geoflora from which AXELROD (1958) suggested that a modern desert vegetation evolved. The Chihuahuan Desert is even now an edaphic patchwork of desert scrub (on limestone, alluvial and sandy soils), mesquite grassland (in arroyos and mesic edges), and montane dry woodlands (pine-juniper-oak) (see Vegetation-Introduction). This was the setting in which AXELROD presumed a modern desert vegetation gradually differentiated and expanded. If the Mapimian subprovince has been in fact a stable climatic region since the Neogene, it might be expected to have a highly diversified and endemic edaphic desert vegetation. The gypsophilous vegetation of three major subdivisions of the desert are summarized in Table 6. The table indicates an exceptionally rich edaphic flora for the Mapimian division.

If parthenogenetic clones of *Cnemidophorus* are removed from consideration, virtually all Chihuahuan endemic herpetofaunal groups are either centered in, or restricted to, the Mapimian subprovince by their distributions. Moreover, a striking majority of these endemics are extremely primitive species, often with mixed character states indicating a nearly transitional condition between a mesic ancestral genus and a more arid adapted genus or species group. In other words, the morphology and evolutionary position of the herpetofauna parallels a transitional differentiation from mesquite-grassland to desert seen in Chihuahuan and especially Mapimian vegetation. Both situations are reflected in the Neogene fossil record. I will document the presence of a primitive endemic herpetofauna with the following chart:

Mapimian Endemic (centered) Species:	Primitive or Transitional Status (Morphology)	Source
1. *Coleonyx reticulatus* (borderline Trans-Pecos/Mapimian)	primitive tropical *Coleonyx*	DAVIS & DIXON, 1958; KLUGE, 1975
2. *Sceloporus merriami* (& *Sceloporus maculosus*)	combined *Sceloporus* and *"Uta"* characters	SMITH, 1939
3. *Uma exsul*	primitive *Uma* almost balanced transitional form between *Uma* and *Callisaurus-Cophosaurus* stocks	SCHMIDT & BOGERT 1947
4. *Elaphe subocularis*	transitional combination of characteristics between *Elaphe* and *Pituophis*	DOWLING, 1957, 1974

The example of *Uma* is particularly favorable to the model of Mexican Plateau origins, since the Chihuahuan species is more primitive than Mohavian and Peninsular congenitors. Other stocks, probably little changed since middle Neogene, include *Gopherus flavomarginatus, Xantusia vigilis* and *X. arizonae*, and *Ficimia cana*. Adjacent mesic portions of the plateau also support primitive endemic stocks, *Phrynosoma orbiculare* being a prime example (see PRESCH, 1969).

The immense size of the plateau, its continued stability, both tectonic and climatic, since Neogene, and its geographical accessibility, all combine to make the region an excellent stage for speciation, radiation, and the survival of relict stocks. Many Madro-Tertiary and Young Northern genera have undergone tremendous radiation in this region, regardless of whether their historical origins were in the plateau proper or on the plains and valleys to the north and west. *Quercus* is an excellent floral example, as are the reptilian genera *Sceloporus, Thamnophis*, and *Crotalus*.

In summary, then, I view the contemporary Chihuahuan Desert as being of Neogene origin. It was a portion of a pre-desert continuum which ran west to the Pacific coast and north to Mohavia, and at times even east into the Kansan plain. The hot, dry central core of the desert, the Mapimian subprovince, still shelters a complex mosaic vegetation and an endemic herpetofauna which appears to be indicative of mid-Neogene stages of differentiation both in ecological organization and in morphology.

Adjacent montane herpetofaunas provide a counterpoint to the development of the desert. No Neogene montane faunas are available for reference. AXELROD (1958), in Neogene vegetation discussion, provided substantial circumstantial evidence of montane paleoecology. I concur with the conclusions of MARTIN & HARRELL (1957), MARTIN (1958), and WAKE & BRAME (1963), namely that montane northern biotas entered the Sierra Madres in the early Neogene (Miocene). Exchanges were achieved when with tropical forest stocks (e.g., *Tantilla cornata – Tantilla rubra*) penetrated the Austro-riparian forests from the southwest via the same corridor. Corridors between eastern forests of the United States and the Sierra Madre Oriental probably crossed the stable uplands of the intervening Edwards Plateau. The snake genus *Storeria* (MARTIN & HARRELL, 1957) and the ancestral stock of tropical plethodontid salamanders (bolitoglossines) probably took this route south. Intriguingly, the plethodontid, *Chiropterotriton prisca*, a primitive member of its genus (WAKE, 1966), is still restricted to the northern Sierra Madre Oriental. To the west, boreal forests (*Pinus ponderosa* and *Pseudotsuga menziesi*) moved south along the southern Rockies, crossed the gap of the present Cochise filter barrier, and formed a continuum with the Sierra Madre Oriental. WAKE (1966) demonstrated forcefully the extremely primitive morphological nature of *Aneides hardyi*. This salamander is almost a linking form between its genus and *Plethodon*. It is restricted to the high *Pseudotsuga* forests of the Sacramento Mountains, now disjunct from the southern Rockies. WAKE suggested that this species was most probably a Miocene relict, not a Pleistocene invader. Similar long term isolation has been hypothesized for relict *Pseudotsuga* forests in the high Chisos, 2000 to about 2500 meters elevation (and for their gastropods) by WELLS (1966) among others. The expansion and union of Arcto-Tertiary forest with Sierra Madrean montane Neotropical Geofloras must be qualified if it is to be viewed as a Neogene event. First, these expansions were primarily Miocene events. Miocene tectonic and volcanic uplift created corridors to the south. In these uplands the climatic trend toward cooling outstripped the trend for drying, resulting in higher P/T ratios and thus dispersal of cool mesic biotas, Geofloras and Old Northern (to the south) and to a much lesser extent Middle American (northeast) herpetofaunas. Mesic Young Northern herpetofaunas probably dispersed (mostly north) at this time as well (e.g., *Thamnophis*). By the beginning of the Pliocene, xeric Madro Tertiary Geofloras expanded and formed a continuum from California through the Mexican Plateau. However, while Miocene dispersal was active it may have operated not only across north-south corridors, but also through the Mexican Plateau. However, while Miocene dispersal was active, it may have operated not only across north-south corridors, but also through the Trans-Plateau Corridor of MARTIN (1958) running west through the Anticlinorium of Arteaga across the central Mexican Plateau to the Sierra Madre Occidental of Durango. It effectively bisects the site of the modern Chihuahuan Desert into northern and southern halves.

Secondly, I consider a definition of the term "montane forest" a crucial qualification to this discussion. These Miocene corridors and resultant relict biotas are confined to very moist or very cool high forests. In the Sierra Madre Occidental-southern Rocky Mountain axis this means high pine, aspen, fir and douglas fir (*Pinus ponderosa, Populus, Abies, Pseudotsuga*) forests above 2000 m and with more than 500 mm annual precipitation. On the eastern axis, eastern United States-Edwards Plateau-Sierra Madre Oriental, northern deciduous hardwood forest biotas merge with montane tropical evergreen hardwoods. Xeric Madro-Tertiary vegetation intervened between the western and eastern forests. Biotas associated with lower, drier, or warmer conditions, particularly with oak savanna or pinyon-juniper woodland, were probably interconnecting throughout the Mexican Plateau and adjacent uplands repeatedly during glacial Maxima (see WELLS, 1966).

Riparian Herpetofauna

Riparian herpetofauna for the Early Pliocene of Kansas have been documented by WILSON (1968). WILSON described modern temperate fish fauna: *Lepisosteus, Notropis, Ictalurus, Fundulus, Micropterus*, and *Lepomis*. A modern amphibian fauna was present, including many species occurring at the same locality today. These include *Ambystoma* sp., *Scaphiopus couchi, Acris, Pseudacris clarki, Hyla gratiosa, Hyla cinerea, Bufo boreas, Bufo cognatus, Bufo marinus*(?), *Bufo pliocompactilis* (a *B. speciosus* ancestor?) and *Rana areolata*. Turtles included *Chrysemys* and *Trionyx*. Possibly riparian lizards included *Gerrhonotus, Ophiosaurus*, and *Eumeces*. Snakes included an erycine (Old Northern) boid *Ogmophis* as well as colubrids *Natrix* and *Thamnophis*. Supplemental evidence from Mio-Pliocene Nebraska (Valentine formation) affirmed the same general ecological and systematic situations (CHANTELL, 1964; TIHEN & ESTES, 1968). VOORHIES (1971) added an *Alligator* to the Valentine Miocene fauna. All authors indicated a slow moving stream with adjacent alluvium (including sand bars) along its banks. These streams were set in relatively level plains. WILSON inferred, especially from mammalian fossils, that the riparian woodland community of the stream was surrounded by warm grassland biota.

The riparian herpetofauna is strikingly modern in its species (especially the amphibians) and in its ecological association and geographical position as well. It differs from the modern Kansan riparian herpetofauna primarily in the presence of warm Old Northern stocks now restricted farther to the southeast, namely *Hyla gratiosa, H. cinerea, Rana aerolata,* and *Alligator*. The latter genus reached south central Oklahoma in the historical past (HIBBARD, 1960) and its presence was thus no radical departure from the present. Vegetation (MACGINITE, 1962) also reflected a warm Arcto-Tertiary aspect with the presence of palms and cypress. The boid is a member of a holarctic subfamily long associated with Arcto-Tertiary and later Madro-Tertiary vegetation (*Charina* and *Lichanura* being extant genera), not Neotropical. Only the presence of *Bufo marinus* (if this identification is valid) indicates a truly tropical stock, at least in terms of geographical derivation.

Late Pliocene Epoch
(including the Plio-Pleistocene Transition)

Paleogeography

Tectonic events of the Late Pliocene made a more dramatic contribution to the geography, and probably the biotic differentiation of North American deserts, than any other episode in the geological history of the continent. The fundamental changes are best summarized by Map 12, which compares Early Neogene conditions to the results of the end Pliocene changes.

Modern coastlines were determined by Late Pliocene events for the whole North American continent, not just the Pacific southwest. SCHUCHERT (1955) indicated that major coast embayments were generally eliminated (or radically reduced) by regional uplifts (and possibly declining sea levels preceding the first glacial pulses). Of particular biological importance was the development of isthmian land bridges (where marine portals had previously passed) at the Bering Straits, the Isthmus of Tehuantepec and the Panamanian Isthmus.

The late Pliocene and early Pleistocene Epochs contained episodes of widespread mountain rejuvenation throughout the Cordilleran southwest. While the Sierra Madres themselves underwent some secondary uplift, more pronounced effects were reported by KING (1948) for isolated ranges in southwest Texas and southern and central New Mexico. COOPER & SILVER (1964) specifically dated the origin of modern uplands in the Cochise filter barrier as Late Tertiary. Coupled with continued temperature deterioration, these events probably terminated the previous Neogene desert (or pre-desert scrub) continuum from Mohavia through the Mexican Plateau, the last link being severed in the region of the present Cochise filter barrier of southeastern Arizona (see Map 12).

Valleys of the Mexican Plateau continued to fill and level with alluvium. Some present alluvial fans, especially along the eastern edge of the plateau, may represent continuous aggregations since Late Pliocene.

Many modern drainage patterns for the modern Chihuahuan Desert may date back to Late Pliocene as well. In particular, the Pecos River had probably already established its drainage pattern and was the Pliocene through Kansan (Middle Pleistocene) headwater of the lower Rio Grande. The upper Rio Grande was evidently still a closed basin drainage, terminating in a series of Chihuahuan (the state of) lakes, such as Lagunas Guzman and Santa Maria. The Neogene history of the upper Rio Grande and alternative models for its expansion were given by STRAIN (1966) and MAXWELL et al. (1967).

MINCKLEY (1969, pp. 48-51) offered an excellent resume of Pliocene-Pleistocene histories for the drainages of the northeastern Mexican Plateau, Cuatro Cienegas, and the Parras Bolson in particular. ARELLANO (1951) suggested that the maximum lake development of the Parras Basin was in the Pliocene. He also indicated that early Pleistocene uplift of the southern or western Parras Basin spilled Lago Mayran to the east. Pliocene to Early Pleistocene drainage patterns from this region may have led east across the Cumbres de Monterrey between Saltillo and Monterrey, eventually joining the lower Rio Grande via the Rio Pesquieria. The Rio Nazas and Rio Aguavanal were probably established by Late Pliocene and already feeding the huge lake or lakes which occupied the Parras Basin. The Parras lakes may have flowed east as ARELLANO suggested, or north-west along the Coahuila-Chihuahua state line until joining the upper Rio Grande via the Rio Conchos as suggested by MEEK (1904). CONANT (1963) suggested a third route of drainage from the Parras Basin (specifically Lago Mayran) to the lower Rio Grande via Cuatro Cienegas. MINCKLEY rejected this third alternative as untenable in light of the structure of the intervening topography. The other two alternatives will be re-evaluated in terms of the

178

affinities of their contemporary herpetofaunas later in this section. A summary of Late Pliocene through Pleistocene drainage patterns and progressive shifts are depicted on Map 13.

Important paleogeographical changes are recounted by ANDERSON (1971). The tectonic basis for these changes appears to be the sliding of the Baja California fault blocks (hypothetically three sections) northward along the southern San Andreas fault. This northward movement apparently began about 25 million years ago. At that time the granitic blocks (or epicontinental plates) of Baja California were wedged into the coastline of western Mexico (Sonora and Sinaloa lowlands). As the Baja California blocks slid north, they collided with deeply rooted Sierra Nevada and San Bernardino mountain blocks to their north (in California). The collision deflected the moving Baja California Blocks to the west. About 6 to 3 million years ago, the end of the Pliocene, this deflection instigated the opening of the Gulf of California and the separation of the Baja California blocks as the distinct modern peninsula. Resulting compression to the north forced the erection of the Transverse Ranges of California at the same time. DURHAM & ALLISON (1960) provided an excellent review of paleogeographical and marine fossil history for Baja California. Apparently, they rejected the importance of Cenozoic epicontinental plate movements in the formation of the Gulf trough and the peninsula. Even so, they concede that the modern Gulf of California may not have occupied the trough continuously until Early Pliocene. Even then, its Pliocene extent may have been quite variable. I accept[1] and apply the model of ANDERSON, but even the alternative view allows extensive land contact between Baja California and the Mainland throughout Early to Middle Neogene. This view contrasts with that of previous biogeographical interpretations, SAVAGE (1960), and especially MACDIARMID (1964), that assumed a relatively stable paleogeography for the Baja California peninsula in Neogene times.

Paleoclimate

Late Pliocene climates in the middle latitudes (25-40 degrees north) were essentially equivalent to those of interglacial periods and to the present. FRYE & LEONARD (1957) indicated that the Late Pliocene Great Plains approached interglacial and modern conditions at the same localities in terms of soils, vegetation, gastropods, and vertebrate faunas. BRATTSTROM (1955) suggested a slightly more mesic climate than present for southeast Arizona. MACGINITE (1958, pp. 70,71) stated directly that the Late Pliocene had a modern climate "...but the outbreaks of polar air so characteristic of the present winters appear to be a development of the Late Pliocene". HIBBARD (1960) rejected this view. He conceded that several lines of evidence indicated a trend toward a cooler, drier climate than during Middle Pliocene: arid caliche soils, the absence of thin shelled tortoises, and the disappearance of swamp dwelling shovel-tusked Mastodons. His rejection rests almost exclusively on the presence of thick shelled large tortoises (*Geochelone*) which presumably could not withstand harsh winters. While this assumption is probably valid, it hardly invalidates all the other lines of evidence favoring the MACGINITE interpretation. It simply means conditions may have been slightly more mesic with milder winters than at present. Using a totally independent line of evidence, marine deposits of the Pacific coast, SAVAGE (1960) indicated

---

[1] I find most recent (1973-76) evidence does not affirm the Atwood-Anderson model.

that the 18°C marine surface isotherm was in approximately the same position[1] in the late Pliocene, interglacial, and present times. In conclusion, I accept the MACGINITE'S statement as justified by available evidence with two slight modifications. Indication of slightly more mesic conditions and milder winters might justify description of the Late Pliocene plains climate as "maritime", in other words, now typical of coast regions at the same latitude. Secondly, HIBBARD (1960, p. 17) did accurately point out dramatic and rapid changes within the Late Pliocene on the Great Plains. Through the interpretation of caliche soil development and disappearance, he reconstructed a pattern that ran "subhumid ... to a semiarid ... to an arid climate. After the formation of the massive caliche there was a return to more moist climate". These rapid changes appeared to include rapid northward advancement and recession of desert conditions, as indicated by the occurrence of caliche. Since the waves of drought must have come from the south, it was the Chihuahuan Desert specifically that participated in these Late Pliocene advances, reaching as far north as perhaps the 40th parallel (in Nebraska).

Geofloras

As indicated in the discussion of Early to Middle Neogene vegetation, modern species were increasingly present in modern associations. These associations included Arcto-Tertiary derivative montane coniferous forest and riparian woodland. Madro-Tertiary derivative associations included mesquite-grasslands, pinyon-oak savannas, chaparral. True desert associations, internally homogenous and geographically expansive, may not have come into existence until End Pliocene (AXELROD, 1950), even though they were edaphically present previously (possibly coalescing into the reticulum previously described). MACGINITE (1962) suggested a similar resolution and expansion of a modern prairie on the Great Plains by the end of Pliocene-Early Pleistocene from a less homogenous Pliocene savanna. Neotropical Geoflora was by this time virtually absent from the middle latitudes of North America, with the exception of very localized relict floras and northern tongues of tubular cactus, thorn scrub, and dry tropical woodland along both coastal lowlands (up to 30 degrees north). At their northern ends the tropical scrubs formed ecotones with Madro-Tertiary Desert and grassland vegetation. From these ecotones, faciations leading to Sonoran Desert and Tamaulipan Acacia-grasslands developed (as defined by SHELFORD, 1963).

Fauna

The Late Pliocene and Early Pleistocene fossil herpetofauna for regions including and adjacent to the Chihuahuan Desert were summarized by GEHLBACH (1965) and AUFFENBERG & MILSTEAD (1963) for North America, and by HOLMAN (1969) for Pleistocene Texas. BRATTSTROM (1955, 1967) provided the only two extensive faunas that pertain directly to this particular time and region. From the region of the Cochise filter barrier in southeastern Arizona BRATTSTROM described a fauna consisting of the salamander *Ambystoma tigrinum*, the toad *Bufo* sp., the chelonians *Kinosternon* and *Geochelone* (*Gopherus* ?), the lizard *Crota-*

---

1. Northward displacement of Pacific borderlands west of the San Andreas fault (as suggested by ANDERSON 1971) could alter this estimation slightly to the south.

phytus, and the snakes *Coluber constrictor, Lampropeltis intermedius* (a *pyromelana* relative, now extinct), *Natrix, Thamnophis*, and *Crotalus lepidus*. BRATTSTROM's (1967) snake fauna from the Late Pliocene Kansas Plains included: *Thamnophis* sp., *Natrix* sp., *Lampropeltis doliata, Lampropeltis getulus, Heterodon nasicus, Heterodon platyrhinos, Elaphe obsoleta, Pituophis melanoleucus, Coluber constrictor, Agkistrodon contortrix, Sistrurus catenatus*, and *Crotalus viridis*. Both faunas indicate modern species living in a climate quite comparable with the present at the same localities. A few indicators of slightly milder (circadian and seasonal) cycles and more mesic conditions are present. At the Arizona locality, the tortoise and the water snake, *Natrix*, served as such indicators. The Kansas snake assemblage has its closest equivalent today in south Texas, in the general vicinity of the Balconian Biotic Province. This 900 kilometer southward displacement may be more due to a fortuitous combination of edaphic habitat conditions (i.e., a combination of riparian, prairie, and rocky environments) than to major differences in climate. Virtually all of the Pliocene snakes still occur in some part of Kansas today.

End Pliocene Differentiation of Modern Desert Biotas

The foregoing account is the functional total of the available fossil record. I am thus forced to employ paleogeographical and geofloral evidence in order to hypothesize the contemporary organization and affinities of the Chihuahuan herpetofauna. The spatial relationships between North American herpetofaunas and their vegetation associations has already been discussed in the last chapter. Explanations of an isomer-like symmetry between eastern and western deserts invoked present climate, topography, latitudes and elevation. However, the common historical source of these now widely disjunct biotas is yet to be explored.

I believe that the sources of both affinity and distinction between major North American Desert biotas are best explained by Late Pliocene events involving physiography, climate, and paleoecology. I consider two particular Pliocene events crucial to the present arrangement and differentiation of desert herpetofaunas (and biotas generally). These are, first, the bisection of the Pliocene Mojavia Desert biotic continuum across the Cochise portal and the Continental Divide (i.e., the Rocky Mountain-Sierra Madre Occidental axis), and second, the enlargement of the Gulf of California in Late Pliocene, which led to northward intervention of an older and more tropical Sonoran Desert biota.

The first event is straightforward in its consequences but deserves some detailed explanation. The presence of a preexisting Pliocene continuum of xeric Madro-Tertiary and Young Northern biotas is strongly affirmed by several lines of circumstantial evidence. I have already substantiated the presence of many modern plant and animal genera and species of this geoflora and faunal element. Many occurred where modern descendant stocks still exist today. In addition, less disruptive topography and milder climatic conditions make the Pliocene desert continuum especially possible. Even now, the 1200 to 1700 meter gap across the Cochise filter barrier is successfully bridged by some desert stocks, *Cnemidophorus tigris* being a well documented example (ZWEIFEL, 1962). The similarity between existing biotas east and west of the Continental Divide indicate a long period of continuous contact between them, something that would have been virtually impossible during the harsh fluctuations of the subsequent Pleistocene. Among the recent Young Northern herpetofauna which reflect this former continuum are: the anurans, *Scaphiopus hammondi, Scaphiopus couchi, Bufo punctatus*; the lizards, *Crotaphytus collaris*, (and siblings), *Crotaphytus wislizeni, Sceloporus magister, Cnemidophorus tigris*, and *Xantusia vigilis*; and the snakes, *Leptotyphlops humilis, Arizona elegans, Hypsiglena torquata, Masticophis taeniatus, Rhinocheilus lecontei, Salvadora hexalepis, Sonora semiannulata, Trimorphodon biscutatus, Crotalus atrox*, and *Crotalus scutulatus*. Other Young Northern snakes, not exclusively desert species, which enforce the continuum model include *Lampropeltis getulus, Masticophis flagellum*, and *Crotalus viridis*. The list is a rather imposing fraction of the total North American desert herpetofauna. Snakes particularly, perhaps because of their lower temperature tolerance and greater motility, have remained intact species across the entire range of the Pliocene continuum despite subsequent disruptions. Many highly distinctive geographical races of these species indicate long evolution *in situ* within sections of the old continuum, indicating long occupation, not recent dispersal from Chihuahuan Desert to Sonoran – or the reverse – across the Cochise filter barrier. Again *Cnemidophorus tigris* offers powerful evidence of the long existence of Sonoran and Chihuahuan races with only secondary contact along the filter barrier (see DESSAUER, FOX & POUGH, 1962).

HIBBARD (1960) documented Late Pliocene fluctuations in desert conditions as far north as Kansas, culminating in cooler more mesic climate. BRATTSTROM (1955) documented the presence of a mesic grassland and pinyon-juniper herpetofauna (*Crotaphytus* and *Crotalus lepidus*) in the Cochise filter barrier at the end of the Pliocene. With this latter direct evidence, I strongly consider the end of the Pliocene as the most likely time of bisection of the desert continuum. The direct evidence is even more convincing when coupled with the timing of local tectonic uplifts and the downswept cooler climates from the north. The Pliocene topography of the Mexican Plateau was essentially modern in arrangement if not in the degree of its relief, and only the last east-west link of the continuum must have passed through this Cochise portal.

The effect of this disruption was to create eastern and western races within members of the old continuum, or to result in sibling species through the classical processes of allopatric speciation. The resulting pairs of species are best represented among the lizard genera *Phrynosoma, Sceloporus, Uma*, and *Cnemidophorus*. However, the toads, genus *Bufo*, and the snake genera *Elaphe, Ficimia, Masticophis* and *Tantilla*, are also represented by east-west pairs of desert sibling species. A complete list has already been provided in Table 23 and the recalculated resemblance values treating siblings as synonyms have been presented in Table 24. The results clearly indicated that a large degree of the difference between eastern and western deserts may be accounted for by the presence of sibling species alone. I believe that lack of consideration of sibling stocks has tended to exaggerate the differences between desert herpetofaunas, especially in quantitative comparisons. This is particularly true of comparisons between Peninsular and Chihuahuan faunas where nomenclatural differences of very closely related stocks, (i.e., *Phrynosoma M'calli, P. modestum, Uma notata-U. exsul*, and *Elaphe rosaliae-E. subocularis*) have effectively obscured striking phyletic affinities (as in SAVAGE, 1960).

The second major event of the Late Pliocene was the expansion of the Gulf of California both geographically and climatically. The semi-isolated Peninsula has remained climatically closer to the Chihuahuan (especially the Mapimian) Desert than the intervening biotic provinces. At the same time, and largely as a result of the same tectonic events, the Sonoran Desert continued to take on modern biu. gical form and expand to the north. The spreading of the Gulf of California opened a wide arid coastal plain to its northeast and possibly enlarged an unobstructed route for summer storm tracks sweeping north. The combination of a semi-arid lowland and regular summer rain storm pattern were fundamental prerequisites to the northern expansion of the Sonoran Desert. Its contemporary biota indicate origins from an ecotone between the Mohavia Desert continuum to the north and a dry tropical thornscrub on the coastal plains to the south (now Sinaloa, Mexico). The mixture of creosote (*Larrea*) and Ocotillo (*Fouquieria*) desert scrub with aborescent tropical thornscrub (tubular cacti) such as *Cereus* and the Palo Verde (*Ceridium*) reflects this ecotonal origin. Its extremely high floral diversity, emphasized by LOWE (1964), is in itself an indication of subtropical conditions.

The Sonoran herpetofauna is also pronounced in its diversity, and a significant portion of its most characteristic and endemic herpetofauna is clearly derived from a tropical Middle American Element. For detailed analysis of the Sonoran herpetofauna see BOGERT & OLIVER (1945), MACDIARMID (1965) and HARDY & MACDIARMID (1969). Just as the Sonoran Desert vegetation differentiated from xeric Madro-Tertiary-Neotropical ecotone, so did its herpetofauna undergo parallel evolution from the compounded semi-arid Middle American Element and xeric Young Northern Element (the Desert and Plains Complex of SAVAGE, 1960).

This Sonoran herpetofauna of early and compound origin may be sorted according to ecological and geographical orientation. The result is two very distinctive assemblages. The first is a set of Sonoran endemics limited by the distribution of reliable summer rains across hot lowland plains. This constitutes the Sonoran endemic assemblage, largely Middle American origin such as *Heloderma suspectum, Micruroides*, and *Procinura*. Modern Middle American species such as *Ctenosaura, Oxybelis, Pternohyla, Agalynchis* and *Boa constrictor* also participate in the southern faciations of the Sonoran Desert. A parallel development (tubular cactus) in the Tamaulipan-Veracruzian interface may have once been connected to the Sonoran, but it has not yet reached true desert aridity. This rain limited Sonoran endemic unit also includes some species clearly derived from the Young Northern Element of the old desert continuum. These include *Bufo retiformis, Sceloporus clarki, Phrynosoma solare, Ficimia quadrangularis, Masticophis bilineatus, Phyllorhynchus browni* and *Crotalus tigris.*

I have identified a second ecogeographical unit of the Sonoran herpetofauna as the California Gulf Arch Assemblage. This includes both genera and species which are restricted to hot flatland desert (in the sense of PIANKA, 1966) forming an arch around the northern edge of the Gulf of California. They are not confined to an area of reliable summer precipitation. They may have originally evolved along the coastal beach fronts of the Baja California protogulf in early Neogene times.[1] That an extensive Mio-Pliocene protogulf existed was well documented by MOORE (1973). The taxa involved are entirely derivatives of the plains and desert complex of the Young Northern Element. They are all lizard and snakes, the entire assemblage being listed in Table 30. The lizard genera *Dipsosaurus* and *Sauromalus*, as

1. These were essentially edaphic deserts of saline exposed dunes and rocky exposures.

well as the snake stocks *Phyllorhynchus*, *Lichanura* and *Chilomeniscus* are generic endemics. The tortoise *Gopherus agassizi* and the snake genus *Chionactis* conform ecologically, and in part geographically, to the Gulf Arch pattern. The lizard *Callisaurus* is both sibling to *Cophosaurus* and also a member of the Gulf Arch Assemblage.

I believe the Gulf Arch fauna to be a well differentiated unit of desert adapted forms. They probably evolved from several different sources: snakes from Old Northern ancestors, *Callisaurus* and *Gopherus* from pre-existing more generalized member of the plains and desert complex of the Young Northern Element, and the primitive iguanids *Dipsosaurus* and *Sauromalus* from the Middle American Element ancestors in the Paleogene thornscrub of the Neotropical Geoflora. The Gulf snake genus *Eridiphas* and the lizards *Sator* and *Anarablyus* might well be added here. The synthetic assemblage, all Young Northern Elements at the generic level of association, has no parallel in higher, cooler Chihuahuan Desert to the east. I assume that even in the early Neogene the Mexican Plateau and its Chihuahuan Desert was too high, and consequently too cool, for these lowlands forms (*Callisaurus* excepted). The absence of all of these stocks from the rather considerable herpetofauna of the Neogene Great Plains tends to support this assumption.

I will now recapitulate the conclusions of the foregoing discussion of Late Pliocene desert biotas. I will then propose and test a biogeographical model of differentiation of biotas based on these conclusions. The conclusions are as follows: (1) a mid Neogene desert continuum or xeric Madro-Tertiary vegetation and desert and plains complex herpetofauna extended from Baja California through the Chihuahuan Desert, even exerting some influence on the more mesic Tamaulipan plain. This belt stretched across North America between 25 and 35 degrees north latitude, with occasional expansions to the 40th parallel across the Kansan Great Plains. Sonoran stock gradually become restricted to the south. (2) The increasingly cool mesic climate which terminated the Pliocene, coupled with regional uplifts, obliterated the Cochise desert portal which had formerly united eastern and western sections of the old desert belt. The result was the *in situ* development of eastern and western subspecies, or in some cases, sibling species pairs. Thus the herpetofauna was already well differentiated before bisection of the continuum took place. (3) The modern western desert biotas, and herpetofaunas in particular, differed from the eastern or Chihuahuan Desert in the presence of two major assemblages: one was the Sonoran Endemic Assemblage, a largely Middle American Element restricted to hot ecotonal lowlands (Neotropical-Madro-Tertiary) with reliable summer rains. The second unit was the California Gulf Arch Assemblage which appears to be a composite of archaic Young Northern stocks (*Dipsosaurus*) and more recently derived (*Chilomeniscus*) genera. This second stock is strongly associated with desert lowlands arched around the San Andreas Fault trough that gave rise to the Gulf[1]. Both of these old stocks expanded to the north as the Gulf took modern form in Late Pliocene, 3 to 6 million years ago.

The spatial biogeographical discussion afforded North American Desert symmetry may now be re-examined. It appears that the cis-trans isomeric symmetry of North American Deserts is due not only to the symmetrical organization of climate and topography, but also to common historical derivation — the old desert

---

1. Also the islands within the Gulf support many insular endemics of these Gulf-Arch stocks (i.e., *Dipsosaurus*, *Sauromalus*, and *Chilomeniscus*).

continuum which constituted the dominant biota of both eastern and western deserts till the Late Pliocene. Essentially, eastern and western fragmented faunas of the old continuum assembled themselves into parallel community assemblages in parallel physical environments. The parallel conformity of communities is a powerful affirmation of the predictive value of well defined communities and their consistent reality, at least in temperate ecosystems. Further support is provided by the disharmonious presence of the Sonoran Desert, not symmetrical in its relationship with other North American Deserts. The Sonoran Desert is an older system superimposed by its relatively recent northern intrusion upon the pattern of the older fragmented desert continuum, as illustrated by Maps 9 and 12.

If the models just discussed are accurate, the differences between North American desert herpetofaunas should be reducible to three major testable factors: the presence of sibling species — especially in pairs east and west of the Continental Divide, the presence of a California Gulf Arch Assemblage differentially clustering the deserts west of the divide together, and third, the presence of the subtropical Sonoran Endemic Assemblage, which makes the herpetofauna of the Sonoran Desert unique among those of North American deserts.

To test the significance of these three factors I designed the following hypothesis: If all sibling species are synonymized, and the California Gulf Arch Assemblage is removed from consideration, all North American desert herpetofaunas will cluster as a single major unit, except for the Sonoran, which will remain somewhat distinctive because of its subtropical endemic assemblage.

The hypothesis was tested by recalculating all resemblance coefficients and coefficient based dendrograms for Mojave, Peninsular, Sonoran, and Mapimian herpetofaunas. The choice of the Mapimian unit was made rather than using all three Chihuahuan subprovinces, or one Chihuahuan Province, because the Mapimian is more comparable in surface area to the other units, and because it contains the complete complement of Chihuahuan endemic and characteristic species, lacking only relictual forms (e.g., *Terrapene*) and parthenogenetic clonal *Cnemidophorus* "species". The sibling pairs synonymized are the same as those listed in Table 23, and the California Gulf Arch forms are the same as those listed in Table 30.

The results of Table 31 are clearly affirmative to the hypothesis and are best interpreted by comparison with the uncorrected resemblance coefficients and dendrogram, Table 15 and Figure 9. A comparison of the derived dendrograms Figures 9 (uncorrected) and 13 (corrected) make the altered relationships particularly graphic. The shifts indicated by the two dendrograms may be summarized by the following table in which group averages are compared.

Comparison	% Faunal Resemblance Uncorrected	% Faunal Resemblance Corrected
1. Between Deserts East and West of Continental Divide	21% C. C. 42% S. C.	41% C C. 71% S. C.
2. Between Chihuahuan, and Mojavian — Peninsular Herpetofaunas	21% C. C. 42% S. C.	41% C. C. 77% S. C.
3. Between Sonoran, and Mojavian — Peninsular Desert.	42% C.C. 70% S. C.	42% C. C. 71% S. C.

Both coefficients indicate a radical increase in faunal resemblance for all units in the comparisons when they have been corrected for the two factors suggested in the hypothesis. Morever, the corrected dendrograms, Figure 13, cluster Mojavian, Peninsular and Chihuahuan herpetofaunas together and set apart the Sonoran fauna based on corrected similarity coefficients.[1] This is precisely what the hypothesis predicted and the exact reverse of the original clustering of faunal units presented in the uncorrected dendogram for S.C. presented on Figure 9 (i.e., two clades, one western including the Sonoran, and the other strictly Chihuahuan).

The relationships indicated by the S.C. based dendrogram on Figure 13 also correspond to Miocene-Middle Pliocene relationships between desert biotas depicted on Map 12, namely that the Mojave, Peninsular, and Chihuahuan Deserts are biogeographical descendant fragments of an older continuum, and that the Sonoran Desert is a new and alien intrusion. Furthermore, these results indicate that sibling species and the Gulf Arch Assemblage are the only factors which make Peninsular and Mojavian herpetofaunas more similar to Sonoran than to the Chihuahuan faunas. The removal of these two units reverses the relative degree of resemblance. I have interpreted these two factors as the results of Late Pliocene tectonic evolutionary and paleoecological changes. This in effect created the sibling species, superimposed an expanded Gulf Arch assemblage in the western deserts, and extended to the north an ecotonal hybrid between desert and tropical thornscrub, the Sonoran Desert. A summary of these relationships is provided by

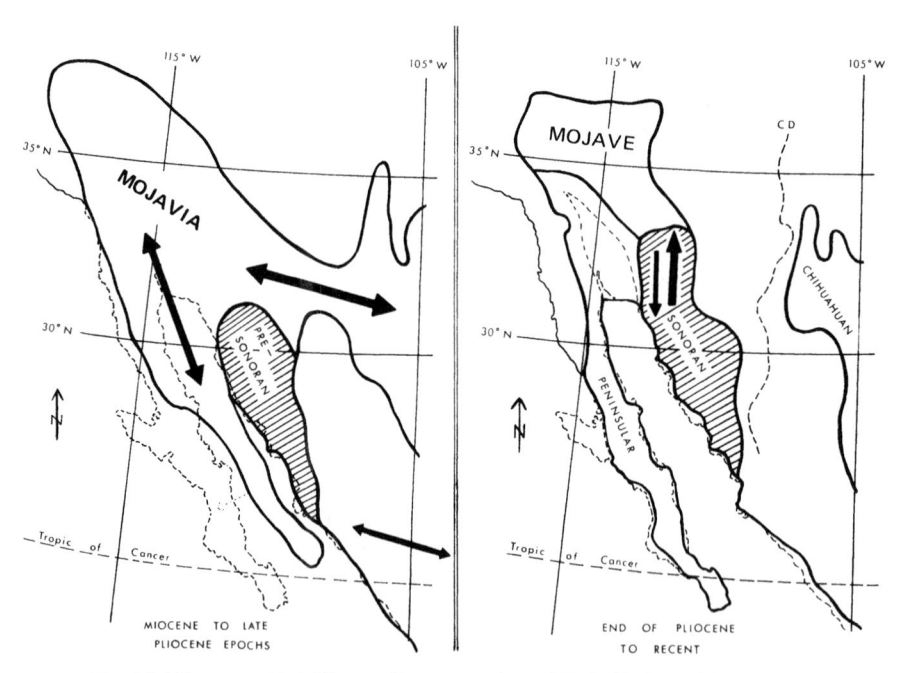

Map 12. Miocene − End Pliocene Fragmentation of Xeric Madro − Tertiary Geoflora.

1. The dendrogram on Figure 13, based on C.C. values is ambiguous, ranks both Sonoran Mapimian as equally distinctive from the others.

Figure 14. The remaining differences between the units of the old Mojavia biota may be attributable to differential local Pleistocene extinctions and extirpations.

The montane herpetofauna of the Late Pliocene Mexican Plateau is virtually unknown from the fossil record. I assume it largely paralleled the modern level of

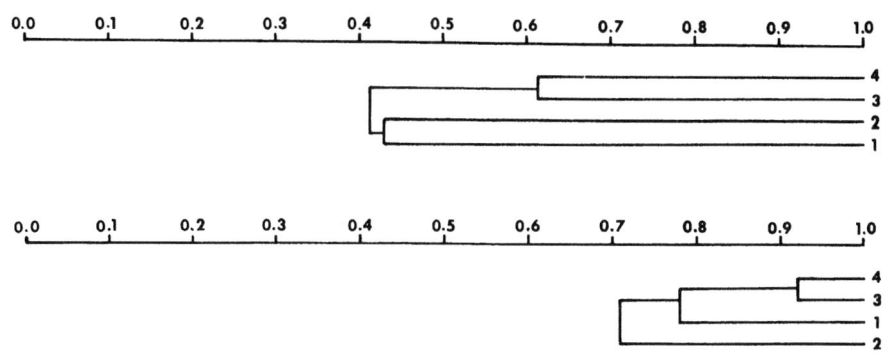

Fig. 13. Faunal Resemblance Dendrogram for the Herpetofaunas of North American Deserts Corrected for Both Siblings and California Gulf Arch Species; Upper Dendrogram: Coefficients of Community; Lower Dendrogram: Coefficients of Similarity; Key: 1. = Mapimian; 2. = Sonoran; 3. = Peninsular; 4. = Mojavian.

differentiation of the plains and desert faunas previously discussed. The derivation of the montane faunas was probably predominantly Young Northern Element-Madrean Complex, the reptile genera *Thamnophis* and *Eumeces* being examples. These stocks are probably of Old Northern derivation of the familial and/or subfamilial level. Old Northern representation on the generic level is provided by the frog genus *Rana* and the snake *Elaphe (triaspis)*. *Gerrhonotus*, judging from its Cretaceous fossil record, is technically a member of the Old Northern Element, but its major radiations are set biogeographically in the woodlands and montane coniferous forest of the Mexican Plateau. Middle American derivates make a small montane contribution, primarily through the leptodactylid frog genera *Eleutherodactylus* and *Syrrhophus*. The plateau centered *augusti* group of *Eleutherodactylus (Hylactophryne)* and *Syrrhophus* almost certainly evolved north of Tehuantepec portal, thus were isolated from other leptodactylids from Eocene through Pliocene.

Table 30. Reptile Species Conforming to the California Gulf Arch Distribution Pattern.

Reptilia:	
Squamata:	
Sauria:	1. *Diposaurus dorsalis*
	2. *Sauromalus obesus*
	3. *Callisaurus draconoides*
Serpentes:	1. *Lichanura trivirgata*
	2. *Chilomeniscus cinctus*
	3. *Phyllorhynchus decurtatus*
	4. *Crotalus cerastes*
	5. *Crotalus mitchelli*
Partial Arch Species:	
Testudinata:	1. *Gopherus agassizi*
Squamata:	
Serpentes:	1. *Chionactis occipitalis*

Table 31. Matrix of Resemblance Coefficients for the Compared Herpetofaunas of Selected North American Deserts Corrected for Sibling Species and the Gulf of California Arch Unit. Coefficients of Community appear in upper right triangle; Coefficients of Similarity in lower left; circled figures indicate number of samples (N).

	Mapimian	Sonoran	Penninsular	Mojavian
Mapimian	㊽	.47	.41	.45
Sonoran	.75	㊽	.35	.49
Penninsular	.73	.53	(51)	.61
Mojavian	.84	.88	.92	㉛

All three of these contributing historical units have a common characterictic, namely that they are relatively old generic stocks, often with a pre-Neogene history. This trend has been interpreted as the result of an upward elevational displacement of older, more mesic stocks, by expanding arid lowland adapted forms. FOWLIE (1965) suggested this model for crotalid snakes in the mountains of the Cochise filter barrier and FINDLAY (1969) independently offered the same model for the same area employing mammalian evidence. Like all too many biological models, it is probably a half truth. I believe that old skink stocks within the genus *Eumeces* and more primitive crotalid snakes (e.g., *Crotalus pricei*) were in fact displaced upward by climate and competing xeric lowland congeners in the early Neogene through the Late Pliocene. But some generic stocks, such as the snakes *Thamnophis*, and the *torquatus* group of the lizard genus *Sceloporus*, are vigorous, well radiated, and not dismissable as mere elevational relicts. Likewise, obvious recent intrusions from the north, such as the *boylei* group of the frog genus *Rana*, may not be dimissed as primitive, though they may be relictual. Extreme low valleys may also be refugia for primitive stocks – e.g., *Coleonyx recticulatus.* The fixation of a Rocky Mountain-Occidental axis and the parallel Edwards Plateau-Oriental axis of mountain faunas was reaffirmed by Late Pliocene faulting and mountain uplift. Doubtless too, the intensifying Chihuahuan Desert between the two axes was becoming an important barrier to cross dispersal – even if a narrow Trans-Plateau dispersal route (MARTIN, 1958) maintained some gene flow.

The Late Pliocene riparian herpetofauna was essentially modern in composition, even on the species level. With few special exceptions it was fundamentally modern in its distribution patterns as well. The end Pliocene hydrographical picture, with subsequent Quaternary modifications is depicted on Map 14. Major patterns have already been discussed in the context of Late Pliocene physiography. Generally, riparian herpetofaunas were probably more complete and extensive across the Mexican Plateau than they are at present. BRATTSTROM (1955) indicated that the water snakes, *Natrix*, extended through the present Cochise filter barrier of Arizona, 600 kilometers northwest of their present limits in the Big Bend of Texas. Isolated populations of *Natrix erythrogaster* in Durango and Zacatecas are well differentiated, and their establishment (but not isolation) could date back to this time. The river systems in which these snakes now dwell are closed drainage systems emptying into ephemeral lakes (Maryan and Viesca) on the western portion of the Parras Basin in Coahuila. CONANT (1963) suggested three alternative routes of former drainage (north, northeast, and east) into the Rio Grande. These were summarized in the last chapter and evaluated quantitatively by

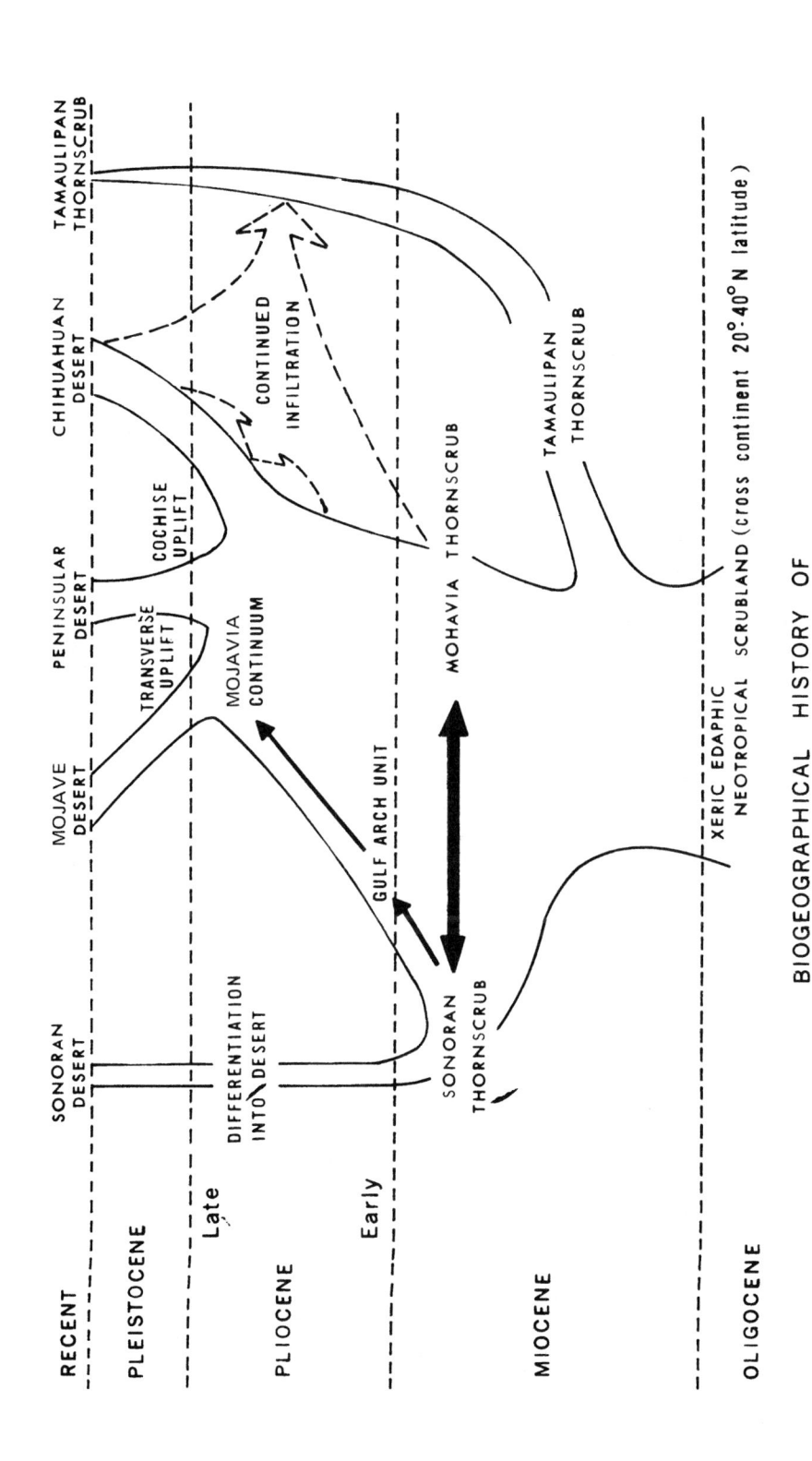

Fig. 14. Biogeographical History of North American Desert Biotas.

BIOGEOGRAPHICAL HISTORY OF
NORTH AMERICAN DESERT BIOTAS

189

the herpetofaunal resemblance coefficients. The results of this analysis – depicted in the dendrogram Figure 12 – strongly indicate that the Nazas-Aguavanal drained north into the upper Rio Grande via the Rio Conchos. This model was suggested by MEEK (1904) and supported by the resemblance of fish faunas. Even now, the intervening gap is pocketed with ephemeral lakes (e.g., near Cebollas and La Perla, Chihuahua) and associated populations of the anuran *Gastrophryne olivacea* and the aquatic turtle *Kinosternon flavescens*. HOWDEN (1963) supported this pattern with distributions of riparian beetles. The hypothetical northeastern drainage of the Nazas-Aguavanal system to the Rio Grande via Cuatro Cienegas and the Rio Salado is not supported by the dendrogram analysis and has also been rejected by MINCKLEY (1969) because of the tectonic history of the intervening region. The third alternative, originally put forth by ARELLANO (1951) suggested an eastern drainage across the Parras Basin and into the lower Rio Grande via the Rio Pesquieria. The dendrogram does not favor this route, but two species do indicate possible prior gene flow between the Nazas system populations and those of hydrogeographical features to its east. In particular, *Natrix erythrogaster* occurs in both the Nazas and Pesquieria, but is absent from the upper Rio Grande and Rio Conchos. In addition, *Kinosternon flavescens* showed an integradation between western and eastern subspecies across the Parras Basin, according to HARTWEG (1938). The eastern form continues through the Rio Pesquieria. As previously noted in the physiographical discussion, much of the Parras Basin sustained standing water and riparian vegetation from Pliocene through Pleistocene times. It is quite possible that the various water systems of the huge basin interconnected and drained away from it by both northwest and eastern routes. The presence of *Thamnophis proximus* in the eastern Parras Basin also indicates its former connection with the Rio Pesquieria to the east.

In conclusion, present herpetofaunal distribution favors a northern drainage of the now closed Nazas and Aguavanal systems, but this evidence does not exclude interconnection with the Rio Pesquieria drainage via the eastern Parras Basin (see Map 13). The tectonic history and geological record indicate both conditions could have existed in Late Pliocene times.

The Pleistocene

A General Description of Pleistocene Physiography, Climate Cycles, and Biogeography.

The Pleistocene contained four episodes of southward expansion of the Laurentide Continental Glacier across North America. These catastrophic cycles of glacial expansions and retreat led to the fixation of modern temperate community assemblages, the extinction of a significant megafauna, including tortoises, and many birds and mammals, and, through the isolation of fragments forged into being the modern biotic provinces of the continent. In this general introduction I hope to establish the time sequence in which Pleistocene events took place, the major tectonic alterations of the north Mexican Plateau, a causal model for the climate of glaciation and its significance for reconstruction of glacial and interglacial climates, and the general biogeographical consequences of glacial-interglacial cycles. Two subsequent sections on the biogeography of Chihuahuan Desert glacial and interglacial intervals will follow, investigating explicit impact on vegetation and herpetofaunas.

The Pleistocene dating sequence of EVERNDEN & EVERNDEN (1970), based on radiometric techniques, has been accepted here. I have adopted their geological dating scale for the entire geological record presented in Table 32. I also concur with their conclusions regarding the initiation of the Pleistocene. If the Plio-Pleistocene boundary is to be drawn strictly according to stratigraphic definition, it must be dated at 1.8 million years, B.P. However, if the more biogeographically relevant definition of first Neogene glaciation is employed, the "biological" Pleistocene began three million years B.P. (evidence for even earlier Neogene glaciation has already been noted).

Several geographically based terminologies exist for the glacial-interglacial sequences in the geological record. The set most geographically applicable to the Chihuahuan Desert was originally designated for the Great Plains directly to its north. The glacial sequence starting with the oldest is: Nebraskan, Kansan, Illinoisan, and Wisconsin. The interglacials in the same progression go: Aftonian, Yarmouth and Sangamon with Recent (or Post-Wisconsin or Holocene) time amounting to a last "interglacial" in its climatic nature.

Physiography

Pleistocene tectonic history for the Mexican Plateau was essentially one of continued development of pre-existing topographical features, largely through additional faulting of blocks in the basin and range landscape. ARELLANO (1951) suggested major uplift along the south edge of the Parras Basin in early Pleistocene. Valley leveling by alluvial aggregation continued from the Tertiary. Thus, the general tectonic trend was to sharpen and reconstruct the earlier features. Major innovations on the earlier landscape were the introductions of new basaltic extrusive flows in central New Mexico (the Malpais region), and the transformation of the upper Rio Grande from a closed basin system to its present drainage from the Rockies to the Gulf of Mexico. As previously discussed in Late Pliocene Physiography, the modern Rio Grande resulted from a cutting of Santa Elena Canyon in the Big Bend region and subsequent connection with the older Pecos-lower Rio Grande drainage, an event which took place between Kansan and Wisconsin Glacials (STRAIN, 1966). Pleistocene uplift probably obliterated any eastern drainage from the Parras Basin, though a northward drainage to the Rio Conchos may have continued through pluvial times (see Map 13). A third major modification to the landscape was probably the appearance and expansion of "pluvial" lakes. Some of these lakes almost certainly have Pliocene origins, the lakes of the Parras Basin and Wilcox Playa of the Cochise filter barrier being documented examples. Nonetheless, great expansions of standing water took place, most of them across the western Trans-Pecos region (New Mexico-northern Chihuahua). A summary of New Mexican lakes was provided by KOTTLOWSKI, COOLEY, and RUHE in WRIGHT & FREY (1965). Major lacustrine basins included the Estancia Basin, Tularosa Basin, Plains of San Augustin, Animas Valley, and Playas Valley in New Mexico. The playas of Santa Maria and Guzman in Chihuahua and the Laguna del Rey, Lago Mayran and Lago Viesca in Coahuila were also included. The karst-like porous topography of the Saladan drainage region (coupled with uplifts) may have inhibited extensive pluvial lake formation on this high portion of the plateau.

For Pleistocene climatic interpretation, I have found STOKES (1955) ocean control theory the most compatible with the biogeographical evidences from the fossil record and from present biotic distributions. His model for glaciation indicated that glacial onset and expansion were the result of warming oceanic climates and the associated increase in evaporation-precipitation. It was the increased precipitation on land masses, especially in continental regions with particularly low temperatures, that atmospheric discharge accumulated as snow, eventually forming glacial masses. As a corollary then, interglacials should have been characterized by cooler climates and drier conditions on the continents. However, TAYLOR (1965) indicated that North American Plains climates were generally more mesic and moderate until late Wisconsin times. The STOKES model suggests that the glacials were not uniform suppressions of the earth's temperatures, but rather an intensification of cold mesic temperate conditions towards the poles and of warmer mesic climates towards the equator, especially in maritime lowlands. A major alternative model was presented by WRY[1] based on reduced marine surface salinities. I have arbitrarily divided glacial maximum North America into a polar glacial zone north of the 30th parallel and a subtropical to equatorial region south of that parallel. North of the 30th parallel, continental climates were under direct glacial influence, either through the presence of ice sheets above the 45th parallel, or their resultant glacial winds and southward displaced cyclonic belts, at about the 35th parallel (according to DILLON, 1956). In this northern division, climate and biota took vertical, downward displacements of 1000 meters (MEHRINGER, 1967). Horizontal displacements across the landscapes toward the south (and also mesic displacement to the west, HIBBARD, 1960) may have been up to 600 kilometers, even between the more moderate 30th and 40th parallels along coastal California (MARTIN, 1958).

South of the 30th parallel climates and associated biotic belts were not nearly so displaced by glacial maxima. There was no direct glacial influence except via occasional northern storms and frosts. The primary shift was toward greater precipitation, which even if accompanied by a slight increase in temperatures (more often summer temperatures were probably reduced), resulted in a raised P/R ratio. HIBBARD's (1960) Kansan glacial Texas biota suggested moderate, mesic, maritime conditions very much compatible with the speculation presented here. HIBBARD also suggested glacials of increasing frigidity — culminating in the Wisconsin. DILLON (1956) undertook an excellent review of North American biogeographical evidence favorable to this model. He suggested that glacial reductions in climate were clinal along a polar-equatorial axis, beginning with 13 C degree suppression of temperatures in the polar regions and reduced to about a 3-5 C degree suppression at about 35 degrees north.

Thus, both vertical and horizontal climatic displacements were very reduced at the latitude of the Chihuahuan Desert, a probability increased by the sheltering rain shadows of the parallel Sierra Madrean ridges around it and the low thermal basin (see Figure 1) in its Mapimian center. Horizontally, significant displacement occurred in the north with a southward shift of cold mesic conditions virtually

---

1. Meteorol. Monogr. 8, 30 (1968).

obliterating the Trans-Pecos Desert (excepting possibly the lowest river valley of the Big Bend of the Rio Grande). Southward, the Mapimian Desert remained essentially *in situ*, but was probably displaced from the foothills of local ranges and those of the Sierra Madres. Its valleys became isolated, but were adjacent mini-refugia, each in its place like a chip of a mosaic. Thus the Mapimian Desert, for both latitudinal and physiographical position, becomes the most logical site for a major glacial refugium. The ample evidence from contemporary biogeographical considerations will be presented later. The Saladan Desert, which was apparently radically uplifted in early Pleistocene, was probably both too cool and too mesic during glacial maxima to support a desert biota. Even a 200 meter downward displacement of the present Saladan vegetation would effectively remove the desert biota from this high portion of the plateau. Furthermore, it is the most southern aspect of the desert, thus most exposed to the intensified mesic climate of the adjacent tropical lowlands (also it is geographically narrower and less insulated by continental rain shadows). In summary, the Chihuahuan Desert underwent a radical glacial compression in response to higher P/T both north and south. However, it probably sustained a stable Mapimian refugial core throughout its Pleistocene perturbations. Map 14 indicates the biotic effects of glacial maximum of the north Mexican Plateau.

The interglacial episodes, at least in the application to Chihuahuan Desert expansions, may have been more significant in their aridity. They lowered P/T ratios rather than raised temperatures, except on the northern borders of the desert where both factors exerted important influence.

Geofloras

The Pleistocene vegetation of North America is just in the process of becoming discovered. Most investigations have been undertaken only during the past decade, and much of the information is restricted to pollen analysis and the aggregations of wood rat (*Neotoma*) middens. The most complete information available is from sites in the Cordilleran southwest or from the Kansan Great Plains. Valuable discussions have been presented by AXELROD (1950, 1958, 1966, 1967), MACGINITE (1962), DILLON (1956), LEOPOLD (1967), MARTIN (1958), MARTIN & HARRELL (1957), MARTIN & MEHRINGER (1965), MEHRINGER (1967) and FLINT (1971). MARTIN & MEHRINGER still stand as the best analysis. Explicit fossil flora of temporal and geographical relevance to the Chihuahuan Desert will be cited under glacial, interglacial, pluvial, and xerothermic headings. These accounts affirm several basic characteristics of Pleistocene vege-tation. First, and most important, temperate Pleistocene vegetation may be generally described in terms of broad geographical associations and dominant species, such as sagebrush scrub, desert scrub, yellow pine forest, grassland, pinyon-juniper woodland, and pine-spruce park-land (MARTIN & MEHRINGER, 1965). LEOPOLD (1967) indicated that major plant extinctions pre-date the Pleis-tocene. All associations, except the last (which may be an ecotonal composite of prairie and northern forest depressed south during Wisconsin glacial maximum), are major modern vegetation types. They constitute the botanical components of the major biotic communities of the southwest today. Since the biotic province concept in the qualitative definition of DICE (1943) is based on a single major geographical community, I conclude that modern communities and biotic provinces became recognizable during the Pleistocene. Obviously, these major units underwent geographical alterations with glacial climatic cycles of displacement and compression. Nonetheless, they were present and extended as broad and apparently homogeneous units. AXELROD (1950) and MACGINITE (1962) strongly suggested that such uniform geographical units did not exist before the Pleistocene, though the same associations of modern species occurred as edaphic local pockets within less distinctive belts in the earlier Neogene. I believe that the transition from edaphic association to geographical community and province was a gradual process of coalescence. This produced modern geographical associations with the advent of modern dry climates in desert and plains regions by Middle to Late Pliocene. MARTIN & MEHRINGER (1965) and RZEDOWSKI (1962) also suggest that distinct deserts may well predate the Pleistocene.

Before leaving the question of Pleistocene vegetation, I wish to comment briefly on the mystery of the warm desert dominant shrub, creosote, *Larrea tridentata*. This species, along with other characteristic warm desert forms, *Koeberlinia spinosa* and *Celtis tala* (MARTIN & MEHRINGER, 1965), also occurs in the deserts of southern South America with no intervening stocks. In the case of *Larrea*, particularly, its fossil presence is unknown until the end of the Wisconsin glacial. When and how it entered the North American desert and became dominant is totally unknown. MARTIN & MEHRINGER noted that it does not have the same insect associates in North and South America. I suspect it was introduced from Patagonia by migratory birds in Late Pliocene or early Pleistocene (GARCIA, *et al.*, 1960). I base this conclusion on its well formed present faunal associations. Whenever established, this vigorous allelopathic sclerophyll must have had a dramatic effect by the reduction of competing species and by thus further differentiation, maintenance and expansion of North American deserts.

Fauna

The best standing discussion of Pleistocene desert herpetofaunas remains that of SAVAGE (1960). AUFFENBERG & MILSTEAD (1965) also have constructed a review of Pleistocene herpetofaunas of the United States, though their models are almost exclusively chelonians of the genera *Terrepene*, *Geochelone*, and *Gopherus*. The general reviews cited for the Late Pliocene herpetofauna record also apply here. GEHLBACH (1965) is particularly valuable. I will recapitulate them here with a critic of several characterizations of the Pleistocene herpetofauna which are generally accepted. These are: (1) expansive speciation of amphibians and reptiles occurred in early to middle Neogene (Pliocene). No new genera are of unquestionable Pleistocene origin. Only a few species appeared, and these were largely amphibian and chelonian mesic (glacial-pluvial) relicts, now isolated to the south or west of their former habitats. (2) Glacial maxima did displace isotherms sufficiently to the south to disrupt previous continuous belts of Tertiary biota and force them into southern glacial refugia. SAVAGE (1960) suggested three such refugia for desert herpetofaunas: one in Peninsular California, one in the Sonora-Sinaloa lowlands, and a last one in the Chihuahuan Desert, centered far south in the present Saladan subprovince. SAVAGE'S model is accepted, but I reject the southward displacement of desert to the Saladan region. Also, a fourth semi-desert thornscrub refugium might be added at the Tropic of Cancer along the Tamaulipan Plain. Its relicts, which probably included ancestors of *Gopherus berlandieri, Crotaphytus reticulatus*, and *Ficmia streckeri*, certainly come close to qualifying as an incipient or antecedent desert herpetofauna. (3) Significant east-west and north-south dispersals and re-invasions took place between the 30th and 40th parallels north during interglacial times. AUFFENBERG & MILSTED (1965) emphasized the eastern dispersal of a xeric plains stock (presumably interglacial) through Pleistocene corridors to the eastern woodlands. They chose *Gopherus* as an illustrative model, largely because of its excellent fossil record. They stated that the modern *Gopherus polyphemus* is a Pleistocene invader from a *G. flavomarginatus*-like stock of the Great Plains and Mexican Plateau. They suggested that its survival in the east was contingent on xeric sand habitats. I accept the model for *Gopherus* as a general model for the eastern Pleistocene invasion of desert and plains complex herpetofaunas. Other members of the complex – also with distinct eastern species living in xeric pockets – include *Scaphiopus holbrooki* and *Pituophis melanoleucus*.

Scaphiopus is known from Miocene Florida. Too little is known of eastern Tertiary biotas to eliminate the probability of Miocene and Pliocene grassland corridors.

Eastern and northern extensions of desert and Great Plains complex herpetofaunas were dismissed by these authors as the result of ecotonal dispersions, not as reliable proof as shifting vegetation or climate. They stated their reservations in particular regarding the northern extensions of the lizards *Cophosaurus texana* and *Phrynosoma modestum* into Sangamon Kansas. I take exception to their interpretation. These lizards survive more often in edaphic desert conditions than in ecotonal transitions, though, admittedly, they occur in both today. They are not primarily ecotonal, however, and are typical of the Chihuahuan Desert core (*Phrynosoma modestum* especially). When their fossil presence in Sangamon is coupled with expansive contemporary caliche desert soils (HIBBARD, 1960), I believe these records make a strong case for the presence of desert conditions in interglacial Kansas.

AUFFENBERG & MILSTEAD (1965) also commented on east-west reconnections between Sonoran and Chihuahuan faunas during interglacials, through the Cochise filter barrier. This was most certainly the case, but their statement that a Sonoran herpetofauna crossed into the Chihuahuan Desert with an inferior reciprocal exchange is without strong biogeographical basis. In fact, none of the true endemic Sonoran stocks (*Heloderma* or *Micruroides*) now penetrate east of the Cochise filter barrier. None of the Gulf Arch assemblages that are so widespread to the west (*Sauromalus, Dipsosaurus, Chionactis, Phylloryhynchus* and *Chilomeniscus*) even reach the barrier. The only desert species that cross the filter barrier are general ubiquitous warm desert forms, *Cnemidophorus tigris Trimorphodon biscutatus*, etc., that have well differentiated eastern and western stocks. I consider them members of the old continuum, fragmented *in situ* at the end of the Pliocene, and no more Chihuahuan in origin than Sonoran. Ecologically speaking, Chihuahuan species or races, adapted to the higher cooler desert would be in better physiological condition to traverse the 1600 m high filter barrier than would be the lowland Sonoran stocks.

(4) The Pleistocene also was characterized by the extinction particularly of large species (and even of large morphs within species) (see MARTIN & WRIGHT, 1967). Poikilotherms suffered less extinction than did birds or mammals. GEHLBACH (1965) reported 4% extinction for Pleistocene fish species and 8% for

amphibians and reptiles, as opposed to 30% for birds and 62% for mammals. Extinction of the giant tortoise, genus *Geochelone*, in North America appeared to be progressive through the glacials, ending with their total disappearance from North America at the end of the Wisconsin in Texas (HIBBARD, 1960). HIBBARD emphasized cold winters as a causal factor, while MARTIN & MEHRINGER (1965) suggested human "overkill" about 12,000 years B.P. They argued that *Geochelone* otherwise should have been able to retreat into the tropics. I don't consider the models in any way mutually exclusive, but rather complementary. Glacial winter chills must have instigated a deteriorating range, probably much as was suggested by AXELROD (1967) for the Wisconsin extinction of the mammalian megafauna. Under such conditions, the remaining populations were probably extremely vulnerable to aboriginal human predation. In this interpretation, I agree with AUFFENBERG & MILSTEAD (1965). The net effect, Pleistocene extinctions, and much more widespread regional extirpations (e.g., *Terrepene carolina* extirpated from the southwestern plains-New Mexico-west Texas in post-Wisconsin times), was to intensify the character and homogeneity of regional biotic assemblages — communities and provinces.

The Glacials

Alterations in Vegetation and Herpetofaunal Distributions

The only synthetic reconstruction of glacial maximum vegetation for the Southwest is that of MARTIN & MEHRINGER (1965). Their analysis, while exceptionally detailed and pioneering in the field, was limited by the accuracy of carbon dating and the questionable accuracy of pollen analysis in reflecting the relative importance of local dominants. In addition, their vegetation map was based on the assumption of 900 to 1200 meter downward displacements of vegetation throughout the region. Since their presentation, several local studies based on wood rat (*Neotoma*) middens have provided greater insight to glacial conditions. WELLS (1966) determined that late Wisconsin pinyon-juniper woodland was displaced about 800 meters, while the yellow pine forest (and its douglas fir stands — *Pseudotsuga mendezii*) was displaced only 600 meters[1]. METCALF (1970), furthermore, indicated differences of several hundred meters in vertical displacement of small mammal faunas on the north versus south sides of mountains in Wisconsin age southern New Mexico. VAN DEVENDER & KING (1971) described a pinyon-juniper woodland displacement of 300 to 600 meters in north-western Arizona at the time of the Wisconsin glacial maximum. The site is now occupied by Sonoran Desert vegetation. MEYER (1973) also indicated moderate displacements (1000 m) leaving the Cuatro Cienegas Bolson floor sclerophyll intact. My conclusions are that the MARTIN & MEHRINGER assumption of 900-1200 meter displacement is not valid for the entire southwest. In conformity with the hypothetical model of DILLON (1956), FLINT (1971)[2] and the evidence of WELLS, VAN DEVENDER & KING, I view displacement as differing with each particular zone. For example, greater proportional descent of pinyon woodland took place than of douglas fir forest, and the downward movement was seriously

1. Displacement does not preclude compression or superimposition.
2. and Gates, 1976 Science 191 (4232): 1138-1144.

affected by northern versus southern exposure. Furthermore, displacement was clinal in degree, being greatest at the north. I will concern myself here only with the pinyon-juniper woodland transition to desert and with only at the glacial maximum of the Wisconsin. This may have been rather brief moment in geological time, according to MARTIN & MEHRINGER (1965), beginning about 20,000 years B.P. and extending only 6,000 years in duration. This pinyon displacement at the maximum was approximately 900 to 1200 meters downward at the 35th parallel north (i.e., Mojave Desert California to Llano Estacado of New Mexico, and perhaps as little as 600 meters in western Arizona). At the latitude of the 30th parallel (i.e., Big Bend, Texas) the displacement of the same interface was 600-900 meters. I have made a rough extrapolation from these declining ranges for the 25th parallel north, and estimate a downward displacement of 300 to 600 meters.[1] This last level would apply to the Mapimian and Saladan subprovinces of the Chihuahuan Desert. The presence of extremely desert adapted herpetofauna and endemic desert vegetation (*Uma exsul* and *Fouquieria shrevei* being examples of each, respectively) on the 1000-1200 meter high valleys of the Mapimian region indicates to me that the desert survived glacial maxima *in situ*. It was never obliterated by the grassland pinyon-juniper woodland that now occurs as relict in adjacent ranges above 1500 meters. If these desert valley biotas were not obliterated (and they could not have been displaced into the 1500-1800 meter Saladan upland to the south), downward displacement could not have been more than 300 meters. Late Pleistocene physiography of the Mapimian was presumably modern and offers no alternative explanation.

On Map 14, I have reconstructed the hypothetical positions of modern biotic provinces and their dominant plant associations at Wisconsin maximum. The map is based on that of MARTIN & MEHRINGER (1965) with additions from other sources previously cited, and HIBBARD (1960). I also considered the present distribution of vegetational and herpetofaunal relicts and endemics. The resulting map is most compatible with all lines of evidence considered. It differs from MARTIN & MEHRINGER primarily in my rejection of the uniform 900-1200 meter displacement of woodlands, and in my evaluation of the importance of pine-spruce (*Pinus ponderosa-Picea*) parklands and sagebrush. Pollen samples, especially when employed without alternative evidence, could easily distort the importance of large wind pollenated species such as conifers. Not only the importance but the vertical and horizontal position of these conifer forests could be easily distorted by blasts of glacial winds off mountain sides to the valleys below. Sagebrush, *Artemisia filifolia*, was doubtless of local importance — especially in cold dry edaphic associations, such as the leeward side of the southern Rockies or in arid sand dune soils. In the Trans-Pecos Desert, the latter is still common in sand dune refugia (i.e., the dunes near Kermit, Texas and the "Medanos" south of Juarez, Chihuahua). However, its importance may have been exaggerated by the edaphic conditions existing at a few New Mexican and Texan sites.

Herpetofaunas probably moved with a good overall coordination relative to their respective plant associations. The result was, in effect, a general expansion of more mesic contiguous biotic provinces at the expense of the Chihuahuan Desert. It was these shifts, the extirpations, differential survival of endemics, and residue of mesic relicts in the post glacial period which resulted in the modern faunal

1. MEYER (1973).

subprovinces of the Chihuahuan Desert. Figure 5 summarized the ecological contributions of this residual mesic relict fauna, all of which were brought to the geographical Chihuahuan Desert by the glacial (and pluvial) expansions depicted on Map 14.

What I do wish to develop here is the derivation of the three Chihuahuan subprovinces in terms of the impact of glacial expansions of adjacent mesic biotas. Because of the limited fossil record, Wisconsin glacial effects will be discussed. Earlier glacials — largely unknown in their impact — were apparently milder and their influence on the present was pre-empted by the Wisconsin.

The Trans-Pecos subprovince was virtually obliterated as a desert. According to WELLS (1966), even the southern tip of this region, Big Bend, was covered by pinyon-juniper woodland down to 700-800 meters elevation. Desert biota might have survived along the canyons and plains of the lowlands carved from the plateau by the Big Bend of the Rio Grande. Present populations of lizards *Sceloporus merriami* and the apparent endemic *Coleonyx reticulatus* in this immediate area — otherwise nearly absent from the Trans-Pecos subprovince — favor this interpretation. Elsewhere in this subprovince, prairie grassland probably covered the lower more mesic valleys while the sclerophytic pinyon woodland occupied the slopes. The woodland and prairie land relicts of Figure 5, groups 1 and 3, probably became established with these vegetation associations expanding from the southern Rockies and northern Sierra Madres. Pinyon woodland belts probably allowed dispersal from the Occidental north-east through the Guadalupes, while the Oriental influence was stronger from the Chisos Mountains and to the east (Edwards Plateau). Prairie and pinyon (Navahonian province) undoubtedly obliterated the Cochise filter barrier as a desert east-west portal. In its place came north-south exchanges between Occidental and southern Rockies-Mogollon Rim, *Thamnophis elegans* and *Phrynosoma douglassi* to the south, and possibly *Crotalus molossus* to the north. Also, the portal became a gateway to a Kansan prairie fauna displaced to the southwest — toward the Sonoran plains. Examples of this movement are the salamander, *Ambystoma tigrinum*, the chelonian, *Terrapene ornata*, the snakes, *Heterodon nasicus* and *Sistrurus catenatus*, and the lizard, *Holbrookia maculata*. The Kansan herpetofauna still has a tremendous influence on the present aspect of the Trans-Pecos subprovince. Even now, Trans-Pecos dune dwellers include the Kansan *Holbrookia maculata* and *Terrepene ornata*, not the Chihuahuan endemic counterparts — *Uma exsul* and *Gopherus flavomarginatus*.

The Wisconsin glacial, however, was not sufficient to re-establish Tertiary connections between high montane forests (*Pinus ponderosa* and *Pseudotsuga*) and their herpetofaunal associates. The salamander *Aneides hardyi* underwent no significant dispersal. MARTIN & MEHRINGER (1965) suggested that this salamander was a glacial relict. They ignored its morphologically primitive nature (almost intermediate between *Plethodon* and *Aneides* — LOWE, 1949, WAKE, 1964). Their interpretation predated the evidence of WELLS (1966) indicating the lack of adequate forest depression for the dispersion of such biota.

The last major glacial influence of the Trans-Pecos was from the southeast, the Tamaulipan province, with its mesquite-grasslands and subtropical Australoriparian woods. The species, now mesic relicts in the geographical Chihuahuan Desert, are depicted by groups 2 and 4 on Figure 5 and listed in the associated discussion of those units. The presence of the eastern coral snake, *Micrurus fulvius*, in such an unlikely spot as a creosote flat of Terrell County, Texas, is just one of the many contemporary evidences for the earlier presence of warm mesic fauna in the region. Mammals, such as peccaries (*Tayassu*), still occur in Big Bend, Texas and add further evidence. That a relatively warm mesic province should expand northwest during glacial times is not surprising considering STOKE'S causal model of warming oceanic conditions. Fossil evidence from the Kansan glacial (HIBBARD, 1960; HIBBARD & DALQUEST, 1966) also indicated the northward expansion of Tamaulipan stocks (*Tayassu, Geochelone, Syrrhophus*) in north Texas.

The Mapimian is fundamentally a glacial refugium. Its position as a thermal cradle has already been established (see Figure 1 and Climate chapter). It contains virtually all of the Chihuahuan endemic herpetofauna and many of these are entirely restricted to this subdivision (*Gopherus flavomarginatus, Uma exsul, Sceloporus maculosus*). Its hot valleys below 1200 meters always sustained a desert. SAVAGE (1960) indicated that the refugium was further south, centered in the modern Saladan subprovince. This proposal does not seem valid. Firstly, the Saladan Desert herpetofauna is virtually without endemics (*Sceloporus cautus*[1] being the single and dubious exception). Secondly, ARELLANO (1951) hypothesized that early Pleistocene faulting considerably uplifted the Saladan region. Thirdly, the present Chihuahuan Desert biota in the Saladan subprovince is entirely a depauperate derivative from the Mapimian to the north (see spatial account). Finally, it is so high, at present, and its Chihuahuan biota so limited to desert valleys, that even a 200 meter displacement of pinyon woodland and acacia grasslands would eliminate the desert biota totally. See the spatial biogeographical account for the relative endemism and distinctiveness of both of these Chihuahuan faunal subdivisions.

The Mapimian Desert was not totally spared glacial mesic intrusions. There is good evidence of a high Sierra Madre Oriental biota only in the Sierra del Carmen and along the high ridges of the Anticlinorium of Arteaga, both peripheral to the Mapimian core. However, pinyon-juniper woodland, residual from Wisconsin and pluvial times, is widespread throughout the ranges of the Mapimian subprovince above 1500 meters. Much of it is receding and rapidly deteriorating, yet it still maintains a saxicolous relictual Sierra Madrean herpetofauna (largely derivative from the Madrean complex of SAVAGE, 1960). These include *Eumeces obsoletus, Gerrhonotus leiocephalus* and *Crotalus lepidus*. The group 3c in Figure 5 is largely responsible for the montane stocks indicated here. This glacial woodland probably occurred as mesic upland reticula across the entire Mapimian Bolson, lacing between the lower refugial desert valleys. I have examined *Crotalus lepidus* from the Sierra Madres above Nueva Delicias in western Coahuila and found them to be intermediate between Occidental and Oriental stocks. The central geographical position of these specimens coupled with its intermediate characters forces me to conclude that a glacial continuum of pinyon herpetofaunas existed across the entire northern Mexican Plateau, not just an extension from one or the other Sierra Madre. This situation would much reduce the significance of Trans-plateau links across the Anticlinorium of Arteaga, except for very high montane biota.

Another mesic distribution of the Mapimian refugium was a corridor of Kansan prairie and its associated herpetofaunas along the western edge of the region. This southwestern corridor apparently extended south of the Rio Conchos along the eastern foothills of the Sierra Madre Occidental. It may have lain along the old

1. specimens from Nuevo Leon indicate integradation with *S. undulatus*

drainage route of the Rio Nazas north to the Conchos. Some of the herpetofauna which still occurs along this route is listed in group 2 of Figure 5. In particular are the prairie mesic and riparian species, *Scaphiopus bombifrons, Bufo woodhousei, Gastrophryne olivacea, Kinosternon flavescens, Sonora episcopa,* and *Tantilla nigriceps.* As noted by CONANT (1963), several Sierra Madrean Occidental *Thamnophis* (e.g., *melanogaster*) penetrate the Mapimian Desert by the way of the Nazas. This too, was probably a condition instigated by glacial extensions of mesic environments.

To the east, riparian-Tamaulipan biotas have made minor intrusions on the Mapimian border. The species involved are noted as group 4 on Figure 5 and listed in the subsequent discussion. Virtually all of these incursions came along river valleys, at Musquiz, Cuatro Cienegas, and the Parras Basin, all in Coahuila. Only Cuatro Cienegas has a large and distinctive fauna. It appears to be a mixture of eastern australoriparian woodland forms such as *Terrepene coahuila* (a derivative of the eastern box turtle *T. carolina,* morphologically no more than a subspecies according to MILSTEAD, 1967), and Tamaulipan stocks such as *Drymarchon* and *Drymobius,* both largely tropical snakes. The superimposition of these mesic subtropical species might be expected during glacial compression of southern biotas. Since lowland conditions at this latitude probably intensified warm mesic conditions during glacials, it is not surprising the fish faunas of Cuatro Cienegas also demonstrate a compounding of temperate and tropical families. Centrarchids, cyprinids, poeciliid, cichlids, and characins all occur sympatrically here. While half of the 18 fish species are endemic, *Terrepene coahuila* is the only endemic reptile. Generally, morphological differentiation is so minor that most forms could have entered during any or all of the pre-Wisconsin glacials, Kansan or Illinoisan being an acceptable estimate (MILSTEAD, 1967). The floor of the Cuatro Cienegas Basin is largely occupied by springs, swamp meadows and salt flats. It is surrounded by a largely separate Chihuahuan Desert biota, which intrudes onto the bolson only along salt flats, bajadas, and dunes. Cuatro Cienegas is best viewed biogeographically as a disjunct glacial isolate of mid-Pleistocene age, derived from the Rio Salado drainage biota on the Tamaulipan Plain.

The Saladan subprovince, as just noted, was totally devoid of desert biota during glacial times, as indicated by COHN (1965) on the basis of modern orthopteran distributions and tectonic evidence. Not only is the present Saladan desert confined to its lowest valleys, but it is virtually undifferentiated and depauperate. Plate 6 effectively illustrates the frail position of the modern desert, wedged below hillsides still dominated by pinyon-juniper woodland.

Based on the assumption of increased rainfall in virtually all parts of glacial North America, it seems unlikely the Saladan region could have supported a desert during any of the glacials. Instead, I envision expansions of more mesic vegetation, primarily acacia grassland from the Trans-Volcanic province from south along the valleys and level western plains of the region, and pinyon-juniper woodland from the Sierra Madre Oriental province. The present biota presents an only slightly dissected picture of this glacial reconstruction (see Map 14). The hypothetical Sierra Madrean area still maintains a chaparral and pinyon biota from Charcas in San Luis Potosi to Concepcion del Oro in Zacatecas. The Oriental lizard *Sceloporus parvus* along with the eastern (or southern) races of *Sceloporus torquatus* and *Sceloporus jarrovi,* as well as an eastern stock of the snake *Salvadora grahamiae,* occupy these relict associations, still widespread above 1700 meters. Other involved species are listed in group 3c of Figure 5. Further west the Saladan landscape is more level and acacia grassland predominates. *Sceloporus torquatus* still commonly occurs along the desert-grassland transition there, where it replaces *Sceloporus poinsetti*. The displacement of these grasslands and woodlands was probably repeated and reversed through several interglacial and subsequent glacial episodes.

The Interglacials

Alternations in Vegetation and Herpetofaunal Distributions

I sense an invisible bias in Pleistocene paleoecological reconstruction against the interglacials. It is assumed that only glacial pulses left mesic relicts in the south. While less likely, it is also conceivable that important xeric relicts have been left in the north as a result of interglacial pulses. Little paleobotanical data are available on interglacial conditions. Most available information comes from gastropods, tortoises and mammal fossils. Even then, the record is largely from the Kansas Great Plains. FRYE & LEONARD (1957) confirmed a general pattern of alternating dry interglacials and wet glacials. FRYE & LEONARD (1957) suggested the Pliocene antiquity of the Pecos River, but virtually all their information on the Pleistocene is from glacial age gastropods. The same is true for METCALF (1967), the only implication for interglacial conditions being that they were dry and inamicable to a diverse snail fauna. AXELROD (1950) indicated that interglacials were progressively severe in their aridity, just as HIBBARD (1960) suggested that glacials were progressive in their temperature (especially winter lows) depressions. HIBBARD (1960) did give a moderately detailed and documented account of interglacials on the southern Great Plains. His most relevant observation was the buildup of caliche (desert) soils during the later portion of interglacials in Kansas. They suggested a northward expansion of Chihuahuan Desert across the southern plains during several of the interglacials.

The general implication of the interglacial records for vegetation is that Chihuahuan Desert scrub expanded at the expense of more mesic associations. At the arid extremes of interglacial times the desert may have stretched from Queretero north to Kansas, and from the Cochise filter barrier in Arizona east through the Edwards Plateau and to Starr County, Texas. The expansion may have been due more to aridity than to any possible increase in temperature. As the desert vegetation expanded it probably displaced more mesic grassland to the north as well as to the east. Undoubtedly, vertical displacements also took place, raising the desert pinyon woodland border from its present 1500 meters at the 30th parallel upward to 1700 or 1800 meters. This upward displacement must have opened portals through the parallel barriers of the Sierra Madre mountains. Across these portals secondary contact between Chihuahuan and Tamaulipan biotas could have been reestablished in the east and between Chihuahuan and Sonoran biotas in the west (through the Cochise portal only). The circumstantial evidence from Chihuahuan gastropod faunas and plains vertebrates from the interglacial make this reconstruction likely, especially with a repetition in the xerothermic period of the post Wisconsin. However, for the present, a convincing paleobotanical record is absent.

Map 15 depicts the reconstruction just undertaken above. The map fails to set a northern limit on desert expansion. Based on HIBBARD'S (1960) observation of caliche soils and ETHERIDGE'S (1960) Sangamon interglacial herpetofauna, the approximate boreal limits are at the 40th parallel north across the Great Plains. ETHERIDGE reported a significant interglacial herpetofauna of the plains. His account included the following lizards in descending order by abundance from Sangamon Kansas: *Cophosaurus texanus, Eumeces obsoletus, Phrynosoma modestum, Crotaphytus collaris, Cnemidophorus sexlineatus*, and *Phrynosoma cornutum*. Only *Eumeces obsoletus* and *Cnemidophorus sexlineatus* are not now typical of the Chihuahuan lizard fauna. The former does occur commonly in rocky refugia or along riparian strips in the Chihuahuan Desert. The latter is extremely similar to the Chihuahuan endemic *C. inornatus* (conceivably the two may not even have been distinct in the Sangamon). The abundance of several desert species, especially *Phrynosoma modestum*, which is now virtually endemic (except for isolated prairie relicts), combined with caliche soils makes a very strong argument for conditions approaching desert in Sangamon Kansas. BRATTSTROM's (1967) analysis of Great Plains snakes is the only other study dealing with a major interglacial herpetofauna. His analysis suggested larger individuals during interglacial times — possibly indicating warmer, drier conditions. The snake fauna does not change character significantly, and was determined to be closest to that of the modern Balconian Province, adjacent to the Chihuahuan Desert. The species involved are largely wide ranging and their provincial affinities are not clear.

Interglacial expansions may have been responsible for some of the extralimital desert relicts discussed in the spatial biogeography chapter (see Peripheral Desert Relict Herpetofaunas). In particular, the extralimital desert relicts in Queretero indicate sufficient differentiation to be the result of some pre-Wisconsin interglacial expansion, *Sceloporus exsul* being an example. For the most part however, these desert relicts are without morphological differentiation from typical desert stocks and are here treated as the result of post-Wisconsin xerothermic expansion, not interglacial.

Just as was the case for vegetation, interglacial desert expansion meant the lateral re-establishment of connection with xeric scrubland biotas in the Tamaulipan Plain and the Sonoran Desert. The former was achieved through lowland gaps in the Coahuila Folded Belt (northern arm of the Sierra Madre Oriental). Abrupt secondary contact probably occurred between Chihuahuan and Tamaulipan stocks along the Rio Grande, at Cuatro Cienegas, and at the Cumbres de Monterey in eastern Coahuila.

To the west, vertical upward displacement of biota opened the Cochise filter barrier to contact between Sonoran and Chihuahuan stocks. Again, there may have been abrupt secondary contact between allied eastern and western races, such as in the case of recent *Cnemidophorus tigris* populations. However, the northward expansion of the Sonoran Desert operated as a second filter barrier, especially during interglacial and xerothermic post-glacials. The biogeographical result is well worth noting. When glacial maxima were in effect, the Cochise Barrier was closed to dispersal or contact between eastern and western desert biota by oak savanna and pinyon woodland (MEHRINGER, 1967; see map 14). Ironically, when the drier interglacial allowed east-west exchanges, the warm mesic lowland Sonoran Desert became a second barrier, this time between the Chihuahuan Desert on the east and its old western partners in the Tertiary Mojave continuum, the Mojave and Peninsular Deserts. As a result, many fragmented sibling species never achieved secondary contact, even in interglacial times. Siblings in groups like the *Elaphe rosaliae* group among snakes, or the lizard genera *Uma* and *Xantusia* and the *Cnemidophorus sexlineatus* species group were all kept in allopatry by the mesic subtropical Sonoran Desert during the only periods of the Pleistocene when secondary contact or dispersal would have otherwise been possible. The contemporary effects of this situation are well illustrated by the sibling species group *Phrynosoma modestum-P. m'calli*. The two simply do not meet at the Cochise filter barrier. In fact, they do not meet at all. *Phrynosoma modestum*, generally confined to the Chihuahuan Desert proper, extends 150 kilometers west of the primary division of the barrier near the New Mexico-Arizona,border. It follows the mesquite grassland ecotone all the way to Nogales, Arizona on the edge of the Sonoran Desert. But even here it fails to make secondary contact with the western sibling, *Phrynosoma m'calli*. The mesic subtropical Sonoran Desert intervenes and forms a complete barrier between the allopatric relatives. *Phrynosoma m'calli* remains confined to the west of the Sonoran Desert proper isolated in the northern dunes of the Peninsular Desert (the Colorado Desert). This Sonoran barrier is minimally 300 kilometers across.

Having established this model, I wish to re-evaluate the suggestion of MARTIN & MEHRINGER (1965) that the lizard *Uma* dispersed across the Cochise filter barrier (presumably west) during post-glacial times. During glacials this barrier was obviously closed to these sand dwelling warm desert lizards by the presence of a cool moist climate and conifer to grassland vegetation. However, during interglacial and post glacial times, this route was also blocked by a warm moist climate and vegetation further west, in other words by the Sonoran Desert. The high degree of morphological difference between eastern and western *Uma* (SAVAGE, MS, indicated that even the fringes of the toes are analogous rather than homologous developments) supports the model of much longer isolation. I hold the eastern and western species to be Mio-Pliocene fragments of the old Mojavia Desert and plains complex. Their absence from intervening sand dune systems, such as those in northern Chihuahua and southern New Mexico, make the post glacial model of dispersion even less tenable.

In addition, as AUFFENBERG & MILSTEAD (1965) suggested, some members of the desert and prairie complex did disperse permanently into the eastern woodlands via xeric corridors and edaphic habitats during Pliocene or interglacial times. Some may have undergone allopatric speciation during subsequent glacials as suggested by SAVAGE (1960) for *Gopherus polyphemus* and *Crotalus adamanteus*. Other examples may include *Sceloporus woodi, Cnemidophorus sexlineatus* and *Crotalus horridus*. While several others failed to speciate, they still may have dispersed by interglacial grassland routes. Some of these include: *Sceloporus undulatus, Coluber constrictor, Masticophis flagellum, Pituophis melanoleucus* and possibly *Lampropeltis calligaster* (perhaps even *L. getulus*).

The most general effect of the interglacials was perhaps not in dispersal of the Chihuahuan herpetofauna, but in the intensification of their xeric commitment physiologically and their increased biogeographical homogeneity as an assemblage. Increasingly intensive interglacials, by extinction and extirpation, reduced the presence of non-desert mesic grassland and woodland species that had survived the

Pliocene and been sustained by the alternating glacials. The fixation of a modern uniform Chihuahuan Desert assemblage was, and in fact still is, taking place through these processes.

The Pluvial Period, Post Wisconsin

Between 15,000 and 12,000 years B.P., the climate of the Wisconsin glacial maximum began to ameliorate. Temperatures gradually rose, glaciers melted into pluvial lakes, and rivers enlarged. Mesic paleoecological conditions continued to be much as they were during glacial times both in degree and extent. I believe that the mesic manifestations of the pluvial were largely the result of montane glacial melts, not increased precipitation. The vegetation of the time, as described by MARTIN and MEHRINGER, WELLS, and VAN DEVENDER *et al.* (all previously cited for the glacial section), was dry cool pinyon-juniper woodland, open ponderosa pine parkland, and very widespread sagebrush scrub. All of these are now indicators of "cold deserts", high plateau steppes and other cold xeric situations. The sagebrush (*Artemisia filifolia*) *Sceloporus graciosus* populations of dunes in southeast New Mexico-southwest Texas are undoubtedly relicts of these conditions (KERFOOT, 1969). *Phrynosoma douglassi*[1] in the adjacent Guadalupe Mountains may be further evidence. Mesic faunas may have advanced along enlarged river systems, but at least in the Trans-Pecos subprovince, these were cold Rocky Mountain stocks such as *Chrysemys picta, Opheodrys vernalis*[1] and *Thamnophis sirtalis*[1]. METCALF (1967) gave some indication of conditions of the sort described here based on pluvial gastropods of the upper Rio Grande. He also (1970) described a small rodent fauna from Late Wisconsin or pluvial eastern New Mexico which indicated that species now typical of Wyoming sagebrush occurred in the area at moderate elevations. This then was the picture for pluvial Trans-Pecos Texas: cool and dry; pinyon, pine and sagebrush, oak, and probably prairie as well widespread with their associated faunas. Drainage patterns were characterized by pluvial lakes trapped in closed basin drainage systems and enlarged cold rivers draining east.

The pluvial of the Mapimian and Saladan subprovinces and adjacent non-desert provinces is largely unknown. The descriptions constructed for the glacials apply here as well. Drainage and lakes were probably still glacial in both pattern and extent of development. In general, the pluvial period is only a slight extension of the Wisconsin glacial. I consider it worthy of description as a discrete episode only because it was the last major cool (and/or mesic) period before the present. The pluvial supercedes all previous glacials in its impact on modern herpetofaunal distributions. The only mesic stocks which could not have dispersed into the Chihuahuan Desert as late as pluvial times are the high montane relicts of douglas fir and high montane pines. Old herpetofaunal relicts such as *Aneides hardyi*, or those modestly differentiated faunas of the Cuatro Cienegas Bolson (e.g., *Terrepene coahuila*, are herpetofaunal examples.

---

1. penetrated south into the Sierra Madre Occidental.

The Xerothermic Period, Post Wisconsin

I have adopted the term "Xerothermic" of AXELROD (1966) over that of the alternative "Altithermic" used by numerous others including most of the authors in WRIGHT & FREY (1965). The former is both more descriptive and more explicit in condition and temporal extent, here estimated at 12,000 to 5,000 years B.P. The Xerothermic is final and exaggerated recapitulation of climatic episodes contained within the latter portions of previous interglacial periods (TAYLOR, 1965). Basically it represents an extreme in Quaternary warm arid conditions and is the most recent in an alternating series of such climate episodes; thus, it supercedes previous trends of the same nature. All of the extralimital Chihuahuan Desert relicts, now clinging to xeric edaphic refugia on cliffs, in flood plains, rock outcrops and sand dunes are the biotic residue of an extended Xerothermic Chihuahuan Desert. Its relicts have been described as peripheral desert faunas in the spatial biogeographical chapter. Its maximum extent is estimated on Map 15. In addition, the northward drainage of the Rio Nazas was permanently severed.

The most conspicuous sustained effect of the Xerothermic was the re-occupation of Trans-Pecos and Saladan lowlands by Mapimian Desert biota at the expense of Kansan prairie, woodland, and the mesquite-grassland ecotone (MEHRINGER, 1967, documented this re-occupation at 7000-7500 years B.P. in San Pedro Valley, Arizona). To the south, the Saladan underwent, and may be still undergoing, the invasion of a depauperate desert biota via the lower valley floors. Dispersal across the eastern two thirds of the Anticlinorium of Arteaga, which separates Mapimian from Saladan subprovinces, is still virtually impossible ecologically for most desert species (ridges in deep pine forest above 1800 meters elevation). Thus dispersal into the Saladan region presumably came through the western border between the two subdivisions, especially along the dissected and leveled river valleys of the Rio Aguavanal (also probably through passes in the adjacent Sierra de Jimulco) in southwestern Coahuila and eastern Durango. This route was also suggested by COHN (1965) for southward desert dispersal.

The general picture of desert re-invasions during the Xerothermic Period, especially for the Saladan subprovince, strikingly parallels BANTA's (1962b) postglacial reconstructed invasion by Mojave herpetofauna into the lowland valleys of another basin and range physiographical province, the Great Basin. In the latter, the desert herpetofauna was also a depauperate derivative of the warmer lowland assemblage. In both, dispersal moved across valleys while relictual pinyon-juniper woodland survived in adjacent ranges. The Rio Aguavanal dispersal route even finds its parallel in the "Sonoran Trailway" of the western Great Basin.

These Xerothermic movements of biota must have taken over huge areas in relatively short periods of time. If WELLS (1966) reconstructions of pluvial pinyon-juniper positions in Big Bend are correct, the Chihuahuan Desert of the Trans-Pecos subprovince had dispersed north at least 600 kilometers up the Pecos River valley. Possibly another 200 kilometers should be added if north Texas-Oklahoma relicts are considered as indicators of the northern limit of the Xerothermic Chihuahuan Desert. SCHULTZ (1967) reported a desert caliche, from Kansas, in excess of 11,000 years old. Even calculating a 600 kilometers expansion over 7000 years, I find that the desert must have averaged a northern movement of nearly one-tenth kilometer per year. This average is much too slow since droughts were episodic. Actual expansions of Chihuahuan scrub vegetation may have been as rapid as three to five kilometers per year, especially up low and accessible river valleys and plains. These rapid movements in vegetation doubtless resulted in broad ecotonal herpetofaunas, often with species densities exacerbated by an edge effect. A lag in Kansan prairie species retreat would superimpose them into an invading Chihuahuan scrub herpetofauna. For certain groups such as *Crotalus* and *Cnemidophorus*, this is exactly the condition found in the Trans-Pecos subprovince and adjacent Cochise filter barrier today, as discussed in the last section of the ecological chapter and depicted on Figures 6 and 7.

The edge effect is particularly apparent in the Cochise filter barrier where desert vegetation has alternately expanded and contracted across the high valleys with each minor shift in P/T ratios. The elevational position of these valleys, at about 1300 to 1600 meters, is right on the transition between desert, prairie and pinyon-juniper woodland. This vertical position, coupled with geographical accessibility to Kansan prairie, to pinyon-juniper woodland of the Sierra Madre and Navahonian provinces, to the Sonoran Desert province, and to the Chihuahuan Desert province, has led to a five way superimposition of expanding and residual faunas. Each assemblage alternately has the local "vegetational rug" pulled out from under it by minor climatic changes, and thus becomes part of the ecotone. In this way such unlikely combinations as the Sonoran *Heloderma*, the Kansan *Terrepene*, the Sierra Madrean *Elaphe triaspis*, and the Chihuahuan *Phrynosoma modestum*, have all been assembled in the same unstable ecotonal grasslands at the edge of the Chiricahua Mountains near Portal, Arizona. To add to the edge effect of these Xerothermic expansions and subsequent perturbations, are the presence of parthenogenetic clonal *Cnemidophorus*. Many of these forms operate as ecological entities equivalent to distinct species. They were both created by hybridization resulting from rapid shifts between desert, grassland, and pinyon woodland. They are at present maintained in ecotonal conditions resulting from these shifts. An excellent analysis of this situation has been provided by WRIGHT & LOWE (1968). The presence of parthenogenetic *Cnemidophorus* is confined to the Cochise filter barrier and the Trans-Pecos subprovince. They are apparently absent from the more stable Mapimian subdivision and from the Saladan region as well. Their absence from the Saladan region may seem surprising since rapid Xerothermic shifts in vegetation have taken place in that region. These shifts are over linear extensions of 200-300 kilometers in the south, and thus have an average rate of displacement of about 0.05 kilometers per year. The edge effect, especially between desert, chaparral, and pinyon-juniper woodland communities, is apparent here, particularly across the Saladan filter barrier of Zacatecas. However, here no hybrid parthenogenetic *Cnemidophorus* occur, apparently because no new mesic upland species were (or are) present to crossbreed with the two forms associated with desert and ecotonal grassland vegetation.

The expansion of Chihuahuan Desert biota and its present ecological and evolutionary consequences are all best attributed to the Xerothermic. The only possible exceptions may be the steep slope creosote patches of Queretero (and other pockets in the Rio Panuco drainage). Here, at least one species, *Sceloporus exsul*, shows sufficient morphological differentiation from its Chihuahuan relative, *S. cautus*, to open the possibility of interglacial rather than the later Xerothermic dispersal (as previously noted).

The Recent, for purpose of definition, will be considered as the past five thousand years. MARTIN & HARRELL (1957) documented a series of climatic perturbations in the Valley of Mexico which might well apply to the Chihuahuan Desert as well. However, resulting biotic alterations were probably too minor, and substantive evidence too geographically distant, to attempt any reconstruction here. Human impact, in the form of introduction, extirpation, and extinction is possibly more significant and better documented. Human introductions of amphibians and reptiles in the Chihuahuan Desert are few, largely peripheral, and of little impact on the pre-existing biota. *Hemidactylus turicus*, the old-world gecko, occurs in human habitations at subtropical riparian villages such as Cuatro Cienegas and Melchor Musquiz, both in Coahuila. (It may extend much further west — expanding along the Rio Grande drainage and possibly even to Camargo, in Chihuahua.) At the former locality it is possible that other tropical species such as *Bufo marinus* and *Drymarchon corias* have been introduced by man directly or inadvertently by the opening of a canal from the Bolson of Cuatro Cienegas to the Rio Salado to the east. Scattered pockets of the bullfrog *Rana catesbeiana* occur across the Trans-Pecos Texas, usually near centers of anglo-saxon population. Because of the latter situation, I suspect these of being introduced as a food frog rather than descending from Australoriparian relicts. In other situations, such as the irrigation of the western Parras basin with the water of the Nazas, pre-existing riparian stocks may have been benefited. Huge choruses of *Gastrophryne olivacea, Bufo cognatus*, and sometimes *Scaphiopus couchi* and *Bufo debilis,* are found in these canals during summer rain season. Some rodent predators, such as the snakes *Pituophis melanoleucus* and *Lampropeltis getulus*, may have also increased densities in agrarian districts. Quantitative evidence to this effect is lacking. I offer these observations only as suggestions for future investigation.

Extirpation by human ecological disturbance had probably been a significant factor in recent Chihuahuan biotic history. The effect of overgrazing in the conversion of grassland ecotones to desert scrub has already been discussed in the vegetation chapter. Radical transformations, altering thousands of acres of land, including soil, drainage, vegetation and microclimate, have taken place throughout the Trans-Pecos subprovince and west through southern Arizona. The effects were graphically demonstrated by LOWE, WRIGHT, COLE & BEZY (1970) with a set of "before and after" photographs of desert grassland taken 67 years apart in Pima County Arizona. They indicated that rapid, radical transformations of this time may instigate further hybridization between sexual and parthenogenetic *Cnemidophorus*. While disturbances of this type have been most severe in the Trans-Pecos as result of cattle grazing, goat grazing (and some cattle) in the Mapimian and Saladan Desert has significantly reduced native grasslands in those areas. Overgrazing may have even accelerated desert expansion in the latter. In addition to disturbing habitat segregation between related faunas, these shifts may extirpate many grassland relicts (i.e., *Ambystoma, Terrapene*) and reduce the primary productivity and efficiency of the indigenous ecosystems.

Man appears to be responsible for the incipient extinction of but a single Chihuahuan species of amphibian or reptile. But what a species it is! *Gopherus flavomarginatus*, the largest terrestrial poikilotherm in temperate North America. It has already been widely expirated by human predation for food. It no longer

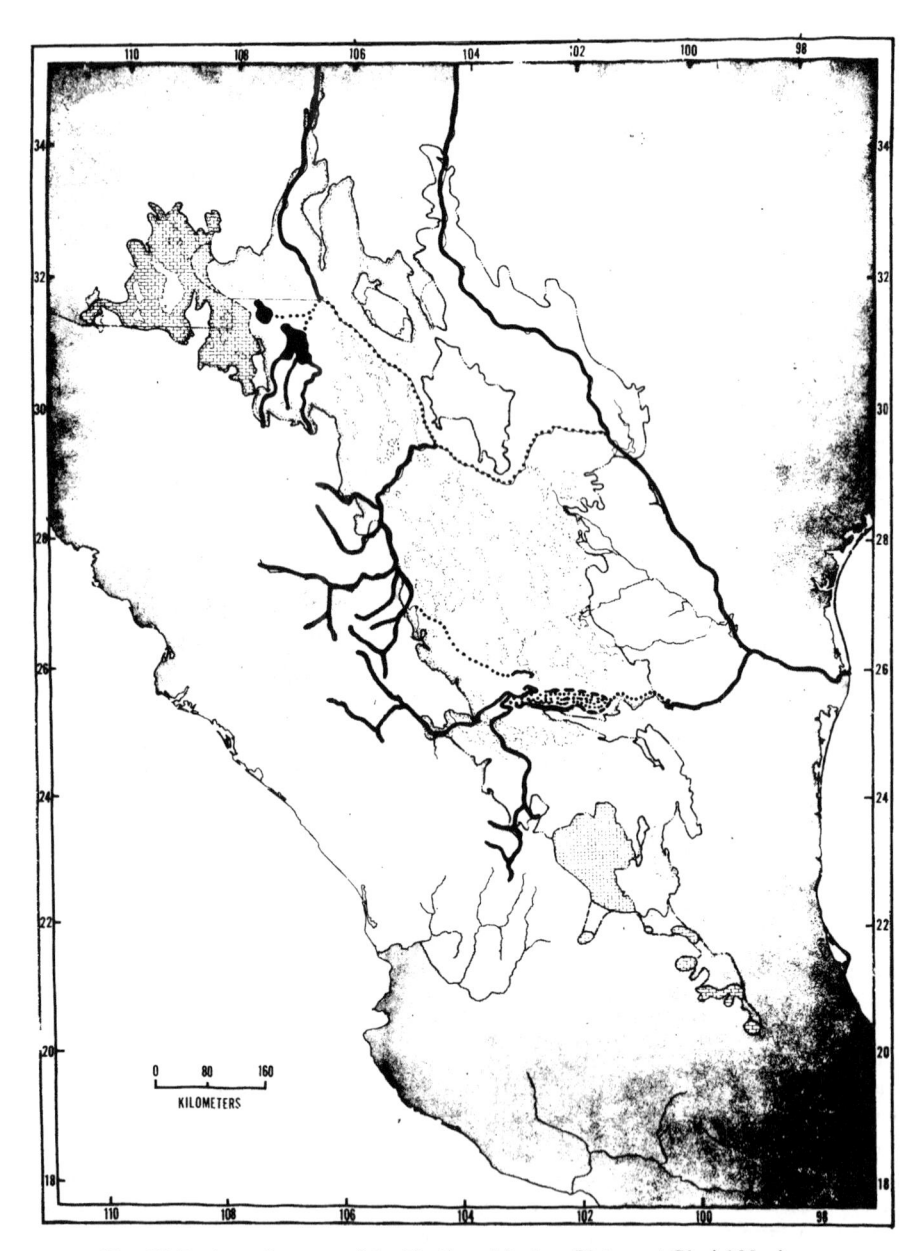

Map 13. Drainage Patterns of the Northern Mexican Plateau at Glacial Maximum.

208

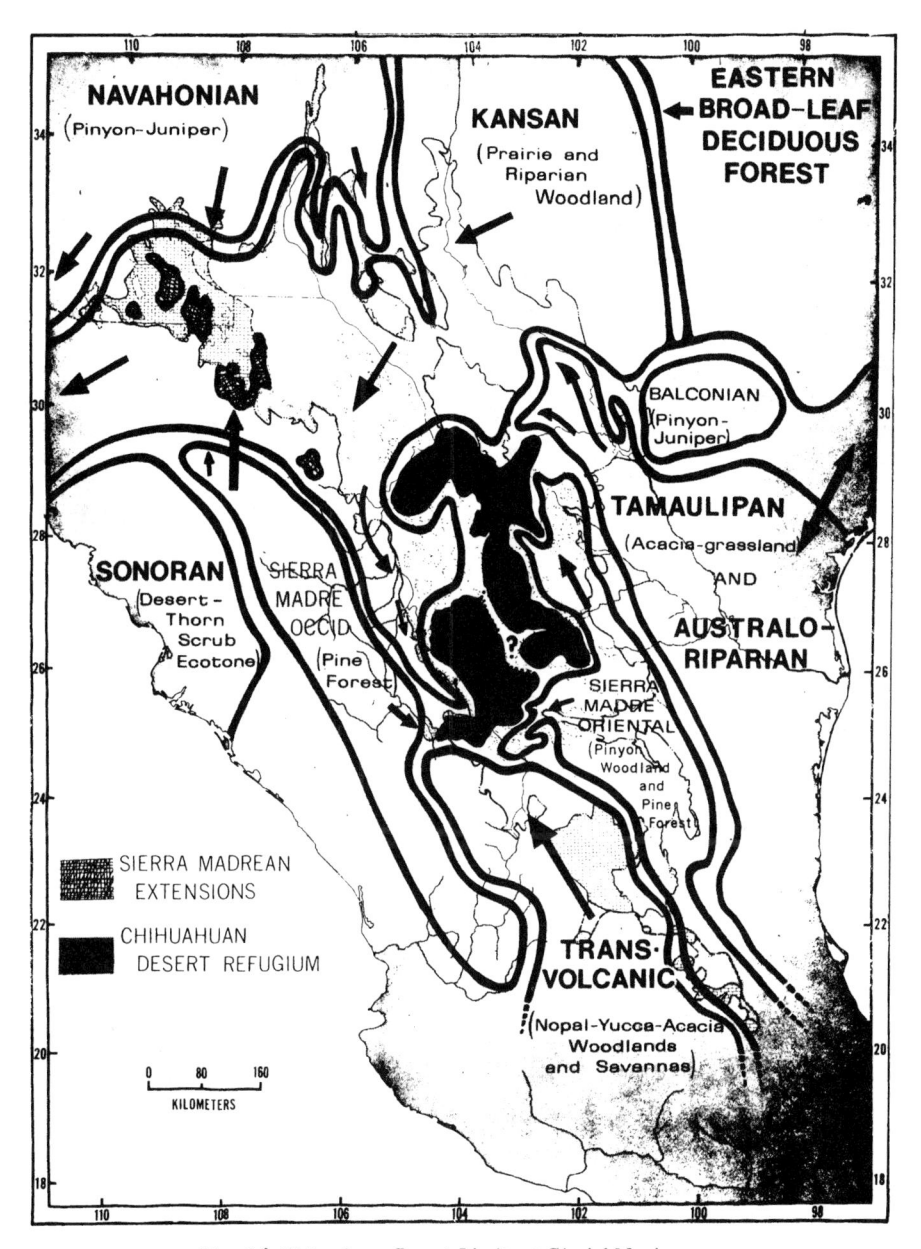

Map 14. Chihuahuan Desert Limits at Glacial Maximum.

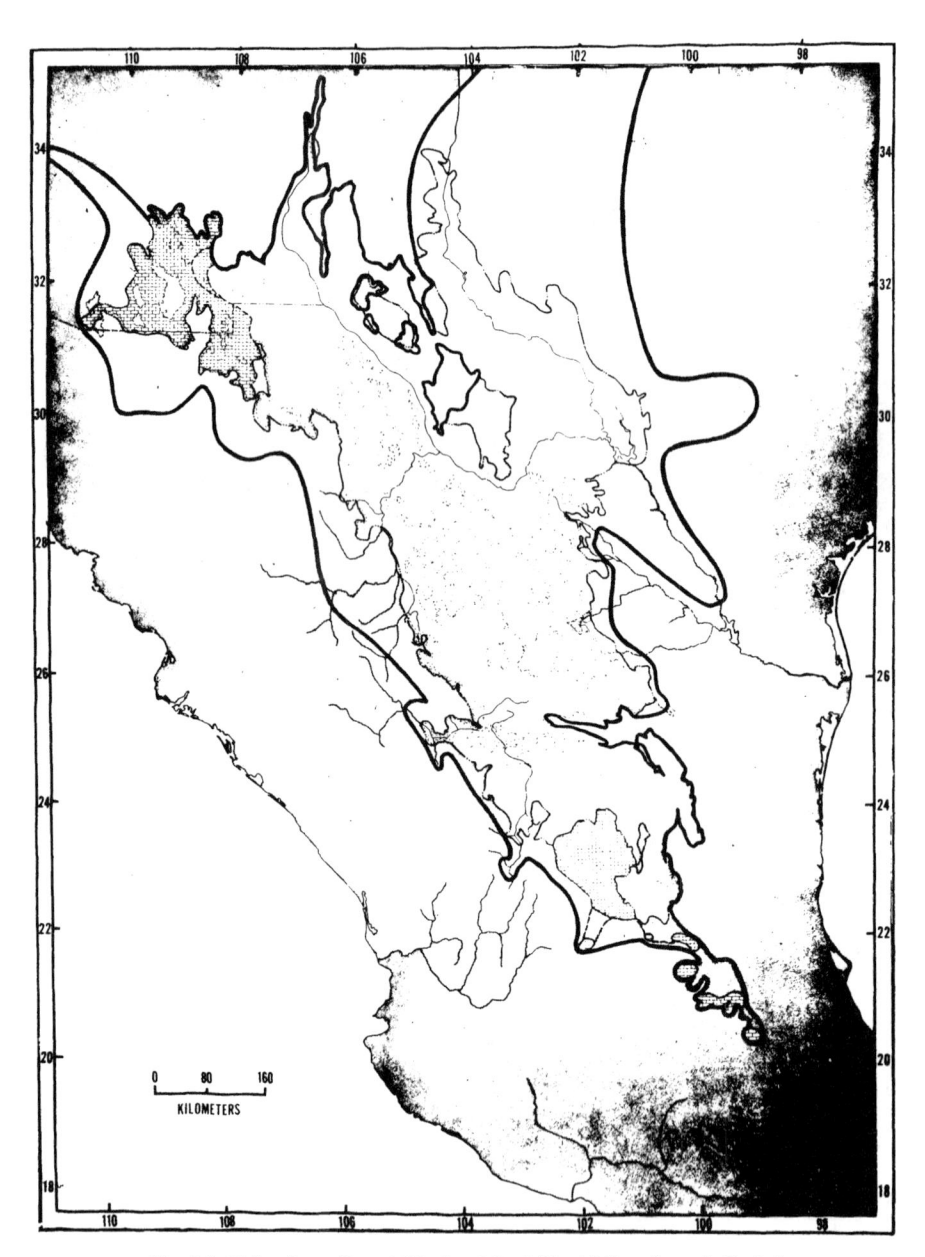

Map 15. Chihuahuan Desert Limits at Post Glacial Xerothermic Period.

210

occurs at or near its type locality at Carillo, Chihuahua, nor is it any longer available as a meat animal in the markets of Cebollas, Escalon, or Conejo. The advent of the Chihuahua Pacific railroad opened much of its range to settlement and exploitation in the 1940's (especially the rail segment between Carillo and El Oro in Coahuila). Freight cars of sprawling tortoises were shipped to Sonoran and Sinaloan towns as a delicacy. Now, drilling for underground water is opening much of the remaining tortoise grasslands to cattle grazing. With cattle will come cowboys, then ejido settlements, followed by more human exploitation, and the devastation of its crucial *Hilaria mutica* grasslands.

It speaks poorly for the values of men in general and herpetologists in particular, that we have spent more time arguing the hypothetical causes of the extinction of *Geochelone* than in seeking alternatives to the impending doom of *Gopherus flavomarginatus*, the largest land chelonian still on our continent. Somewhere, even now, in silent agony one more giant dies. Somewhere in the valleys of the Bolson de Mapimi lies an archaic desert grassland, of unspoken beauty and unmeasured value. Several living fragments of the history of this continent still survive in those valleys. I hope mankind will have the chance to see their beauty and explore their mysteries before they too lie in silenced agony, overgrazed, exploited, and devastated beyond all recall.

Table 32. Paleontological Calendar of the Evolution of the Chihuahuan Desert and Its Extant Herpetofauna.

Geological Time Scale	Age[1] (years) B.P.	Paleoclimate	Physiography	Geoflora	Herpetofaunal Elements	Differntiation and Dispersal of Herpetofauna
Mesozoic Era; Cretaceous Period	135 to 65 x 10[6]	Subtropical to warm temperate (ESTES, 1970) wet	Low plains deeply erod- ed old ranges re- ceding epi- continental seas; Lara- mide Orogeny	Neotrop- ical with some Arcto- Tertiary broad- leafs	Predominantly Neotropical with signifi- cant Old Northern representation. Mesozoic Archo- saurians still present	Modern families present except colubrids; some modern Old Northern genera; expand with sea recession
Cenozoic Era; Paleocene- Oligocene Epochs	65 to 25 x 10[6]	as above, gradual deteriora- tion of temp. and ppt.	terrestrial with local embayment at east edge Rockies, Sierra Madres and basin range est. low	as above, edaphic Madro- Tertiary pockets	as above, first Young Northern genera. Total ex- tinction of Archosaurian terrestrial fauna completed.	as above and some Old Northern genera added from Eurasia
Neogene Period Early Miocene Epoch	25 to 18 x 10[6]	temp. deter- ioration and further drop in ppt.	as above	as above Madro- Tertiary expan. acceler- ates	Southern restriction and/or ex- tinction, of S. American Ele- ment and other wet tropic	Gradual expansion of Young Northern genera?
Middle and Late Miocene	18 to 12 x 10[6]	Increase in temp. & ppt. followed by sharp drops	Extensive orogeny and volcanism in Sierra Madre,	Radical expansion of Madro- Tertiary	Young North- ern genera increase in kind; older	Rapid and major expan- sion of Young Northern genera, recession and fragmentation

Epoch	Time	...in both, P/T ratios drop	Occidental and Plateau	...esp. elements grassland	...elements highly restricted	...of Old Northerns and Middle American
Pliocene Epoch	12 to 6x10^6	Cont. temp. & ppt. (esp. summer) deterioration after initial reverse	Further orogeny esp. in Sierra Madre Oriental followed tectonic stability & alluvial aggregations	Early reversal followed by continual expansion	Young Northern increase; Old Northern and Middle more restricted to riparian & montane habitat	Early reversal followed by continuation of Miocene trend. Modern species groups
Late Pliocene	6 to 3x10^6	Initiation of first glacial (Nebr.). Gradual increase in aridity	Fut. orogeny of Sierra Madre Occidental	Madro-Tertiary but peripheral Arcto-tertiary expansion by end of epoch. Modern species present	Still Young Northern but increase in Old Northern Element	Young Northern sibling species formed by isolation from western stocks. Old Northern invasion via grassland and riparian habitats
Cenozoic Era[2] Pleistocene Epoch A. Glacial	1.8 x 10^6 to 15,000	N. of 30th parallel — cool & mesic (high P/T) S of 30th — warmer & mesic (still high P/T)	Generally stable & mod. topographie, aggreg. alluvium fills valleys; local uplift of fault bl. basalt flows in New Mexico	Madro-Tertiary mesic grassland & woodland dominate Arcto-Tertiary expansion in montane & riparian habitats -modern species now form. mod. communities	Mesic (Madrean Complex) of Young Northern favored with Old Northerns in riparian & montane environments	Mesic Young Northern & Old Northerns expand Southward & lower in elevation at expense of xeric Young Northern modern species present racial patterns and community assemblages taking form. Some major extinctions (i.e., *Geochelone*)

Table 32. Paleontological Calendar of the Evolution of the Chihuahuan Desert and Its Extant Herpetofauna (continued).

Geological Time Scale	Age[1] (years) B.P.	Paleoclimate	Physiography	Geoflora	Herpetofaunal Elements	Differentiation and Dispersal of Herpetofauna
B. Inter-glacial[2]	1.8 x 10⁶ 70,000	N. of 30th parallel warmer and more xeric (low P/T ratio) S. of 30th cooler (?) and xeric (low P/T)	as for glacial but caliche soil favored	Xeric Madro-Tert. desert scrub expands Arcto-Tert. retreat	Xeric Young Northerns dominant recess., frag., extirpation of mesic Young Northerns and Old Northerns	as above but with reversal of expansion, xeric Young Northern extend up in lat. & elev.
Recent[2] (Post Glacial) A. Pluvial	15,000 to 12,000	Cool & mesic high P/T, ppt. and ice melt ext.	Gen. topo. stable & mod. & river discharge (Nazas, Santa Maria) sufficient for open	as for glacial but less extreme	as for glacial but less extreme	as for glacial; contemporary riparian and montane assemblages established
B. Xero-ther-mic[2]	12,000 to 7,000	Much warmer more xeric than pluv. or pres., low P/T	as above, calcifica-tion, soils favored, discharge reduced to form closed drainage systems	as for inter-glacial but more extreme; first evidence of *Larrea*	as for inter-glacial	Rapid exp. of xeric Young Northerns from glacial-pluvial refugia along river valleys in all directions; frag., & extirpation of mesic faunas
C. Contem-porary Condi-tions	7,000 to present	as for xero-thermic, but cooler & more mesic	Riparian sys-tems altered, reconnected, ext. by irri-	as for inter-glacial but fur-	as for inter glacial; few human intro-ductions	as above but frag. of desert biotas along periphery (i.e., river canyons,

| (slightly) higher P/T | gation; further reduction of podzol soils by overgrazing livestock | ther expansion des. sclerophyll at expense of grassland | (i.e., *Hemidacytus*) | flood plains, xeric edaphic patches within Kansan grassland. Possible local expansion of mesic stocks with agriculture |

1. Radiometric dating of Cenezoic from EVERNDEN & EVERNDEN (1970).
2. Terminal dates from MARTIN & MEHRINGER (1965).

XI. SUMMARY

1. This biogeographical analysis is directed at the fulfillment of three major objectives, in the following sequence: (1) a definition of the Chihuahuan Desert in explicit and biologically relevant parameters; (2) the testing of this definition by a quantification of the conformity of herpetofaunal distributions to the previously defined Chihuahuan Desert; and (3) comparative spatial and historical analyses of the Chihuahuan Desert herpetofauna, or more specifically, the present composition and organization of herpetofauna relative to its paleogeographical and paleoecological history.

2. The biogeographical definition of the Chihuahuan Desert synthesized here is based on a correlation between arid climatic conditions, physiography, and vegetation. The following environmental factors define the Chihuahuan Desert: Climate: average annual temperatures 16-22°C, average annual precipitation 75-300 mm, P/T ratio 7-21, average annual solar radiation (in Langleys) 500+; Physiography: Sierozem soil, often a caliche with horizon under 0.5 meters deep, elevation 600-1500 m north of the 25th parallel, 1500-1800 m south of the 25th parallel, hydrogeographical patterns — east of the Continental Divide — all of the desert proper in Rio Grande or closed basin drainage, general topography — basin and range with fault blocks running northwest, basins filling with alluvium; Vegetation: Sclerophyll scrub — the following in combination — *Larrea tridentata*, *Flournesia cernua* and *Parthenium incanum*, also *Agave lechiguilla*, *Parthenium argentatum*, *Euphorbia antisyphylitica*, *Fouquieria splendens* and *Bouteloua gracilis*. *Larrea*, *Fouquieria* and *Bouteloua* are not endemic.

The definition provided above was translated into a map which served subsequently as a base map throughout the text.

3. Using the base map as center of reference, a checklist of all amphibian and reptile species occurring within 25 kilometers of the previously defined Chihuahuan Desert was constructed. Approximately 170 species were discovered to occur within this area.

4. Of these 170 species occurring geographically in the Chihuahuan Desert it was determined that only 34% actually participate reliably in the Chihuahuan Desert scrub ecosystem. The remainder are relict or peripheral, or riparian forms which are not part of the ecologically participant herpetofauna of the desert. Low primary productivity of the desert scrub was suggested as a cause of the low percentage.

5. An ecological distribution of herpetofaunas by plant association was undertaken. The Chihuahuan Desert scrub herpetofauna was broken down into four edaphic associations: alluvial, barrial, dune, and saxicolous. Dendrograms, based on progressive group averages of coefficients of community and of similarity coefficients, both set the saxicolous association apart from the others in having the most distinctive herpetofauna. This association also had the highest percentage of herpetofauna totally restricted to a single edaphic unit.

Barrial and dune associations appear to have high faunal resemblance values. The barrial had the highest species density of the four edaphic situations.

6. In addition to the edaphic associations within Chihuahuan Desert scrub, five

non-desert associations occurring within the geographical limits of the Chihuahuan Desert were also defined and analyzed. The five are desert riparian vegetation, mesquite grassland, chaparral, juniper-piñon woodland, and coniferous forest. Of the five, desert riparian herpetofauna was generally indicated as the most distinctive by faunal resemblance analysis and C.C. and S.C. based dendrograms. It also had the highest percentage of species totally restricted to 31% of its environment. Coniferous forest, with 26% of its herpetofauna restricted, tied with the Chihuahuan Desert scrub as the second most distinctive. Quantitative faunal analysis indicated strongly that mesquite grassland is an ecotone. Mesquite grassland lacks a restricted fauna (except for parthenogenetic hybrid clones). It exhibits an edge effect by sustaining the highest number of species of any assocation examined, and appears to be almost perfectly balanced in the derivation of its herpetofauna, two-third shared with the Chihuahuan Desert scrub and two-thirds with the juniper-piñon woodland.

7. Ecological shifts of mesic relicts to refugial habitats and micro-habitats within the Chihuahuan Desert are analyzed. In general, these shifts are toward riparian habitat, relict grasslands and woodlands, and saxicolous situations within the desert proper.

8. Through the illustrative examples of selected genera, *Sceloporus, Crotalus* and *Cnemidophorus*, habitat segregation is briefly evaluated. The various interactions of edge effect, latitude, elevation, species density, and habitat segregation are discussed for each genus. No one correlation pattern explains latitudinal gradients in species density for all of the genera, some of which are the reverse of others. Possible historical factors are discussed later.

9. A spatial biogeographical analysis of the Chihuahuan Desert herpetofauna was undertaken. It involves the resolution of a primary area, the correlation of this area with the previously constructed base map to test its predictive value, the quantification of the herpetofauna of the primary area to determine its status as faunal province, the comparison and quantified evaluation of subdivisions within the primary area and finally a comparison with adjacent faunal units and other North American Desert faunas.

10. To determine the primary area a grid crucifix of 6400 kms^2 squares had been set across the wide and long axes of the northern Mexican Plateau with tangential axes drawn across potential filter barriers. Indices of faunal change were calculated from specific range maps for all indigenous herpetofauna for each square in the grid. From the maximum IFC values the optimal limits of a primary area were constructed. These faunal limits were then tested against the limits predicted by the original base map (constructed without faunal information). The correlation between predicted and actual herpetofaunal maximum change was significant in accuracy. Thus the base map borders were validated in their general positions.

11. The participant herpetofauna of the Chihuahuan primary area was then re-evaluated against other previously determined faunal provinces and found to be distinct at the provincial level from all adjacent provinces and from non-adjacent deserts with which it was compared. Provincial distinction was set at a coefficient of community of 60% or lower.

12. The Chihuahuan herpetofaunal province was found to be internally homogeneous at the province level, but three distinctive subprovinces could be discriminated on the basis of the IFC value grid as well as two major filter barriers.

The subprovinces are from north to south as follows: the Trans-Pecos Subprovince, the Mapimian Subprovince and Saladan Subprovince. Two major filter barriers, areas of broad ecotonal transition, were the Cochise Filter Barrier of southeastern Arizona and the Saladan Filter Barrier in Zacatecas. The former involved transitions between Sonoran, Kansan, Navahonian, Sierra Madrean Occidental, and Chihuahuan herpetofauna. The latter was simply between Chihuahuan and Trans-Volcanic Province stocks.

13. Extralimital desert relict herpetofaunas, surviving in edaphic xeric pockets were also discussed. The southern Kansan Province, the Mogollon rim of Arizona, pockets in the Tamaulipan Province, and canyon slopes (Rio Panuco drainage) in the Trans-Volcanic Province harbor such relict herpetofaunas.

14. Similarities between the Chihuahuan herpetofauna and those of other provinces are evaluated. The greatest faunal resemblance between the Chihuahuan fauna and that of a non-desert province lies with the prairie Kansan Province.

15. Between desert faunal provinces, striking quantitative and qualitative parallels were discovered between Chihuahuan, Mojave, and Peninsular herpetofaunas and even dominant vegetation. A 1000 kilometer transect across the Continental Divide between these three deserts indicates a reverse symmetry — a cis/trans isomeric reflection between the Chihuahuan Desert biota (especially vegetation) and that of the Peninsular and Mojave Deserts. In addition the central and southern Peninsular Desert shows a latitudinally parallel geography to the Chihuahuan. Many herpetofaunal stocks are represented by sibling species in both deserts which are absent from the intervening Sonoran. Climatic, topographical, and latitudinal considerations are discussed as causal sources of these symmetrical arrangements of closely related biotas. The Sonoran herpetofauna also has a significant shared unit with Chihuahuan, but not as many sibling species in common. It also differs more in climate and vegetation.

16. A historical biogeographical interpretation of modern distributions was developed using paleographical, paleobotanical, and direct fossil evidence. See Table 32 for a complete résumé.

17. Modern families of amphibians and reptiles appeared in the Late Cretaceous record of central North America. A Cretaceous riparian woodland flora and fauna, some modern even at the genus level, may have given rise to the modern Chihuahuan riparian biota directly.

18. The Paleocene Period of Cenozoic North America contained the development of two major Geofloras, Arcto-Tertiary and Neotropical. Associated with them was the parallel development of two herpetofaunal elements, Old Northern and Middle American respectively. By late Paleogene times a third Geoflora, expanding from xeric edaphic ecotones between the previous two Geofloras, came into being. This was the Madro-Tertiary Geoflora. With it developed an arid adapted "pre-desert" herpetofauna that was to become the New Northern Element.

19. The Neogene Period marked the beginning of a continuous late Cenozoic climatic deterioration, with drier, cooler, and more fluctuating conditions replacing old subtropical steamy warm wet environments. With this trend toward modern climates, recent genera of amphibians and reptiles appeared. Older tropical groups retreated or became extinct.

20. By Pliocene times high cold Tertiary forest were already isolated and some montane herpetofauna with them. By Middle to Late Pliocene times modern amphibian and reptile species appeared. By late Pliocene times, a xeric Madro-

Tertiary vegetation had developed from originally edaphic pockets, gradually coalescing into a geographical continuum from Peninsular California to the Mojave east to the Chihuahuan Desert and the southern Great Plains. In this continuum many ubiquitous desert herpetofaunas evolved as a subdivision of the desert and plains faunal complex.

21. End Pliocene events, particularly the depression of cooling isotherms from the northeast and the opening of the Gulf of California, broke up the old Mojavia into modern fragments, namely the Peninsular, Mojave and Chihuahuan Deserts. It was suggested that greater faunal resemblances between deserts west of the Continental Divide was due to the presence of a Gulf of California Arch unit and sibling species pairs fragmented by the cooling and uplift of Cochise filter barrier. It was further suggested that the Sonoran Desert was older than others biotically with a more pronounced Middle American Element. However, the Sonoran Desert represents a relatively recent northward intrusion into the old Mojavia derivatives. As a test, resemblance coefficients and resulting dendrograms were re-calculated synonymizing sibling species and deleting the Gulf Arch faunal unit. The results indicated that these two unit had in fact determined the faunal similarity between the three western deserts. When these two units were eliminated, the new dendrogram clustered all the members of the old desert continuum together (Chihuahuan, Mojavian, Peninsular) and set the Sonoran Desert apart as a separate entity.

22. The Pliocene glacial maximum compressed the Chihuahuan Desert both horizontally and vertically. Horizontal displacement in the north virtually eliminated the Trans-Pecos subprovince, as retreated before cooler more mesic conditions. The central Mapimian subprovince remained as a stable refugium with the most endemics, some of which bear primitive character states typical of hypothetical Tertiary stocks.

23. The Saladan subprovince, due to Pleistocene uplift and intensified warm mesic climates in the subtropical latitude of glacial North America, was probably devoid of desert biota during glacial times. Vertical downward displacement of the desert pinyon-juniper, interface was probably 900-1200 m at the 35th parallel, reduced to 600-900 m at the 30th parallel, and 300-600 m at the 25th parallel, during glacial maxima.

24. The Pleistocene Interglacials were generally characterized by drying climate and in the north a warming climate as well. The Chihuahuan Desert biota expanded at this time, stretching perhaps from Kansas to Queretero on the north-south axis, and from eastern Arizona to south central Texas on the east-west axis.

25. The post glacial pluvial period was essentially a deteriorating continuation of Wisconsin Glacial conditions. This was the last point at which some mesic riparian stocks were still continuous across the Chihuahuan Desert or that montane pinyon-juniper woodland formed a continuous reticulum across the desert.

26. The Xerothermic Period which followed the pluvial operated much as an interglacial in its climatic and geographical effects on desert biota. As arid conditions advanced both north and east, mesic corridors became and still remain disrupted. The Chihuahuan Desert may have expanded at this time very rapidly, averaging several tenths of a kilometer per year. The results of this rapid expansion are seen in herpetofaunal ecotones, edge effects, and hybrid parthogenetic clones. The Xerothermic desert may have extended nearly as far as the

interglacial desert. It probably reached its northern limit in Kansas.

27. Recent desert limits have been set by minor retreats from the Xerothermic maximum. Human induced alterations include irrigation, overgrazing of mesquite-grassland favoring a sclerophyll expansion, and the incipient extinction of *Gopherus flavomarginatus*, North America's largest reptile, through human exploitation for food and land abuse.

LITERATURE

ADDICOTT, W.O. 1969. Tertiary climatic changes in the marginal north-eastern Pacific Ocean. *Science*, 165:583-586.

ANDERSON, D.L. 1971. The San Andreas Fault. *Sci. Amer.*, 225(5): 52-68.

ANDERSON, J.D. 1962. A new subspecies of the ridge-nosed rattlesnake *Crotalus willardi* from Chihuahua, Mexico. *Copeia*, 1962(1): 160-163.

ANDERSON, J.D. & W.Z. LIDICKER, 1963. A contribution to our knowledge of the herpetofauna of the Mexican state of Aguascalientes. *Herpetologica*, 19(1): 40-51.

ARELLANO, A.R.V. 1951. Research on the continental neogene of Mexico. *Amer. J. Sci.* 249:604-616.

ATWATER, T. 1970. Implications of plate tectonics for the Cenozoic tectonic evolution of western North America. *Geol. Soc. Amer. Bull.*, 81 (12):3513-3535.

ATWOOD, W.W. 1940. The physiographic provinces of North America. Ginn and Co., Boston. 536 pp.

AUFFENBERG, W. 1955. A reconsideration of the racer *Coluber constrictor* in the eastern United States. *Tulane Stud, Zool.*, 2(6):89-155.

AUFFENBERG, W. 1963. The fossil snakes of Florida. *Tulane Stud. Zool.*, 10(3):131-216.

AUFFENBERG, W. 1969. Tortoise behavior and survival. *Florida State Mus.* Patterns of Life Series. Rand McNally and Co.

AUFFENBERG, W. & W.W. MILSTED, 1965. Reptiles in the Quaternary of North America in (Wright, H.E. & D.G. Frey, Ed.) the Quaternary of the United States, Princeton Univ. Press, Princeton, New Jersey. 922 pp.

AXELROD, D.I. 1950. Studies in Late Tertiary paleobotany. Carnegie Inst. Washington, Publ. 590:1-323, Pls. 1-19, Figs. 1-4.

AXELROD, D.I. 1956. Mio-Pliocene floras from west-central Nevada. *Univ. Calif. Publ. Geol. Sci.*, 33:1-322, Pls. 1-32, Figs. 1-18.

AXELROD, D.I. 1958. Evolution of the Madro-Tertiary geoflora. *Bot. Rev.*, 24:433-509, Figs. 1-10.

AXELROD, D.I. 1962. A Pliocene *Sequoiadendron* forest from western Nevada. *Univ. Calif. Publ. Geol. Sci.*, 39:195-268.

AXELROD, D.I. 1964. The Miocene Trapper Creek flora of southern Idaho. *Univ. Calif. Publ. Geol. Sci.*, 39:1-148.

AXELROD, D.I. 1966a. The Eocene Copper Basin flora of Northeastern Nevada. *Univ. Calif. Publ. Geol. Sci.*, 59:1-125.

AXELROD, D.I. 1966b. The Pleistocene Soboba flora of southern California. *Univ. Calif. Publ. Geol. Sci.*, 60:1-109.

AXELROD, D.I. 1967. Quaternary extinctions of large mammals. *Univ. Calif. Publ. Geol. Sci.*, 74:1-42.

AXELROD, D.I. 1968. Tertiary floras and topographic history of the Snake River Basin, Idaho. *Geo. Soc. Amer. Bull.*, 79:713-734.

BANDY, O.L., R. CASEY, & C. WRIGHT, 1970. Late Neogene planktonic zonation, magnetic reversals and radiometric dates, Antarctic to the tropics. *Antarctic Res. Ser. 15* (Antarctic Oceanology I):1-24.

BANTA, B.H. 1962a. The amphibians and reptiles from the state of Aguascalientes, Mexico in the collections of the California Academy of Sciences. *Wasmann Journ. Biol.*, 20(1):99-105.

BANTA, B.H. 1962b. Preliminary remarks upon the zoogeography of the lizards inhabiting the Great Basin of western United States. *Wasmann Journ Biol.*, 20(2):253-286.

BEAMAN, J.H. & J.W. ANDERSON, 1966. The vegetation floristics and phytogeography of the Summit of Cerro Potosi. *Amer. Midl. Nat.*, 75(1): 1-33.

BERGGREN, W.A. 1969. Cenozoic chronostratigraphy, planktonic foraminiferal zonation and radiometric time scale. *Nature*, 224:1072-1075.

BEZY, R.L. 1967. A new night lizard. (*Xantusia vigilis sierrae*) from the southern Sierra Nevada in California. *J. Ariz. Acad. Sci.*, 4(3):163-167.

BEZY, R.L. 1972. Karyotypic variation and evolution of lizards in the family *Xantusiidae*.

Contrib. Sci. Nat. Hist. Mus. Los Angeles Co. No. 227:1-29.

BLAIR, F. 1950. The biotic provinces of Texas. *Texas J. Sci.*, 2(1):93-117.

BLUMER, J.C. 1909. On the plant geography of the Chiricahua Mountains. *Science*, 30:720-724.

BOGERT, C.M. 1960. The influence of sound on the behavior of amphibians and reptiles. p. 137-317. In Langon, W.E. & W.N. Taualga, (Eds.) Symposium on Animal Communication, *Amer. Inst. Biol. Sci.*, Washington.

BOGERT, C.M. & W.G. DEGENHARDT, 1961. An addition to the fauna of the United States; the Chihuahuan ridge-nosed rattlesnake in New Mexico. *Amer. Mus. Novit.*, 2064:1-15.

BOGERT, C.M. & J. OLIVER, 1945. A preliminary analysis of the herpetofauna of Sonora. *Bull. Amer. Mus. Nat. Hist.*, 83(6):303-425.

BOSE, E. & O.A. CAVINS, 1927. The Cretaceous and Tertiary of southern Texas and northern Mexico. *Univ. Tex. Bull.*, 2748:7-142.

BRAND, D.D. 1936. Notes to accompany vegetation map of northwest Mexico. *Univ. New Mexico Bull. Biol. Ser.*, 4(4):1-27.

BRANSCOMB, B. 1958. Shrub invasion of a southern New Mexico desert grassland range. *J. Range Mgmt.*, 11:129-132.

BRATTSTROM, B. 1955. Pliocene and Pleistocene amphibians and reptiles from southeastern Arizona. *Jour. Paleo.*, 29:150-154.

BRATTSTROM, B. 1967. A succession of Pleiocene and Pleistocene snake faunas from the high plains of the United States. *Copeia*, 1967(1):188-202.

BRAY, W.L. 1905. The vegetation of the sotol county in Texas. *Univ. Texas Bull.*, 60, Sci. Ser., 6:1-24.

BUFFINGTON, L.C. & C.H. HERBEL, 1965. Vegetational changes on a semi-desert grassland range. *Ecol. Monogr.*, 35(2):139-164.

CARPENTER, C.C. 1967. Display patterns of the Mexican iguanid lizards of the genus *Uma*. *Herpetologica*, 23(4):285-293.

CARR, A.F. 1952. Handbook of Turtles. Cornell Univ. Press 542 pp.

CHANEY, R.W. 1938. Ancient forest of Oregon: a study of earth history in western America. *Carnegie Inst. Wash. Publ.* 501:631-648.

CHANEY, R.W. 1940. Tertiary forests and continental history. *Geol. Soc. Amer. Bull.*, 51:469-488.

CHANEY, R.W. 1947. Tertiary centers and migration routes. *Ecol. Monogr.* 17:140-148.

CHANTELL, C.J. 1964. Some Mio-Pliocene hylids from the Valentine formation of Nebraska. *Amer. Midl. Nat.*, 72(1):211-225.

CHEW, R.M. & A.E. CHEW, 1965. The primary productivity of a desert shrub (*Larrea tridentata*) community. *Ecol. Monogr.*, 34(4):355-375.

CLEMENTS, F.E. 1936a. Plant indicators. *Carnegie Inst. Wash. Publ.*, 290:1-388.

CLEMENTS, F.E. 1936b. The origin of the desert climax and climate. In Essays in geobotany in honor of William Albert Setchell. *Univ. California Press*. 87-140.

COHN, T.J. 1965. The arid land Katydids of the North American genus *Neobarettia* (Orthoptera: Tettigonidae): their systematics and a reconstruction of their history. *Misc. Pub. Mus. Zool., Univ. Michigan* No. 126, 180 pp.

CONANT, R. 1975. A field guide to the reptiles and amphibians of the eastern and central North America. Houghton-Mifflin, Cambridge, Mass.

CONANT, R. 1963. Semi-aquatic snakes of the genus *Thamnophis* from the isolated drainage system of the Rio Nazas and adjacent areas in Mexico. *Copeia*, 1963:473-479.

CONANT, R. 1965. Miscellaneous notes and comments on toads, lizards and snakes from Mexico. *Amer. Mus. Novit.* (2205):1-10.

COOPER, J.G. 1859. On the distribution of the forests and trees of North America with notes on its physical geography. *Ann. Rep. Smithsonian Inst.*, 1858:246-280.

COOPER, J.R. & L.T. SILVER, 1964. Geology and ore deposits of the dragoon quadrangle Cochise County, Arizona. *Geol. Survey Prof. Paper* 416:1-196.

CORRELL, D.S. & M.C. JOHNSTON, 1970. Manual of the vascular plants of Texas. Texas Research Foundation. Renner, Texas. 1881 pp.

DARROW, L. 1944. Arizona ranges resources and their utilization. I Cochise County. *Univ. Arizona Agr. Exp. Sta. Tech, Bull.*, 103:311-366.

DAVIS, W.B. & J.R. DIXON, 1956. A new *Coleonyx* from Texas. *Proc. Bio. Soc. Washington*, 71:149-152.

DEGENHART, W. & W. MILSTEAD, 1959. Notes on a second specimen of the snake *Tantilla cucullata* Minton. *Herpetologica*, 15(3):158-159.

DEGENHARDT, W.G. & J.L. CHRISTIANSEN, 1974. Distribution and habitats of turtles in New Mexico. *Southwestern Nat.*, 19(1):21-46.

DEGENHARDT, W.G., T.L. BROWN, & D.A. EASTERLA, 1975. The taxonomic status of *Tantilla cucullata* and *Tantilla diabola*. *Texas J. Sci.*, 27: (in press).

DENTON, G.H. & R.L. ARMSTRONG, 1969. Miocene-Pliocene glaciations in southern Alaska. *Amer. J. Sci.*, 267: 1121-1142.

DESSAUER, H.V., W. FOX & F.H. POUGH, 1962. Starch gel electrophoresis of transferreins, esterases, and other plasma proteins in hybrids between two subspecies of whiptailed lizard (genus *Cnemidophorus*). *Copeia*, 1962:767-774.

DICE, L.R., 1943. The biotic provinces of North America. Univ. Mich., pp. viii-79.

DICE, L.R. & P.M. BLOOM, 1937. Studies of mammalian ecology in south-western North America, with special attention to the colors of desert mammals. Carnegie Inst. Wash. Publ. no. 485:1-iv, 1-129.

DILLON, L.S. 1956. Wisconsin climate and life zones in North America. *Science* 123(3188):167-176.

DIXON, J.R. 1969. Taxonomic review of the Mexican skinks of the *Eumeces breviorstris* group. *Contrib. Sci. Nat. Hist. Mus. Los Angeles Co. No.* 168:1-30.

DIXON, J.R. & C.S. LIEB, & C.A. KETCHERSID, 1971. A new lizard of the genus *Cnemidophorus. (Teiidae)* from Quretaro, Mexico. *Herpetologica*, 27(3):344-354.

DIXON, J.R., C.A. KETCHERSID & C.S. LIEB, 1972a. The herpetofauna of Queretaro, Mexico, with remarks on taxonomic problems. *Southwestern Nat.*, 16(3-4):225-227.

DIXON, J.R., 1972b. A new species of Sceloporus (Undulatus group: Sauria, Iguanidae) from Mexico. *Proc. Biol. Soc. Wash.* 84(38):307-312.

DORF, E. 1942. Upper Cretaceous floras of the Rocky Mountain region. II. Flora of the Lance formation at its type locality, Niobrara County, Wyoming. *Carnegie Inst. Wash. Publ.*, 508:1-168, 3 Figs., 17 pls.

DORF, E. 1960. Climatic changes past and present. *Amer. Sci.*, 48:341-364.

DOWLING, H.G. 1957. A taxonomic study of the ratsnake genus *Elaphe* Fitzinger. V. The *rosilae* section. Occ. Paps. Mus. Zool. Michigan, 583:1-22.

DOWLING, H.G. 1958. Pleistocene snakes of the Ozark Plateau. *Amer. Mus. Novit.*, Herp. Information Search Systems. (1882):1-9.

DOWLING, H.G. 1974. Yearbook of Herpetology. ed. H.G. Dowling. Vol. 1, 167-170. Herp. Information Search Systems Publications.

DREWES, H. 1968. New and revised stratigraphic names in the Santa Rita Mountains of southeastern Arizona. *Geol. Sur. Bull.*, 1274-C:1-15.

DUELLMAN, W.E. 1960. A taxonomic study of the Middle America snake, *Pituophis deppei. Univ. Kansas Publ. Mus. Nat. Hist.*, 10(10):399-610.

DUELLMAN, W.E., 1970. The Hylid frogs of Middle America. Monogr. Mus. Nat. Hist. Univ. Kans. Lawrence.

DUNN, E.R., E.R. HALL, C.L. HUBBS, E. MAYR, G.G. SIMPSON, C.M. BOGERT & W.F. BLAIR, 1943. Criteria for vertebrate subspecies, species and genera. *Ann. New York Acad. Sci.*, 44(2):105-188.

DURHAM, J.W. 1950. Cenozoic marine climates of the Pacific coast. *Geol. Soc. Amer. Bull.*, 61:1243-1264, Figs. 1-3.

DURHAM, J.W. & E.C. ALLISON, 1960. The geological history of Baja California and its marine fauna. *Syst. Zool.*, 9(24):47-91.

EARDLEY, A.J., 1951. Structural geology of North America. New York.

EARLE, A.M. 1962. The middle ear of the genus *Uma* compared to those of other sand lizards. *Copeia* (2):185-188.

ESCOTO, J.A.V. 1964. Weather and climate of Mexico and Central America. In Handbook of Middle American Indians (West, R.C., Vol. Ed.). Univ. Texas Press, Austin, 1:187-215.

ESTES, R. 1964. Fossil vertebrates from the late Cretaceous Formation of eastern Wyoming. *Univ. California Pub. Geol. Sci.*, 49:1-180.

ESTES, R. 1970. Origin of the recent North American lower vertebrate fauna: an inquiry into the fossil record. *Forma et Functio*, 3:139-164.

ETHERIDGE, R.E. 1960. Additional notes on the lizards of the Cragin Quarry Fauna. *Michigan Acad. Sci.*, 55:113-117.

EVERNDEN, J.E. & R.K.S. EVERNDEN, 1970. The Cenozoic time scale. *Geol. Soc. Amer.*, Special Paper, 124:71-90.

FINDLAY, J.S. 1969. Biogeography of southwestern boreal and desert mammals. In Contributions in Mammology (Jones, J.K., Jr. Ed.). *Univ. Kansas, Mus. Nat. Hist., Misc. Publ.,* 51:113-128.

FLINT, R.F. 1971. Glacial and Quaternary geology. John Wiley and Sons, Inc. New York, 892 pp.

FOUQUETTE, M.J. & H.L. LINDSAY, 1955. An ecological survey of reptiles in parts of northwestern Texas. *Texas J. Sci.* 7(4):402-421.

FOUQUETTE, M.J. & F.E. POTTER, 1961. A new black-headed snake (*Tantilla*) from southwestern Texas. *Copeia,* 1961(2):144-148.

FOWLIE, J. 1965. The snakes of Arizona. Azul Qunita Press, Fallbrook, California. 164 pp.

FOX, W. 1948. Effect of temperature on development of scutellation in the garter snake *Thamnophis elegans atratus. Copeia,* 1948(4):252-262.

FRYE J.C. & A.B. LEONARD, 1957. Ecological interpretations of Pliocene Pleistocene stratigraphy in the Great Plains region. *Amer. J. Sci.,* 255:1-11.

GARCIA, E., C. SOTO & F. MIRANDA, 1960. (1961). *Larrea* y clima. *Anales Inst. Biol.,* 31:133-171.

GARDNER J.L. 1951. Vegetation of the creosote bush area of the Rio Grande valley in New Mexico. *Ecol. Monogr.,* 21:379-403, Figs. 1-7.

GEHLBACH, F.R. 1965. Amphibians and reptiles from the Pliocene and Pleistocene of North America: a chronological summary and selected bibliography. *Texas. J. Sci.,* 17(1):56-70.

GEHLBACH, F.R. 1967. Vegetation of the Guadalupe escarpment, New Mexico-Texas. *Ecology,* 48(3):404-419.

GEHLBACH, F.R. 1971. Lyre snakes of the *Trimorphodon biscutatus* complex: a taxonomic resume. *Herpetologica,* 27(2):200-211.

GEHLBACH, F.R. & J.A. HOLMAN, Paleoecology of amphibians and reptiles from Pratt Cave, Guadalupe Mountains National Park, Texas. *Southwestern Nat.,* 19(2):191-198.

GILMORE, C.W. 1928. Fossil lizards of North America. *Mem. Nation. Acad. Sci.,* 22:ix + 201 pp., 196 Figs., 27 Pls.

GOLDMAN, E.A. & R.T. MOORE, 1946. The biotic provinces of Mexico. *J. Mammal.,* 126(4):347-360.

HAGMEIER, E.M. 1966. A numerical analysis of the distributional patterns of North American mammals. II. Re-evaluation of the provinces. *Syst. Zool.,* 15:279-299.

HAGMEIER E.M. & C.D. STULTS, 1964. A numerical analysis of the distributional patterns of North American mammals. *Syst, Zool.,* 13:125-155.

HARDY, L.M. 1970. Systematic revisions of the genera *Pseudoficimia, Gyalopion* and *Ficimia* (Serpentes: Colubridae), 207 pp. Unpubl. Dissertation. Univ. New Mexico. Univ. Microfilms.

HARDY, L.M. & R.W. MCDIARMID, 1969. The amphibians and reptiles of Sinaloa, Mexico. *Univ. Kansas Mus. Nat. Hist.,* 18 (3): 39-252.

HARTWEG, N. 1938. *Kinosternon flavescens steinergeri,* a new turtle from northern Mexico. *Univ. Michigan Occ. Papers Mus. Zool.,* 371:1-5.

HASTINGS, J.R. 1959. Vegetation change and arroyo cutting in southeastern Arizona. *J. Arizona Acad. Sci.,* 1:60-67.

HECHT M.K. 1956. A new xantusiid lizard from the Eocene of Wyoming. *Amer. Mus. Novit.,* (1774):1-8.

HERNANDEZ, X.E. & H.M. GONZALEZ, 1959. Los pastizales de Chihuahuan. *Secr. Agric. Ganaderia, Circular la Campana* (3):1-48.

HEYER R.W. & J.M. SAVAGE, 1967. Variation and distribution in the tree-frog genus *Phyllomedusa* in Costa Rica, Central America. *Beiträge zur Neotropischen Fauna* α,5(2).

HIBBARD C.W. 1960. The president's address: an interpretation of Pliocene and Pleistocene climates in North America. *Michigan Acad. Sci., Arts and Letters,* 62nd Ann Rept., pp. 5-30.

HIBBARD, C.W. & W.W. DALQUEST, 1966. Fossils from the Seymour formation of Knox and Baylor Counties, Texas, and their bearing in the late Kansas climate of that region. *Contrib. Mus. Paleont. Univ. Michigan* 21(1):1-66.

HIBBARD, C.W, & D.W. TAYLOR, 1960. Two late Pleistocene faunas from southwestern Kansas. *Univ. Mich., Contr. Mus. Paleontology,* 16(1):1-223.

HINDE, R.B. 1843. The regions of vegetation, in Edward Belcher: Narrative of a voyage round the world, performed in her majesty's ship Sulphur, during the years 1836-1842, 2:325-460.

HOLDRIDGE, L.R. 1967. Life zone ecology. Trop. Sci. Center, San Jose, Costa Rica. 206 pp.

HOLMAN, J.A. 1966. A small Miocene herpetofauna from Texas. *Quart. J. Florida Acad. Sci.*, 24:267-275.

HOLMAN, J.A. 1969. The Pleistocene amphibians and reptiles of Texas. *Publ. Mus. Michigan State Univ.*, 4(5):161-192.

HOWDEN, H.F. 1963. Speculations on some beetles, barriers and climates during the Pleistocene and Pre-Pleistocene periods in some non-glaciated portions of North America. *Syst. Zool.*, 12:178-201.

HUMPHREY W.E. 1956. Tectonic framework of northeast Mexico. *Trans. Gulf Coast Assn. Geol. Soc.*, 6:26-35.

IMLAY, R.W. 1936. Evolution of the Coahuila Peninsula, Mexico. IV. Geology of the western part of the Sierra de Parras. *Geol. Soc. Amer. Bull.*, 47:1091-1152, 10 pls.

JACCARD, P. 1902. Lois de distribution florale dans la zone alpine. *Bull. Soc. Vaudoise Sci. Nat.*, 38:69-130.

JAEGER, E.C. 1957. The North American deserts. Stanford Univ. Press. 308 pp.

JENKINS, O.P., 1951. Geologic Guidebook of the San Francisco Bay Counties Calif. *State Division of Mines Bull.*, 154:1-392.

JOHNSTON, I.M. 1941. Gypsophily among Mexican desert plants. *J. Arnold Arboretum*, 22(2):145-170.

JOHNSTON, I.M. 1943. Plants of Coahuila, eastern Chihuahua and adjoining Zacatecas and Durango. I-II. *J. Arnold Arboretum*, 24:306-339.

JOHNSTON, I.M. 1944. Plants of Coahuila, eastern Chihuahua and adjoining Zacatecas and Durango. III-V. *J. Arnold Arboretum*, 25:47-83, 113-183, 431-453.

JOHNSTON, M. 1971. Revision of *Colubrina* (Rhamnaceae). *Brittonia*, 23(10):2-53.

JONES, W.R., R.M. HERNON & S.L. MOORE, 1967. General Geology of Santa Rita Quadrangle, Grant County, New Mexico. U.S. *Geol. Survey Prof. Paper*, 555:1-144.

KERFOOT, C.W 1969. Geographic variability of the lizard, *Sceloporus graciosus* Baird and Girard in the eastern part of its range. *Copeia*, 1969(1):139-159.

KIESTER, A.R. 1971. Species density of North American amphibian and reptiles. *Syst. Zool.*, 20:127-137.

KING, P.B. 1935. Outline of the structural development of Trans-Pecos Texas. *Bull. Amer. Assoc. Petrol. Geol.*, 19:221-261.

KING, P.B. 1948. Geology of the southern Guadalupe Mountains in Texas. *U.S. Geol. Surv. Prof. Paper* 215:1-183, Pls. 1-19, Figs. 1-24.

KING, P.B. 1958. Evolution of the modern surface features of western North America, in (Hubbs, C.L., Ed.) Zoogeography. Washington: *Amer. Assoc. Adv. Sci. Publ.*, 51:3-60, Figs. 1-11.

KING, P.B. 1959. The evolution of North America. Princeton, N.J.: Princeton Univ. Press. 189 pp., 96 figs., 1 map.

KLAUBER, L.M. 1945. The geckos of the genus *Coleonyx* with descriptions of the new subspecies. *Trans. San Diego Soc. Nat. Hist.*, 10:1330216.

KLAUBER, L.M., 1972. Rattlesnakes: their habits, life histories and influence on mankind. Second edition. University of California Press. Berkeley, 2 vols. XXX, XVI, 1533 pp. ill.

KLUGE, A. 1962. Comparative osteology of the eublepharid lizard genus *Coleonyx* Gray. *J. Morphol.*, 110(3):289-332.

KLUGE, A.G. 1975. Phylogenetic relationships and evolutionary trends in the eublepharine lizard genus *Coleonyx. Copeia*, 1975:24-35.

KOTTLEWSKI, F.E. 1958. Lake Otero-second phase in formation of New Mexico's gypsum dunes. *Geol. Soc. Amer. Bull.*, 69:1733-1734 (abstr.).

LEGLER, J.M. 1959. A new tortoise, genus *Gopherus*, from north central Mexico. *Univ. Kansas Publ., Mus. Nat. Hist.*, 13(3):73-84.

LEGLER, J.M. 1960. A new subspecies of slider turtle. (*Pseudemys scripta*) from Coahuila, Mexico. *Univ. Kansas Publ., Mus. Nat. Hist.*, 11(5):335-343.

LEONARD, A.B. & J.C. FRYE, 1962. Pleistocene molluscan faunas and physiographic history of Pecos Valley in Texas. *Univ. Texas Buy. Econ. Geol. Rep. Inv.*, 45:1-45.

LEOPOLD, A.S. 1950. Vegetation zones of Mexico. *Ecology*, 31:507-518.

LEOPOLD, A.S. 1967. Late Cenozoic patterns of plant extinctions. In *Pleistocene Extinctions*

(P.S. Martin & H.E. Wright, editors) Yale University Press, New Haven. pp. 203-246.

LE PICHON, X. 1968. Sea-floor spreading and continental drift. *J. Geophys. Res.* 73:3661-3697.

LE SUEUR, H. 1945. The ecology of the vegetation of Chihuahua, Mexico, north of parallel twenty-eight. *Univ. Texas Publ.*, 4521:1-92.

LITTLE, E.L. & J.G. KELLER, 1937. Amphibians and reptiles of Jornada experimental range, New Mexico. *Copeia*, 1937(4):216-222.

LOWE, C.H. 1955. The eastern limit of the Sonoran Desert in the United States with additions to the known herpetofauna of New Mexico. *Ecology*, 36:343-345.

LOWE, C.H. 1956. A new species and a new subspecies of whiptailed lizards (genus *Cnemidophorus*) of inland southwest. *Bull. Chicago Acad. Sci.*, 10(9):137-150.

LOWE, C.H. (Ed.) 1964. The vertebrates of Arizona. Univ. Arizona Press, Tuscon, 259 pp.

LOWE, C.H. & J.W. WRIGHT, 1966. Evolution of parthenogenetic species of *Cnemidophorus* (whiptail lizard) in western North America. *J. Arizona Acad. Sci.*, 4(2):81-87.

LOWE, C.H. & J.W. WRIGHT, 1968. Weeds, polyploids, parthenogenesis, and the geographical distribution and ecological distribution of the all female species of *Cnemidophorus. Copeia*, 1968(1):128-139.

LOWE, C.H., WRIGHT, J.W., C.J. COLE & R.L. BEZY, 1970. Natural hybridization between the teiid lizards *Cnemidophorus sonorae* (parthenogenetic) and *Cnemidophorus tigris* (bisexual). *Syst. Zool.*, 19(2):114-127.

LUNDELIUS, E.L. 1967. Late-Pleistocene and Holocene faunal history of central Texas. In (P.S. Martin & H.E. Wright, editors) *Pleistocene Extinctions*, Yale University Press, New Haven. pp. 287-320.

LYNCH, J.D. 1968. Genera of leptodactylid frogs in Mexico. *Univ. Kansas Publ. Mus. Nat. Hist.,* 17(11):503-515.

LYNCH, J.D., 1970. A taxomic revision of the leptodactylid frog genus *Syrrhophus* Cope. *Univ. Kansas Publ. Mus. Nat. Hist.,* 20:1-45.

MACGINITE, H.D. 1953. Climate since the late Cretaceous. In (Hubbs, C.L., Ed.) Zoogeography. *Amer. Assoc. Adv. Sci.*, 51:61-79. Washington.

MACGINITE, H.D. 1958. Fossil plants of the Florissant beds, Colorado. *Carnegie Inst. Washington Publ.*, 599:1-198.

MACGINITE, H.D. 1962. The Kilgore flora. A late Miocene flora from northern Nebraska. *Univ. California Publs. Geol. Sci.*, 35:67-158.

MALDONADO-KOERDELL, M. 1964. Geohisfory and paleogeography of middle America, Handbook of Middle American Indians, vol. 1. Univ. Texas Press, Austin, Texas.

MARROQUIN, J.S., G. BORJA, R. VELAZQUES & J.A. LA CRUZ, 1964. Estudio ecologico dasonomico de las zonas aridas del norte de Mexico. *Spec. Publ. Inst. Nacional de Investigaciones Forestales*, (2):1-166.

MARSHALL, J.T., Jr. 1957. Birds of the pine-oak woodland in southern Arizona and adjacent Mexico. *Pacific Coast Avifauna*, 32:1-125.

MARTIN, P.S. 1958. A biogeography of the reptiles and amphibians in the Gomez Farias region, Tamaulipas, Mexico. *Misc. Publ., Mus. Zoo., Univ. Michigan*, 101:1-102.

MARTIN, P.S. & B.E. HARRELL, 1957. The Pleistocene history of temperate biotas in Mexico and eastern United States. *Ecology*, 38(3):468-479.

MARTIN, P.S. & P.J. MEHRINGER, 1965. Pleistocene pollen analysis and biogeography of the southwest, in (Wright, H.E. & Frey, D.G., Eds.) The Quaternary of the United States Part I: 433-452. Princeton Univ. Press.

MARTIN, P.S. & H.E. WRIGHT, Jr. 1967. Pleistocene Extinctions. Yale University Press. New Haven. 453 pp.

MAXWELL, R.A., J.T. LONSDALE, R.T. HAZZARD & J.A. WILSON, 1967. Geology of the Big Bend National Park, Brewster County, Texas Univ. Texas Publ., 6711:1-320.

MAYR, E. 1942. Systematics and the origin of species. Columbia University Press., New York, 334 pp.

MAYR, E., E.J. LINSLEY & R.L. USINGER, 1953. Methods and principles of systematic zoology. McGraw-Hill Book Company, Inc., New York. 336 pp.

MAYR, E., 1970. Populations, Species and Evolution. The Belknap Press of Harvard University Press, Cambridge. 453 pp.

MCCOY, C.J. 1970. A new alligator lizard (genus *Gerrhonotus*) from the Cuatro Cienegas, Coahuila, Mexico. *Southwestern Nat.*, 15(1):37-44.

228

MCCOY, C.J. & W.L. MINCKLEY, 1969. *Sistrurus catenatus* (Reptilia: Crotalidae) from the Cuatro Cienegas Basin, Coahuila, Mexico. *Herpetologica*, 25(2):152-154.

MCDIARMID, R., 1966. A study in biogeography: The Herpetofauna of the Pacific Lowlands at Western Mexico. Masters of Science Thesis. University of Southern California. 308 pp.

MCDIARMID, R.W. 1968. Variation, distribution and systematic status of the blackheaded snake *Tantilla yaquia* Smith. *Bull. So. California Acad. Sci.*, 67(3):159-178.

MCDOWELL, S.B. 1964. Partition of the genus *Clemmys* and related problems in the taxonomy of the aquatic Testudinidae. *Proc. Zool. Sco. London,* 143, pt. 2.

MEEK, S. 1904. The fresh waterfishes of Mexico north of the Isthmus of Tehauntepec. *Field Columbian Mus. Zool. Ser.*, 5:1xiii-252.

MEHRINGER, P.J. Jr. 1967. The environment of extinction of the Late-Pleistocene megafauna in the arid southwestern United States. In (P.S. Martin & H.E. Wright, editors) Pleistocene Extinctions Yale University Press, New Haven, pp. 247-266.

METCALF, L. 1967. Late Quaternary mollusks of the Rio Grande Valley. *Univ. Texas El Paso, Sci. Ser.*, 1:62.

METCALF, A.L., 1970. Quaternary Surfaces, Sediments and Mollusks: Southern Mesilla Valley, New Mexico and Texas. New Mexico Geo. Soc. 20th Field Conference:, 158-164.

MEYER, E.R. 1973. Late-Quaternary paleoecology of the Cuatro Cienegas Basin, Coahuila, Mexico. *Ecology*, 54(5)982-995.

MILSTEAD, W.W. 1960. Relict species of the Chihuahuan Desert. *Southwestern Nat.,* 5(2):75-88.

MILSTEAD, W.W. 1965. Changes in competing populations of whiptail lizards (*Cnemidophorus*) in southwestern Texas. *Amer. Midl. Nat.*, 73(1):75-80.

MILSTEAD, W.W. 1967. Fossil box turtles (*Terrapene*) from central North America, and box turtles of eastern Mexico. *Copeia*, 1967(1):168-179.

MILSTEAD, W.W. 1969. Studies on the evolution of box turtles (genus *Terrapene*). *Bull. Florida State Mus. Bio. Sci.*, 14(1):1-113.

MINCKLEY, W.L. 1969. Environments of the bolson of Cuatro Cienegas, Coahuila, Mexico. *Univ. Texas, El Paso, Sci. Ser.*, 2:1-65.

MINTON, S.A. 1956. A new snake of the genus *Tantilla* from west Texas. *Fieldiana Zool.*, 34:449-452.

MINTON, S.A. 1959. Observations on amphibians and reptiles of the Big Bend region of Texas. *Southwestern Nat.*, 3:28-54.

MOLL, E.O. & S.M. LEGLER, 1971. The life history of a Neo-tropical slider turtle *Pseudemys scripta* (Schoepff) in Panama. *Bull. Los Angeles Co. Mus. Nat. Hist.*, (11):1-102.

MOLL, E.O., K.L. WILLIAMS, & H.M. SMITH, 1963. Herpetological explorations on the Rio Conchos, Chihuahuan, Mexico. *Herpetologica*, 19(3):205-215.

MOORE, D.G. 1973. Plate edge deformations and crustal growth, Gulf of California structural province. *Geol. Soc. Amer. Bull.* 84:1883-1906.

MORAFKA, D.J. 1977. Is there a Chihuahuan Desert? Wauer, R.H. and Riskind, D.H. (eds.) 1975. Transactions of Symposium on the Biological Resources of the Chihuahuan Desert Region, U.S. and Mexico. National Park Service. Washington, D.C.

MULLER, C.H. 1937. Plants as indicators of climate in northeast Mexico. *Amer. Midl. Nat.*, 18:986-1000.

MULLER, C.H. 1939. Relations of the vegetation and climatic types in Nuevo Leon, Mexico. *Amer. Midl. Nat.*, 21:687-720.

MULLER, C.H. 1940. Plant succession in the *Larrea-Flourensia* climax. *Ecology*, 21:206-212.

MULLER, C.H. 1947. Vegetation and climate of Coahuila, Mexico. *Madrono*, 9:33-57.

ODUM, E.P. 1971. Fundamentals of ecology. Third Edition. W.B. Saunders Co., Philadelphia, London, Toronto.

PENNOCK, L. 1966. A karyotype study of parthenogenetic species in the Teiid lizard genus *Cnemidophorus* from south-western United States. Dissertation. Univ. Colorado. Univ. Microfilm 675999. 83 pp.

PETERS, J.A. 1955. Use and misuse of the biotic province concept. *Amer. Nat.*, 89(844):21-28.

PETERS, J.A. 1971. A new approach in the analysis of biogeographic data. Smith. Contrib. Zool., 107:1-28.

PIANKA, E.E. 1966. Latitudinal gradients in species diversity: a review of concepts. *Amer. Nat.*, 100(910):33-46.

PIANKA, E.E. 1967. Lizard species diversity. *Ecology*, 48(3): 333-357.

PIANKA, 1970. Comparative autecology of the lizard *Cnemidophorus tigris* in different parts of its geographics range. *Ecology*, 51(4):703-720.

POUGH, F.H., 1966. (copeia crstalus). Ecological relationships of rattlesnakes of Southeastern Arizona with notes on other species. *Copeia*, (4):676-683.

PRESCH, W.F. 1969. Evolutionary osteology and relationships of the horned lizard genus *Phrynosoma* (Family Iguanidae). *Copeia*, 2:250-276.

PROVINE, J., (1969). A Biogeography of the Herpetofauna of Val Vevele County, Texas. Unpublished Master of Science Thesis, University of Texas, Arlington.

RAUN G.G. 1959. Terrestrial and aquatic vertebrates of a moist, relict area in central Texas. *Texas J. Sci*, 11(2):158-171.

RAUN, G.G. & F.R. GEHLBACH, 1972. Amphibians and reptiles in Texas. *Dallas Mus. Nat. Hist. Bull.*, 2:1-61 (140 maps).

RIEMER, W.J. & H.G. DOWLING, (editors). 1963 et. seq. Catalogue of American Amphibians and Reptiles, *Amer. Soc. Ich. and Herp.*

ROBINSON, M.D. & T.R. VAN DEVENDER, 1973. Miocene lizards from Wyoming and Nebraska. *Copeia*, 1973:698-703.

RZEDOWSKI, J. & G.C. RZEDOWSKI, 1957. Notas sobre la flora y la vegetacion del estado San Luis Potosi. V. La vegetacion a lo largo de la carretera San Luis Potosi-Rio Verde. *Act. Cient. Potosina*, 1:7-68.

RZEDOWSKI, J. & F. MEDELLIN LEAL, 1958. El limite sur de distribution geographica de Larrea tridentata. *Act. Cient Potosina*, 2(2):133-147.

SAVAGE, J.M. 1960. Evolution of a peninsular herpetofauna. *Syst. Zool.*, 9:184-212.

SAVAGE, J.M. 1966. The origins and history of the Central American herpetofauna. *Copeia*, 4:719-766.

SCHMIDT, K.P. 1940. Notes on Texas snakes of the genus *Salvadora*. *Field Mus. Nat. Hist. Zool. Ser.*, 24(12):143-150.

SCHMIDT, K.P. & C.M. BOGERT, 1947. A new fringe-footed sand lizard from Coahuila, Mexico. Amer. Mus. Novitates, no. 1339, 9 pp.

SCHMIDT, K.P. & D.W. OWENS, 1944. Amphibians and reptiles of Northern Coahuila, Mexico. *Zool. Ser. Field Mus. Nat. Hist.* 29(6):96-116.

SCHMIDT, T.F. & J. SMITH, 1944. Amphibians and reptiles of the Big Bend region of Texas. *Field Mus. Nat. Hist. Zool. Ser.*, 29:75-96.

SCHUCHERT, C. 1935. Historical geology of the Atillean-Caribbean region or the lands bordering the Gulf of Mexico and the Caribbean Sea. John Wiley and Sons, New York.

SCHUCHERT, C. 1955. Atlas of paleogeographic maps, John Wiley and Sons, New York. 84 charts.

SCHULTZ, G.E. 1967. Four superimposed Late-Pleistocene vertebrate faunas from southwest Kansas in (P.S. Martin & W.E. Wright, edits.) *Pleistocene Extinctions* pp. 321-336.

SEGERSTROM, K. 1962. Geological investigations on Mexico. Chapter C. Geology of south-central Hidalgo and north-eastern Mexico. *U.S. Geol. Surv. Bull.*, 1104-C:87-162.

SEIFERT, W. & R.W. MURPHY, 1972. Additional specimens of *Coleonyx reticulatus* (Davis and Dixon) from the Black Gap wildlife management area, Texas. *Herpetologica*, 28(1):24-27.

SHREVE, F. 1942a. The desert vegetation of North America. *Bot. Review*, 8(4):195-246.

SHREVE, F. 1942b. Grassland and related vegetation in northern Mexico. *Madrono*, 6:190-198.

SHREVE, F. 1951. Vegetation of the Sonoran Desert. *Carnegie Inst. Washington, Publ.*, 59:1-192.

SIMPSON, G.G. 1960. Notes on the measurement of faunal resemblance. *Amer. Sci.*, Bradley 258-A:300-311.

SINGEWALD, Q.D. 1936. Evolution of the Coahuila Peninsula. Mexico V. Igneous phenomena and the geologic structure near Mapimi. Geol. Soc. Amer. Bull., 47:1153-1176.

SMITH, H.M. 1939. The Mexican and Central American lizards of the genus *Sceloporus*. *Field Mus. Publ. No. 445, Zool. Ser.*, 26:1-397.

SMITH, H.M. 1940. An analysis of the biotic provinces of Mexico, as indicated by lizards of genus of *Sceloporus*. *Sobretiro de los Anales de la Escuela Nacional de Ciencias Biologicas (Mexico)*, 2(1):95-110.

SMITH, H.M. & E.H. TAYLOR, 1941. A review of the snake genus *Ficimia*. *J. Washington Acad. Sci.*, 31(8):356-368.

SMITH, H.M. 1947. Subspecies of the Sonoran toad (*Bufo compacactilis*). *Herpetologica*, 4(1):7-13.

SMITH, H.M. 1949. Herpetogeny in Mexico and Guatemala. *Ann. Assoc. Amer. Geographers,* 39(3):219-238.

SMITH, H.M. & H.K. BUECHNER, 1947. The influence of the Balcones Escarpment on the distribution of amphibians and reptiles in Texas. *Bull. Chicago Acad. Sci.,* 8(1):1-16.

SMITH, H.M., C.W. NIXON, & S.A. MINTON, 1949. Observations on constancy of color and pattern in soft-shelled turtles. *Trans. Kansas Acad. Sci.,* 52(1):92-98.

SMITH, H.M. & E.H. TAYLOR, 1966. Herpetology of Mexico. Eric Lundberg, Ashton, Maryland. (Bound Collection).

SMITH, H.M. & J.E. WERLER, 1969. The status of the northern red blackheaded snake, *Tantilla diabola* Fouquette and Potter. *Herpetologica.* 3(3-4):172-173.

SMITH, H.M., K. WILLIAMS & E.O. MOLL, 1963. Herpetological explorations on the Rio Conchos, Chihuahua, Mexico. *Herpetologica,* 19(3):205-215.

SOKAL, R.R. & R. SNEATH, 1963. Principles of Numerical Taxonomy. W.H. Freeman and Co., San Francisco. 359 pp.

STEBBINS, R.C. 1954. Amphibians and reptiles of western North America. McGraw-Hill Book Co., New York. 528 pp.

STEBBINS, R.C. 1958. A new alligator lizard from the Paramint Mountains, Inyo County, California. *Amer. Mus. Novit.,* 1883:1-28.

STEBBINS, R.C. 1966. A field guide to western reptiles and amphibians. Houghton-Mifflin Co., Cambridge, Mass., 279 pp.

STEJNEGER, L. & T. BARBOUR, 1943. A checklist of North American amphibians and reptiles (5th edition). *Bull. Mus. Comp. Zool.,* 93. 195 pp.

STEVENS, M.S., J.B. STEVENS & M.R. DAWSON, 1969. New early Miocene formation and vertebrate local fauna, Big Bend National Park, Brewster County, Texas. Tex. Mem. Mus. (Austin). The Perace-Sellards Ser., 15:1-53.

STEVENS, R.L. 1964. The soils of Middle America and their relations to Indian peoples and cultures, in (West, R.C., Vol. Ed.) Handbook of Middle American Indians 1(8):265-315.

STIRTON, R.A. 1963. Time, Life, and Man. John Wiley and Sons, Inc., New York. 558 pp.

STOKES, N.L. 1955. Another look at the Ice Age. *Science,* 122 (3174):815-821.

STRAIN, W.S. 1966. Blancan mammalian fauna and Pleistocene formation, Hudspeth County, Texas. *Texas Mem. Mus. Bull.,* 10:1-55.

STUART, L.C. 1964. Fauna of Middle America, in Handbook of Middle American Indians, Vol. 1, Univ. Texas Press, Austin, Texas., 570 pp.

TAMAYO, J.L. 1962. Geografia General de Mexico (2nd Ed.). Institutio Mexicano de Investigaciones Economicas, Mexico.

TAMAYO, J.L. & R.C. WEST, 1964. The hydrography of Middle America, in Handbook of Middle American Indians, Vol. 1. Univ. Texas Press, Austin, Texas. 570 pp.

TANNER, W.W. 1966. A re-evaluation of the genus *Tantilla* in the southwestern United States and northwestern Mexico. *Herpetologica,* 22(2):134-153.

TAYLOR, D.W. 1965. The study of Pleistocene mollusks in North America, in (Wright and Frey eds.) The Quaternary of North America. Princeton Univ. Press, Princeton, p. 597-612.

TAYLOR, E.H. 1949. A preliminary account of the herpetology of the state of San Luis Potosi, Mexico. *Univ. Kansas Sci. Bull.,* 33(2):169-215.

TAYLOR, E.H. 1950. Second contribution to the herpetology of San Luis Potosi. *Univ. Kansas Sci. Bull.,* 33(11):441-457.

TAYLOR, E.H. 1952. Third contribution to the herpetology of the Mexican state of San Luis Potosi, Mexico. *Univ. Kansas Sci. Bull.,* 34(13):793-815.

TAYLOR, E.H. 1953. Fourth contribution to the herpetology of San Luis Potosi. *Univ. Kansas Sci. Bull.,* 35(2): no. 13.

THORNTHWAITE, C.W. 1931. The climates of North America according to a new classification. *Geogr. Rev.* 21:663-655.

TIHEN, J.A. 1949. The genera of gerrhonotine lizards. *Amer. Midl. Nat.,* 41(3):580-601.

TIHEN, J.A. 1964. Tertiary changes in the herpetofauna of temperate North America. *Senck. Biol.,* 45, 3(5):265-277.

TIHEN, J.A. & R. ESTES, 1964. Lower vertebrates from the Valentine formation of Nebraska. *Amer. Midl. Nat.,* 72(2):265-279; 453-472.

TINKLE, D.W. 1959. Observations of the lizards *Cnemidophorus tigris, Cnemidophorus tesselatus* and *Crotaphytus wislizeni. Southwestern Nat.,* 4(4):195-200.

TRUEB, L.T. 1968. Variation in the tree frog *Hyla lancasteri. Copeia,* (2):285-299.

TWENTE, 1952. Pliocene lizards from Kansas. *Copeia* (2):70-73.

UNITED STATES DEPARTMENT OF COMMERCE. 1968. Climatic atlas of the United States. Environmental Data Service, U.S. Governmental Printing Office, Washington, D.C. 80 pp.

UVARDY, M.D.F. 1969. Dynamic zoogeography. Van Nostrand Reinhold Co., Reinhold Co., New York. 445 pp.

VAN DEVENDER, T.R. & J.E. KING. 1971. Late Pleistocene vegatational records in the western Arizona. *J. Arizona Acad. Sci.,* 6(4):240-244.

VANZOLINI, P.E. & E.E. WILLIAMS, 1970. South American anoles: the geographic differentiation and evolution of the *Anolis chrysolepis* species group (Sauria, Iguanidae). *Arquivos Zool.,* 19(1-2):1-24.

VIVO, J.A. & J.E. GOMEZ, 1946. Climatologia de Mexico. Instituto Panamericano de Geographia e Historia Direccion de Geografia, Meteorologia e Hidrologia. Pub. NO. 19.

VOORHIES, M.R. 1971. Paleoclimatic significance of Crocodilian remains form the Ogallalla Group (Upper Tertiary) in northeastern Nebraska. *J. Paleontology,* 45(1):119-121.

WAKE, D.B. 1964. Comparative osteology and evolution of the lungless salamanders, Family Plethodontidae. Unpublished Dissertation, University of Southern California.

WAKE, D.B. 1966. Comparative osteology and evolution of the lungless salamanders, Family Phethodontidae. *Mem. So. California Acad. Sci.,* 5:1-111.

WAKE, D.B. & A.H. BRAME,1963. The salamanders of South America. *Contrib. Sci. Nat. Hist. Mus. Los Angeles Co.,* (65):1-12.

WALKER, W.J. 1966. Subspecies of *Cnemidophorus gularis.* Unpublished Dissertation, Univ. Microfilm, 87 pp.

WARNOCK, B.H. 1970. Wildflowers of the Big Bend County, Texas. Sul. Ross State Univ., Alpine, Texas. 157 pp.

WEAVER, J.E. & F.E. CLEMENTS, 1938. Plant ecology, McGraw-Hill Book Co., New York.

WEAVER, J.E. & M. ROSE, 1967. Systematics, fossil history and evolution of the genus *Chrysemys. Tulane Stud. Zool.,* 14(2):63-73.

WEBB, W.L. 1950. Biogeographic regions of Texas and Oklahoma. *Ecology,* 31(3):426-433.

WEBB, R.G. 1960. A new softshell turtle (genus *Trionyx*) from Coahuila, Mexico. *Univ. Kansas Sci. Bull.,* 40(2):21-30.

WEBB, R.G. 1962. North American recent soft-shelled turtles (Family Trionychidae). *Univ. Kansas Publ. Mus. Nat. Hist.* 13(10):429-611.

WEBB, R.G. 1965. A new night lizard (*Xantusia*) from Durango, Mexico. *Amer. Mus. Nov.,* 2231:1-66.

WEBB, R.G. 1970. Another new night lizard (*Xantusia*) from Durango. *Sci. Nat. Hist. Mus. Los Angeles Co., Contrib.* 194:1-10.

WEIDIE, A.E. & G.E. MURRAY, 1967. Geology Parras Basin and adjacent areas of northeastern Mexico. *Amer. Assn. Petrol. Geol. Bull.,* 51:678-695.

WEIGMAN, J., 1960. The Flora of Baja California. *Syst. Zoo.* 9:189-234.

WELLS, P.V. 1966. Late Pleistocene vegetation and degrees of vegetation and degrees of pluvial climatic change in the Chihuahuan Desert. *Science,* 152:970-975.

WERMUTH, H. & R. MERTENS, 1961. Schildkroten-Krokodile Bruckenechsen. Veb Gustave Fischer Ver 422 pp.

WEST, R.C. 1964. Natural environment and early cultures Vol. 1 in, Handbook of Middle Americans (Wauchope, R., Gen. Ed.). Univ. Texas Press Austin. 570 pp.

WOLFE, J.A. 1971. Tertiary climatic fluctuations and methods of analysis of Tertiary floras. *Palaeogeography, Palaeoclimatol, Palaeoecol.,* 9:27-57.

WILLIAMS, K.L., P.S. CHRAPLIWY, & H.M. SMITH, 1959. A new fringe-footed lizard (*Uma*) from Mexico. *Trans. Kansas Acad. Sci.,* 62(2): 66-172.

WILSON, R.L. 1968. Systematics and faunal analysis of Lower Pliocene vertebrate assemblage from Trego County, Kans. *Univ Mich. Contrib. Mus. Paleontology,* 22(7):75-126.

WRIGHT, A.H. & A.A. WRIGHT, 1957.Handbook of snakes. Comstock, 2 vol. 1107 pp.

WRIGHT, H.E. & D.G. FREY, 1965. The Quaternary of the United States. Princeton Univ. Press, Princeton, New Jersey 922 pp.

WRIGHT, J.W. & C.H. LOWE, 1967. Evolution of the alloploid parthenospecies *Cnemidophorus tesselatus* (Say). *Mammalian Chromosomes Newsletter,* 8(2):95-96.

WRY, 1968. Meteorol. Monogr. 8. 30.

ZWEIFEL, R.G. 1962. Analysis of hybridization between two subspecies of the desert whiptail, *Cnemidophorus tigris. Copeia,* 1962(4):749-774.

ZWEIFEL, R.G. & W.E. DUELLMAN, 1962. A synopsis of the lizards of the *sexlineatus* group (Genus *Cnemidophorus*). *Bull. Amer. Mus. Nat. Hist.,* 123(3):159-210.
232

APPENDIX I: BIOGEOGRAPHICAL FEATURES

Plate 1. Chihuahuan Desert scrub. Edaphic gypsophilous dune association. Dominants are *Fucquieria shrevei*, an endemic *Sporobolus*, and various halophytic shrubs. *Uma exsul* occurs here. North edge of Laguna del Rey, Coahuila. 22 kms S. of El Oro.

Plate 2. Chihuahuan Desert scrub. Edaphic saxicolous association. Dominant vegetation includes *Fouquieria splendens, Agave, Jatropha, Yucca,* and *Hectia.* On Coahuila 30, 50 kms NE San Pedro de las Coloniae, Coahuila, Elevation 1120 M.

Plate 3. Riparian woodland. Trees consist of *Salix* and *Populus.* Sedges and seep willow (*Baccharis*) are also common. *Chrysemys scripta* and *Natrix erythrogaster* occur here. Rio Nazas at La Goma, Durango. Elevation 1200 M.

Plate 4. Mesquite-grassland ecotone. Note drained slopes support oak woodland (encinal) while grassland valleys are pocketed with advancing scrub (*Prosopis, Acacia* and a cylindro-opuntia). This deteriorating grassland has probably suffered overgrazing. Ecotonal conditions affirmed by the presence of hybrid clonal *Cnemidophorus* (*C. uniparens* and *C. exanguis*) in the general region. On Mex. 45 kms W. Ricardo Flores Magon Grandes, Chihuahua. Elevation 1660 M.

Plate 5. Anticlinorium of Arteaga. Separating the Mapimian and Saladan subprovinces, it is photographed here from the Lower Mapimian Desert floor. The wall is 300-600 M high. The location here is 25 km W. Paila, Coahuila at 1100 M elevation.

Plate 6. Saladan Desert. Note the invasion of desert sclerophyll (*Larrea*) across the valleys, receding patchy stands of piñon-juniper woodland in the foothills (mixed with *Yucca carnerosana* and *Nolina*) and a continuous piñon woodland along the higher ridges. Desert invasion of valleys is probably a result of Xerothermic Period expansions. The location is 10 kms E. of Concepcion del Oro, on the road to Mazapil, Zacatecas. Elevation 2500 M.

Plate 7. Cis-trans symmetry of the North American deserts. Joshua tree woodland. This is essentially an arborescent *Yucca* (*brevifolia* in the Mojave, *filifera* in the Chihuahuan Desert) faciation of a *Larrea tridentata* desert scrub association. This photo corresponds to points 4 on the thousand kilometer transect discussed in the text. Upper: Chihuahuan Desert, 80 kms SW of Saltillo, Coahuila (near Rocomonte on Mex. 54), Elevation 1700 M. Lower: Mojave Desert, Pipes Wash, 11 kms N. of Yucca Valley, San Bernadino Co., California. Elevation 1000 M.

Plate 8. The Cochise filter barrier. Note the disruptive basin and range topography that allows for the differential and intermittent dispersal of eastern and western desert faunas. Photo taken from Chiracahua Mountains north of Barefoot Park, Cochise County, Arizona, facing north. Elevation 2000 M. Taken in montane coniferous (pine) forest.

Plate 9. Southern edge of the Saladan filter barrier. At this elevation, the Chihuahuan Desert biota becomes both depauperate and stunted. Stunted succulents and *Larrea* only 20 cm high grow in this area. A summer rain storm moves across the northeast. Substrate is an impoverished residual karst caliche. *Holbrookia maculata* and *Lampropeltis mexicana* occur here. The location is in San Luis Potosi, 80 kms W. of San Luis Potosi. Elevation 2290 M.

Plate 10. Edaphic desert relict of the Rio Panuco drainage. Desert vegetation of *Larrea, Fouquieria splendens, Agave lechiguilla, Hectia,* etc. occur as an edaphic slope scrub above the Rio Extorax. *Ficimia cana* and *Sceloporus exsul* occur in this region. The photograph was taken 60 kms NE. Tisquisquiapan, Queretero. Elevation 2025 M.

Plate 1

Plate 2

234

Plate 3

Plate 4

Plate 5

Plate 6

236

Plate 7

Plate 8

Plate 9

238

Plate 10

APPENDIX II. SPECIES DISTRIBUTION MAPS

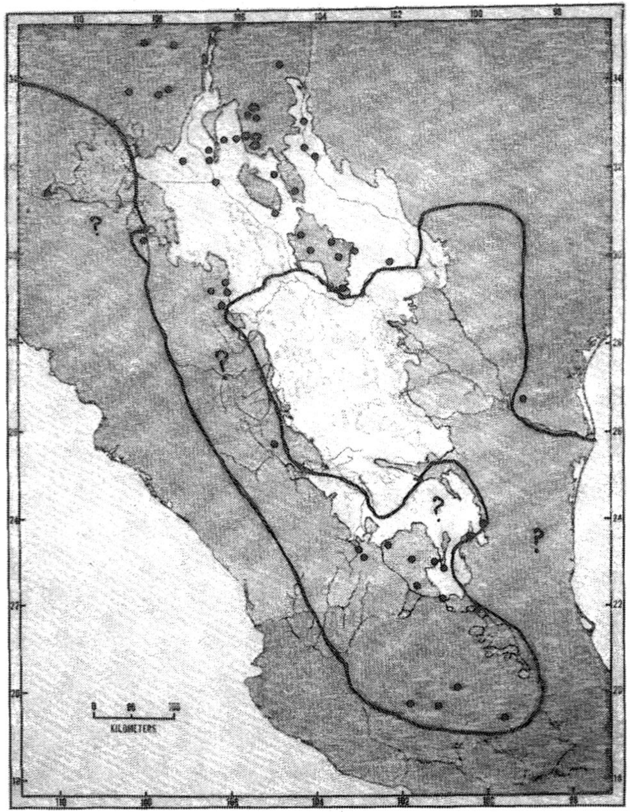

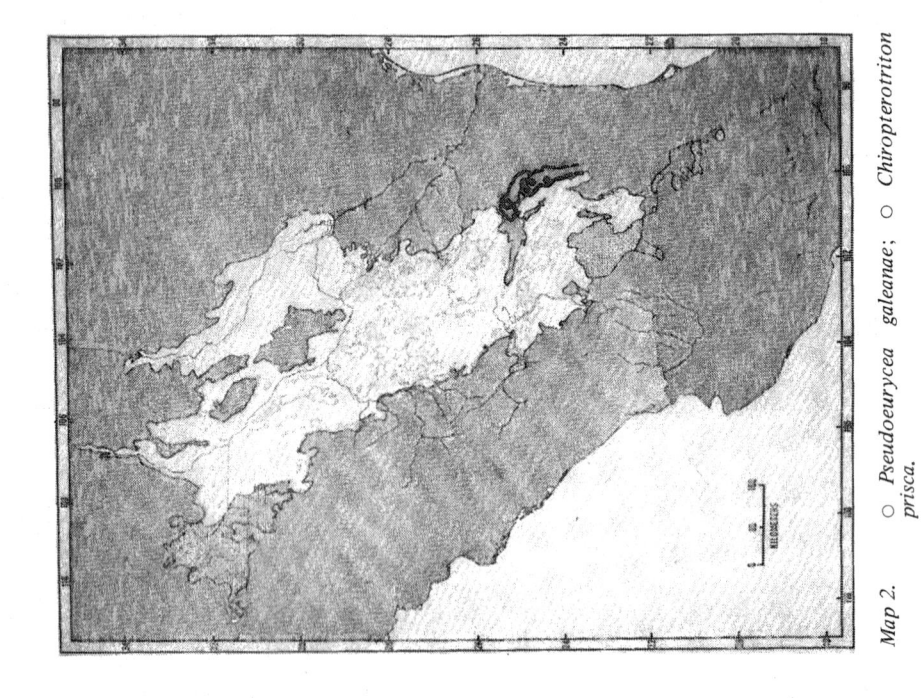

Map 1.   ● *Ambystoma tigrinum* (range limits shown);   ○ *Aneides hardyi* (range limits not shown).

Map 2.   ○ *Pseudoeurycea galeanae*;   ○ *Chiropterotriton prisca*.

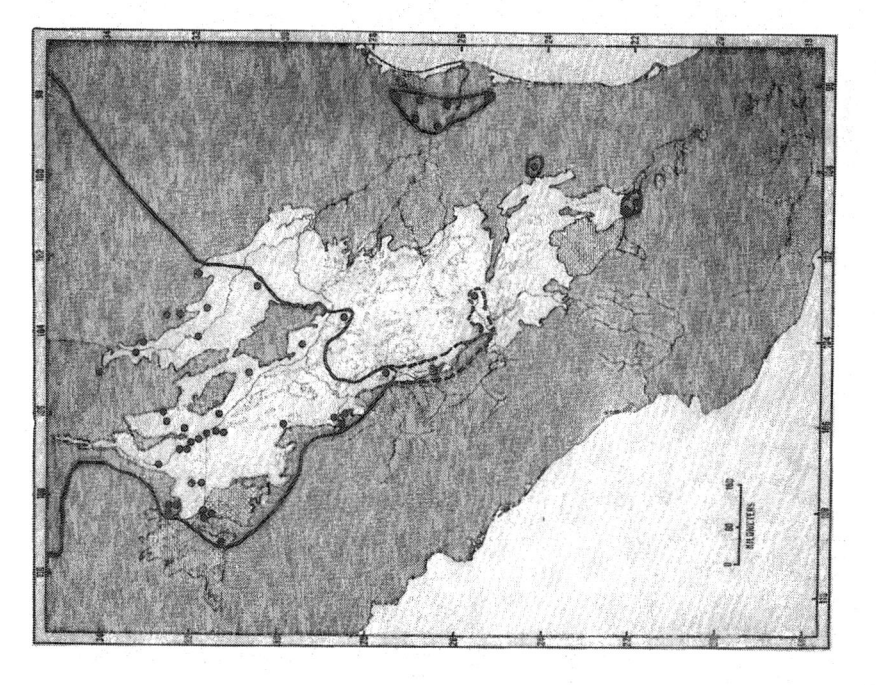

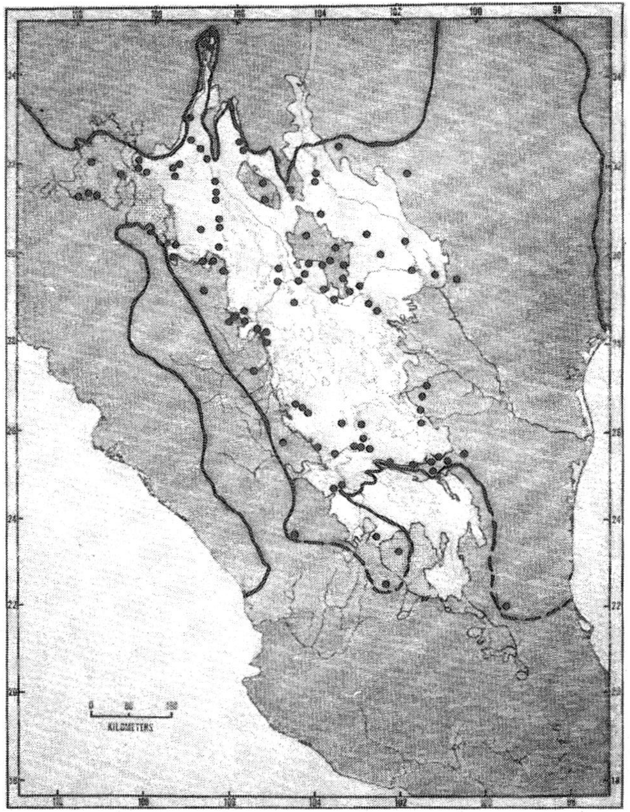

Map 4. *Scaphiopus couchi.*

Map 3. *Scaphiopus bombifrons.*

243

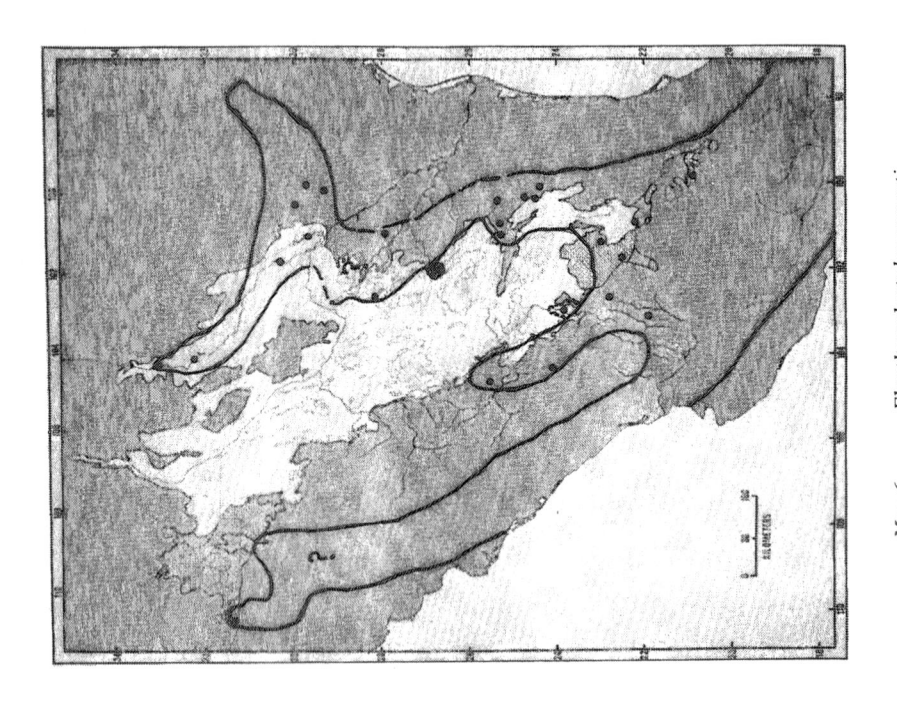

Map 6. *Eleutherodactylus augusti.*

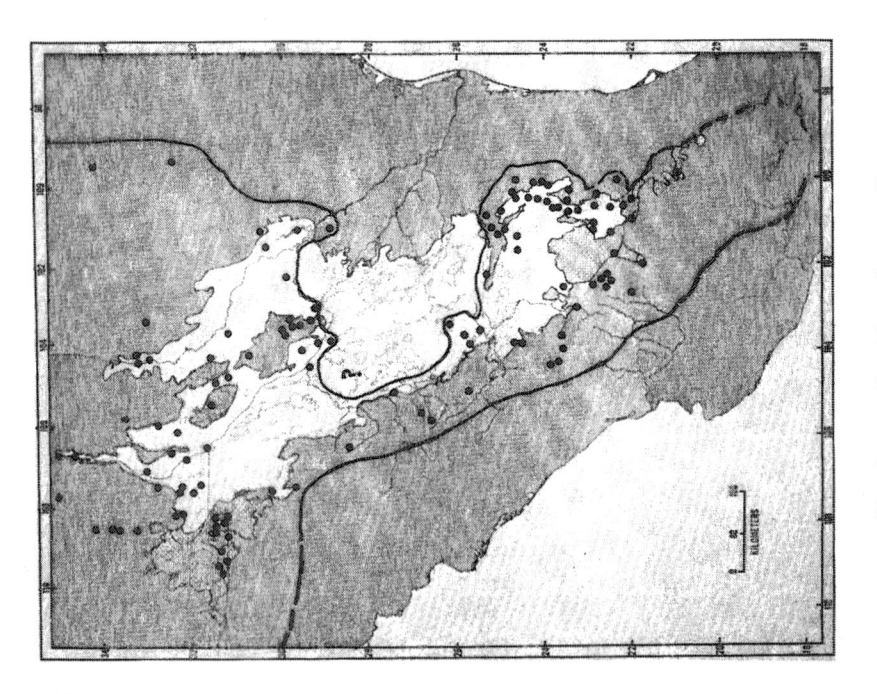

Map 5. *Scaphiopus hammondi.*

244

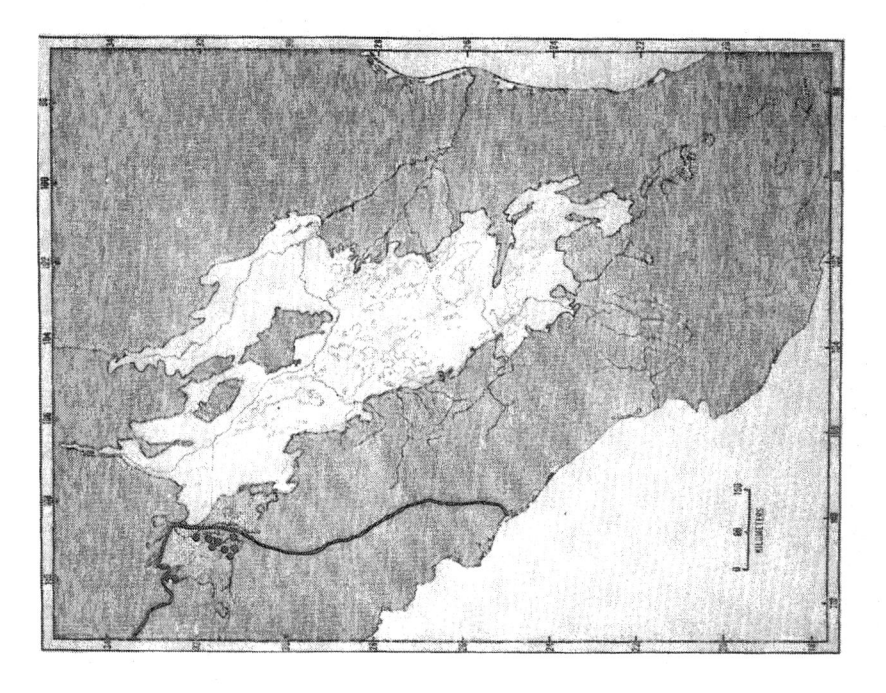

Map 8.   *Bufo alvarius.*

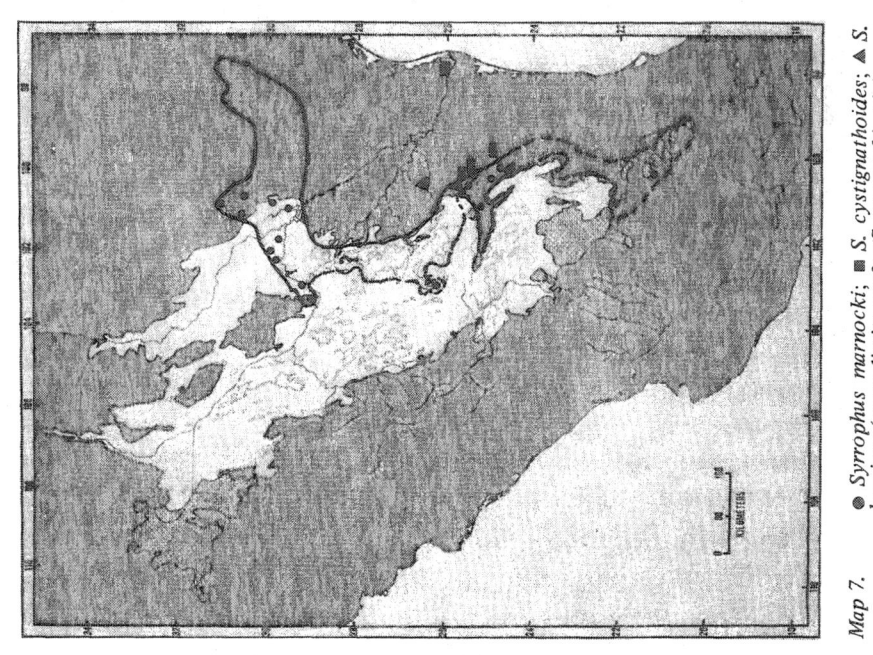

Map 7.   ● *Syrrophus marnocki;* ■ *S. cystignathoides;* ▲ *S. longipes* (range limits set for *S. marnocki* only).

245

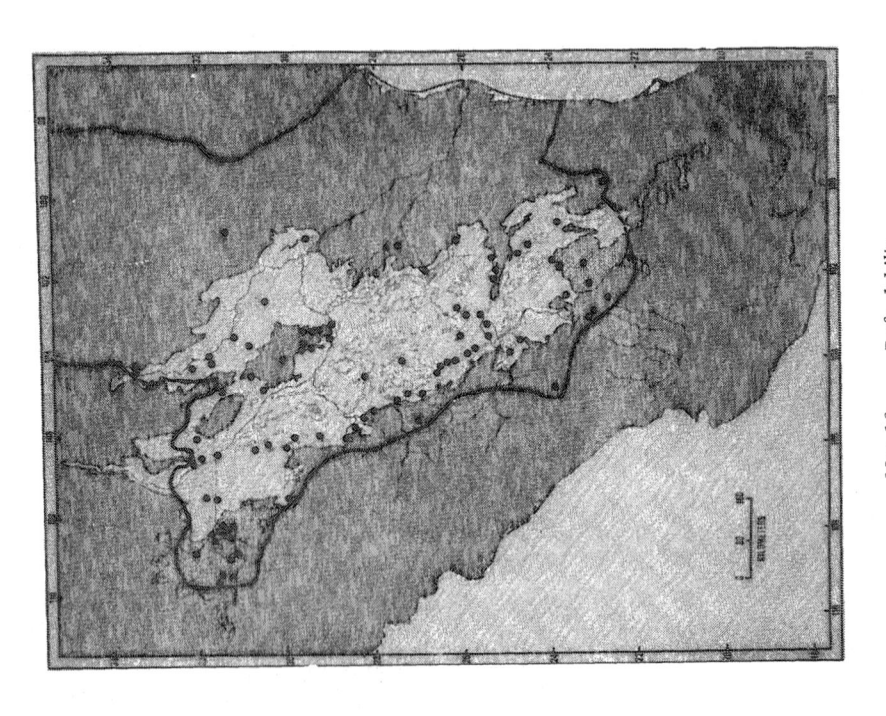

Map 10.   *Bufo debilis.*

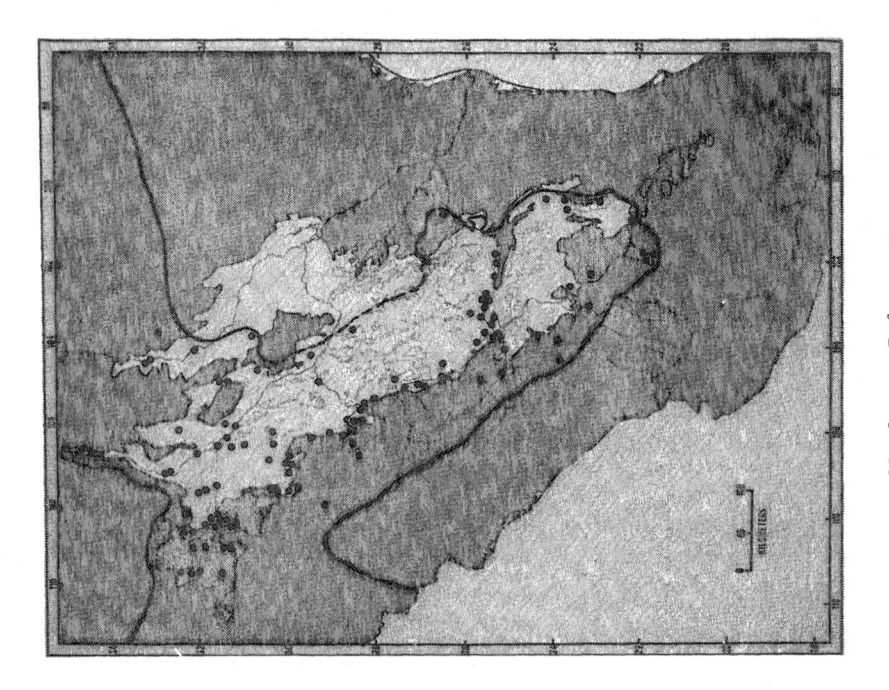

Map 9.   *Bufo cognatus.*

246

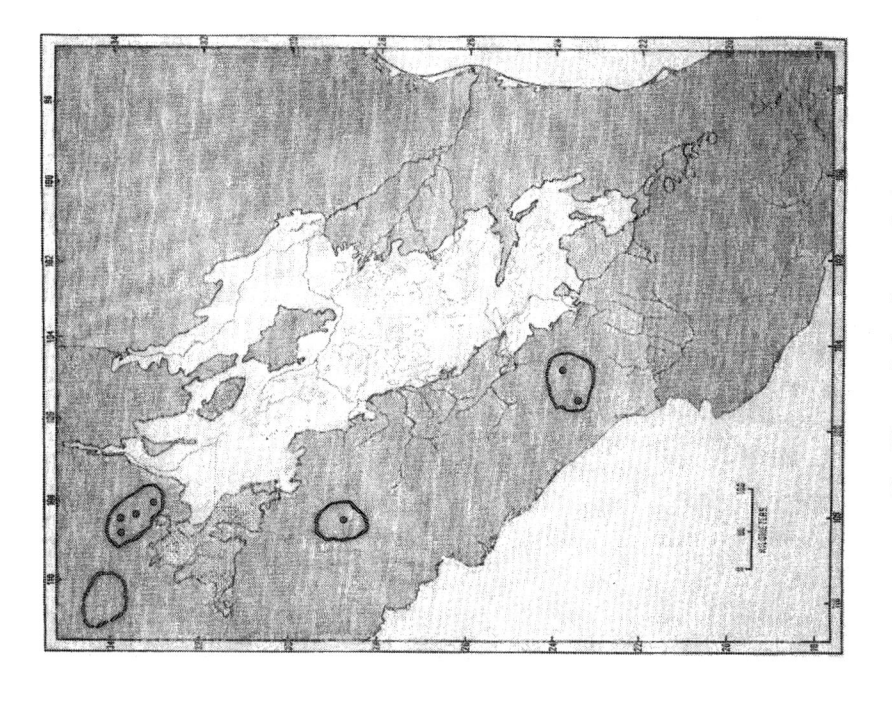

Map 12.   *Bufo microscaphus.*

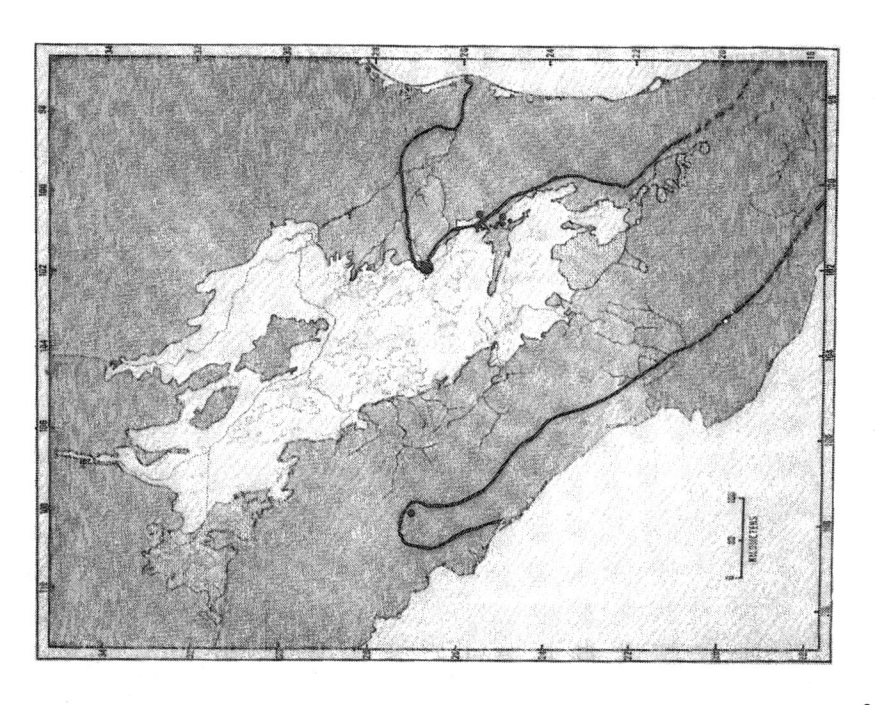

Map 11.   *Bufo marinus.*

247

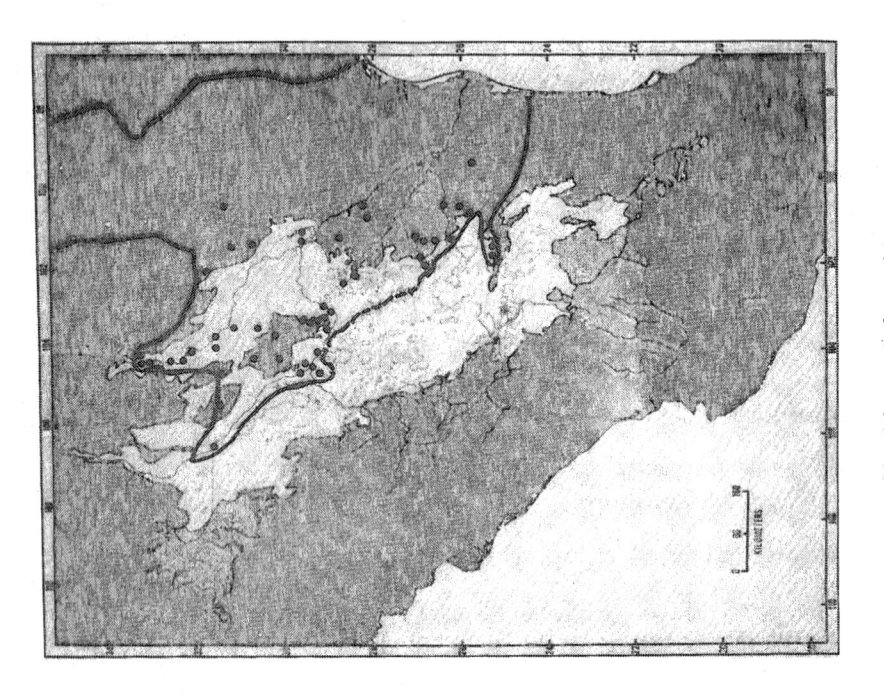

Map 14. *Bufo speciosus.*

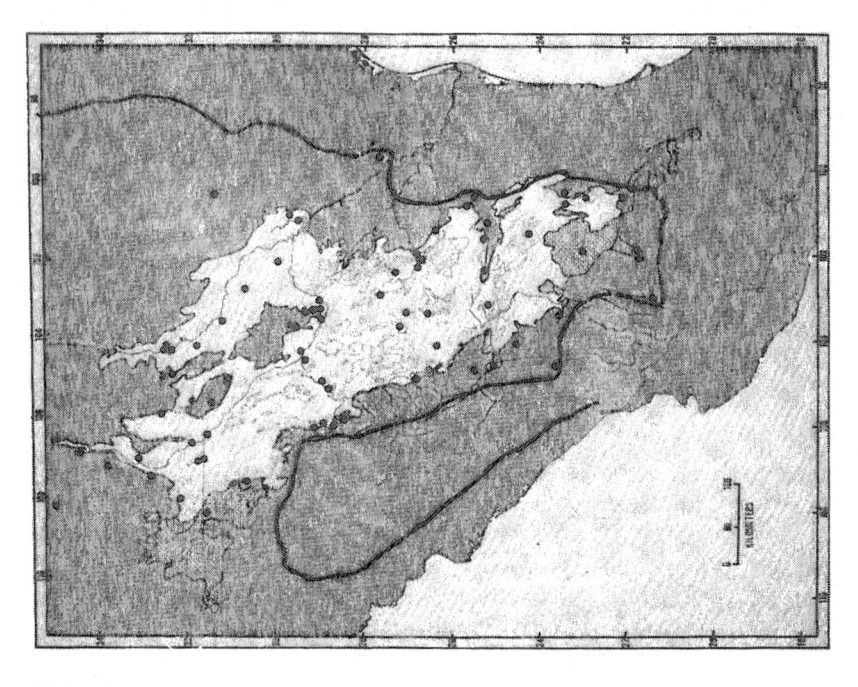

Map 13. *Bufo punctatus.*

248

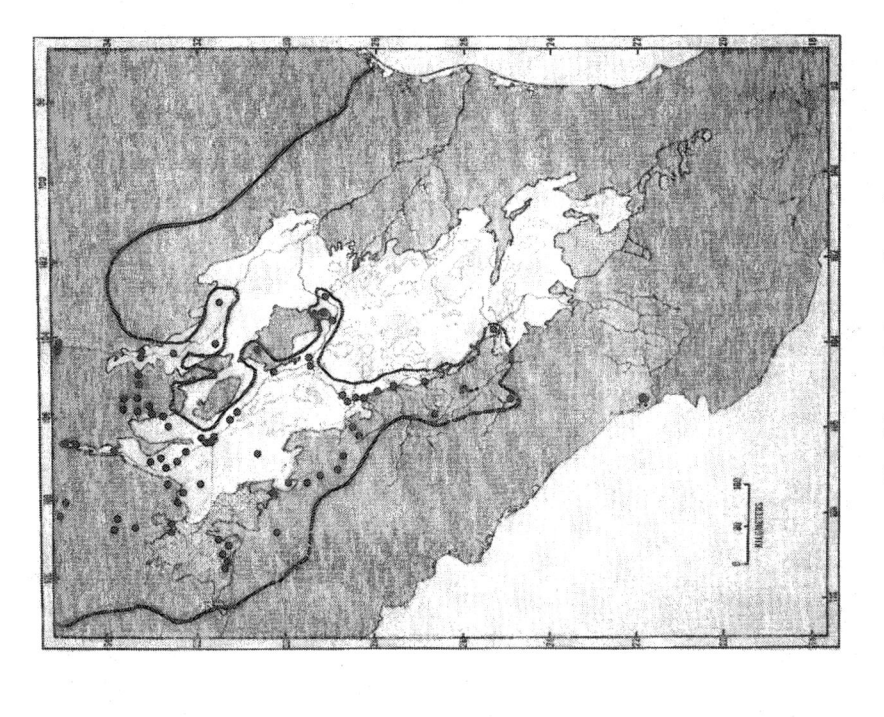

Map 16.   Bufo woodhousei.

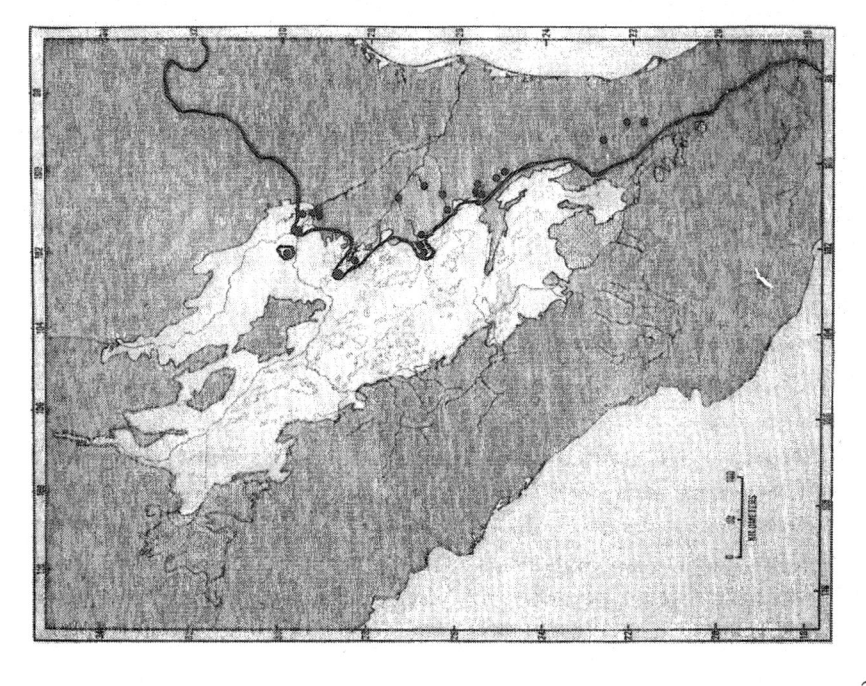

Map 15.   Bufo valliceps.

249

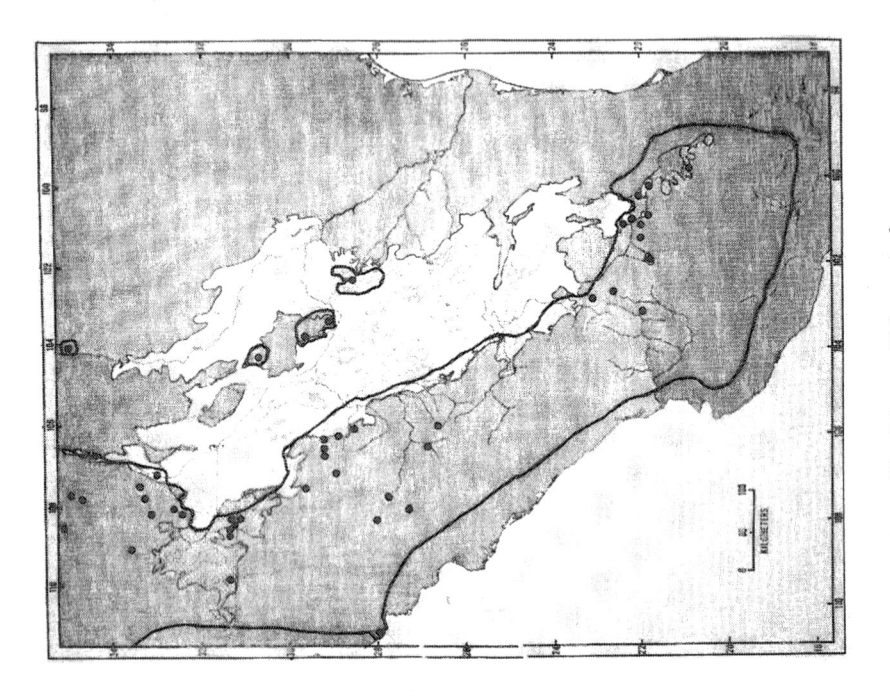

Map 18. *Hyla arenicolor.*

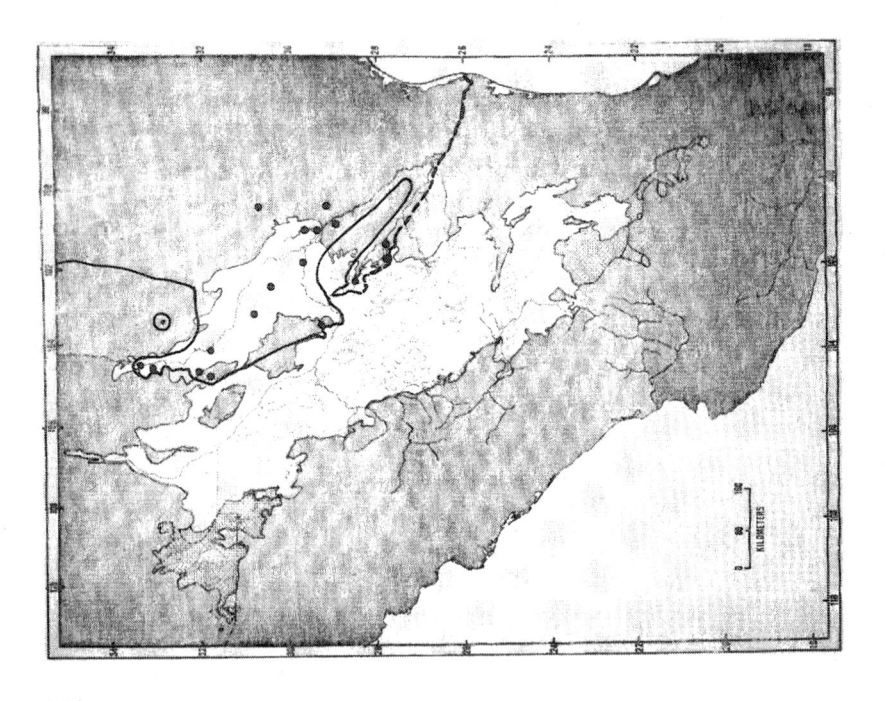

Map 17. *Acris crepitans.*

250

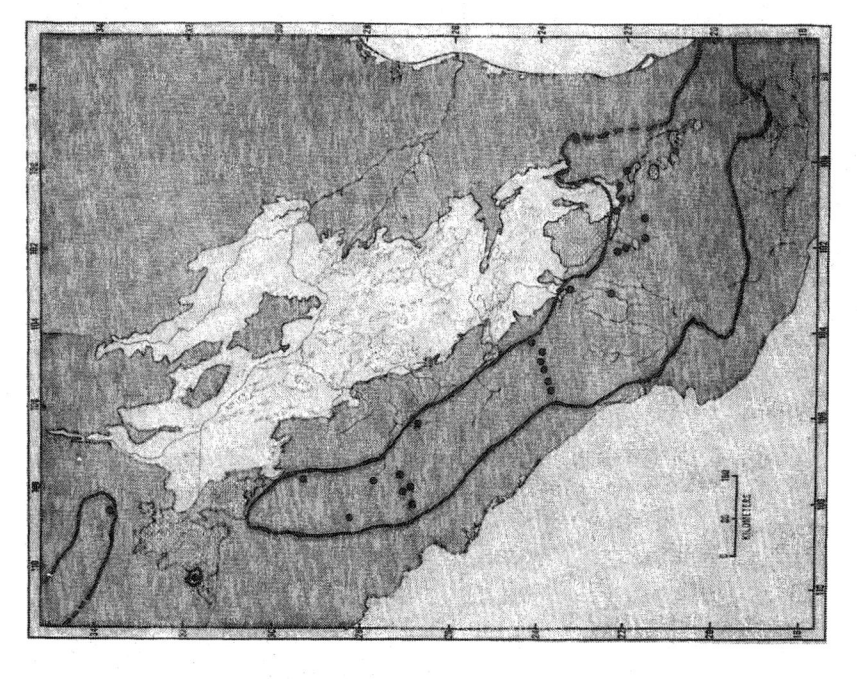

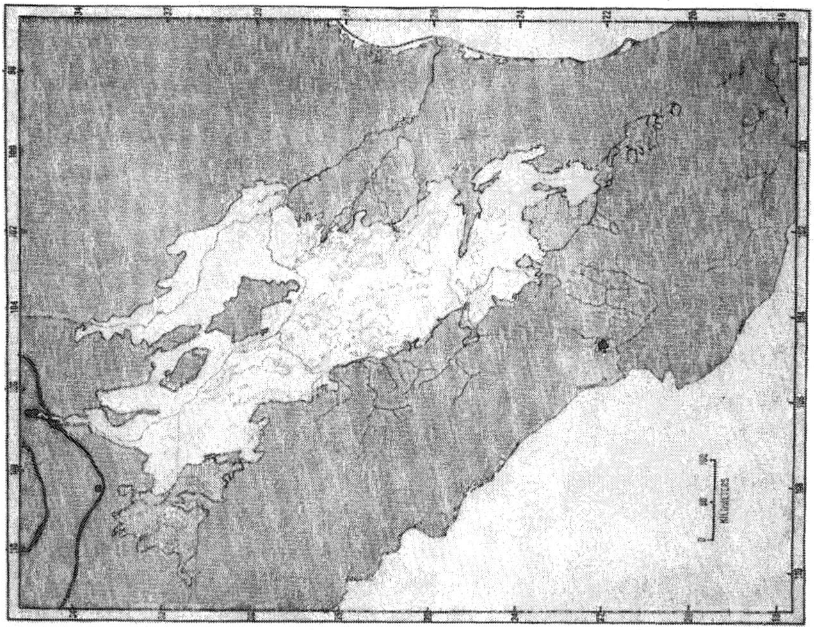

Map 20. *Pseudacris triseriata.*

Map 19. *Hyla eximia.*

251

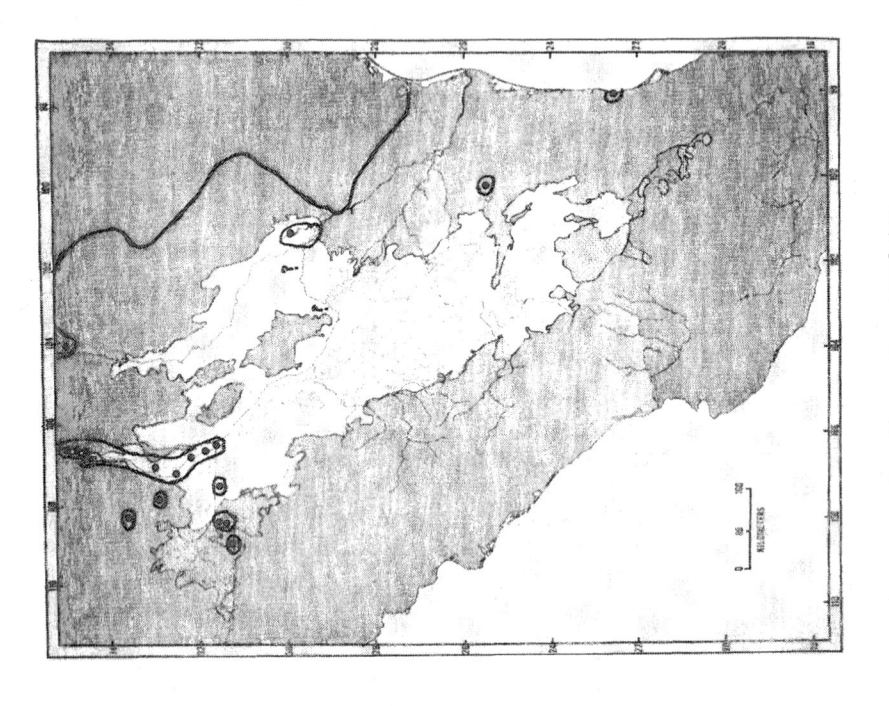

Map 22. *Rana catesbeiana.*

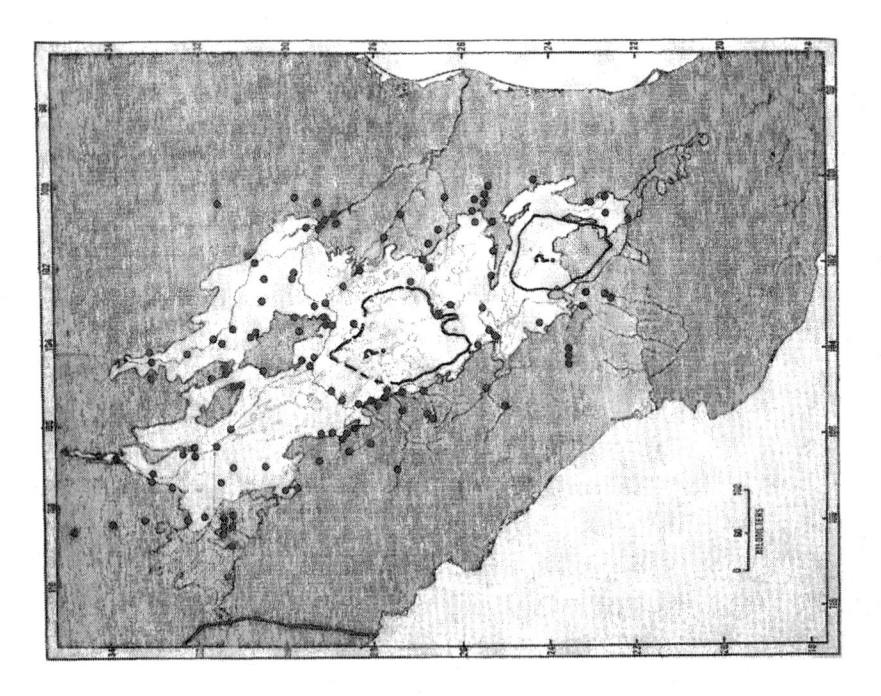

Map 21. *Rana pipiens.*

252

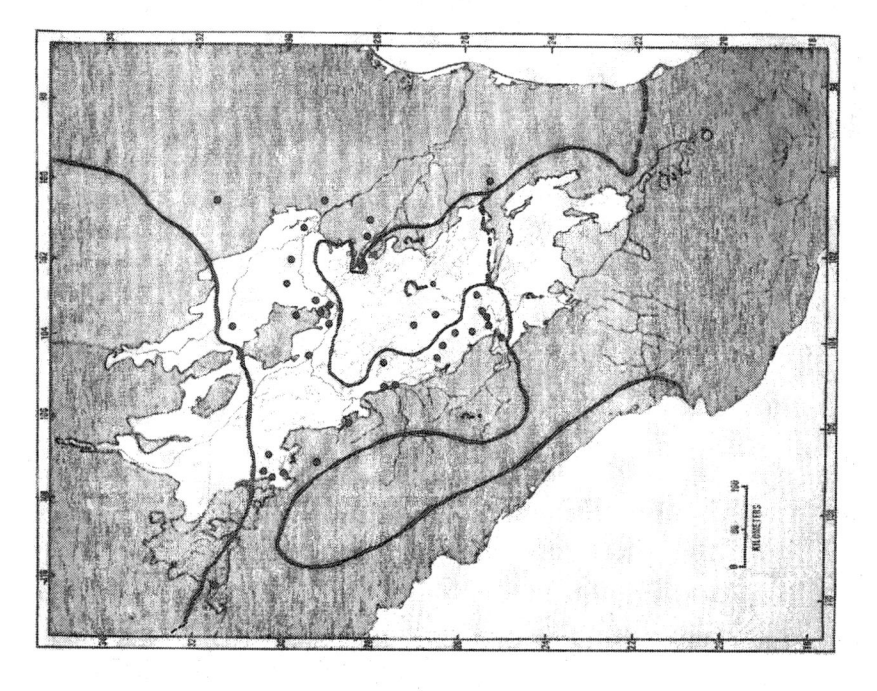

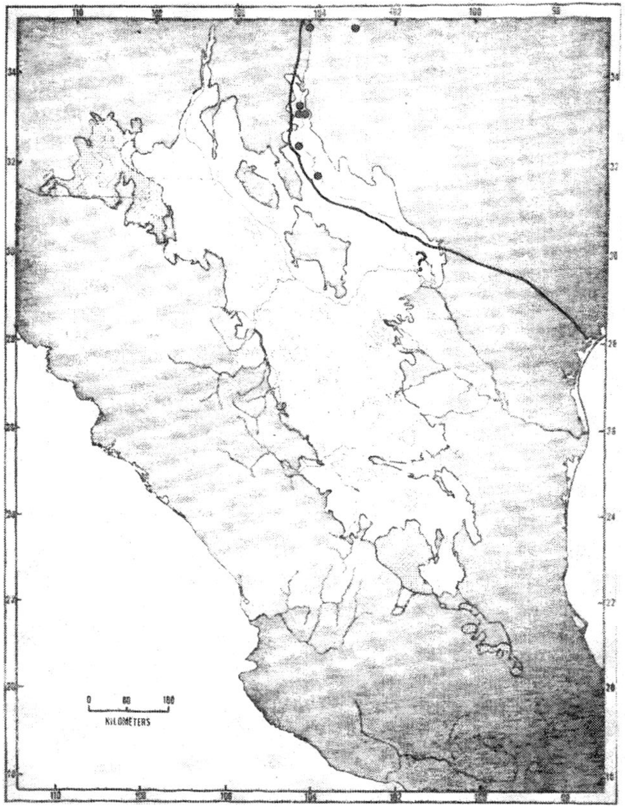

Map 24.   *Chelydra serpentina.*

Map 23.   *Gastrophryne olivacea.*

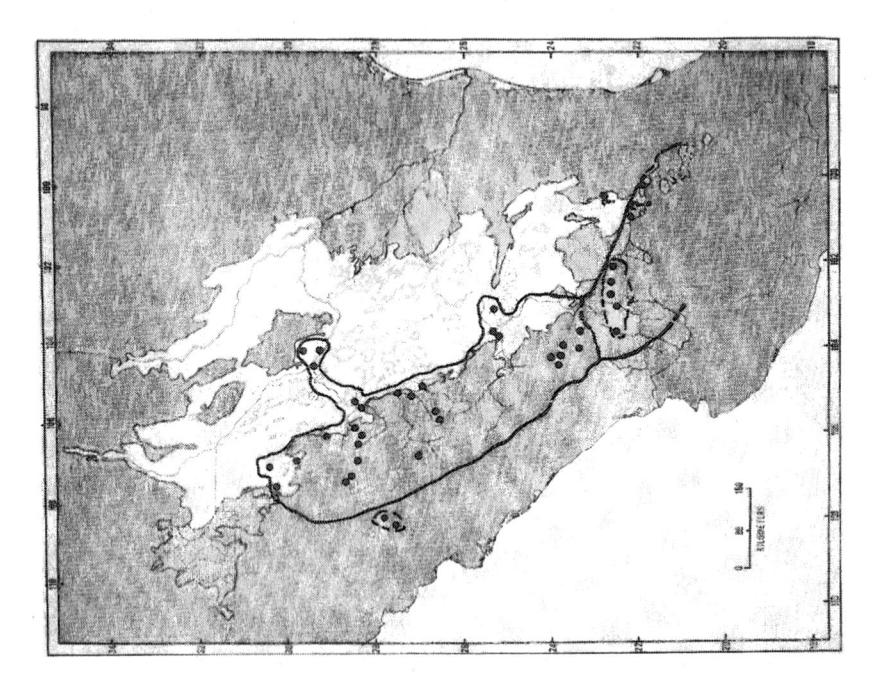

Map 26. *Kinosternon hirtipes.*

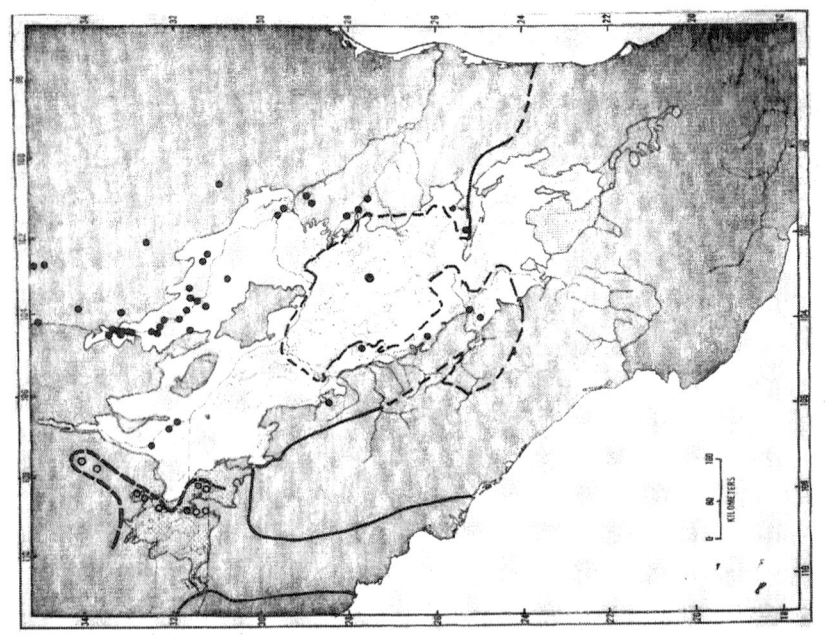

Map 25. ○ *Kinosternon sonoriense;* ● *Kinosternon flavescens,* (range limits set for *K. flavescens* only).

254

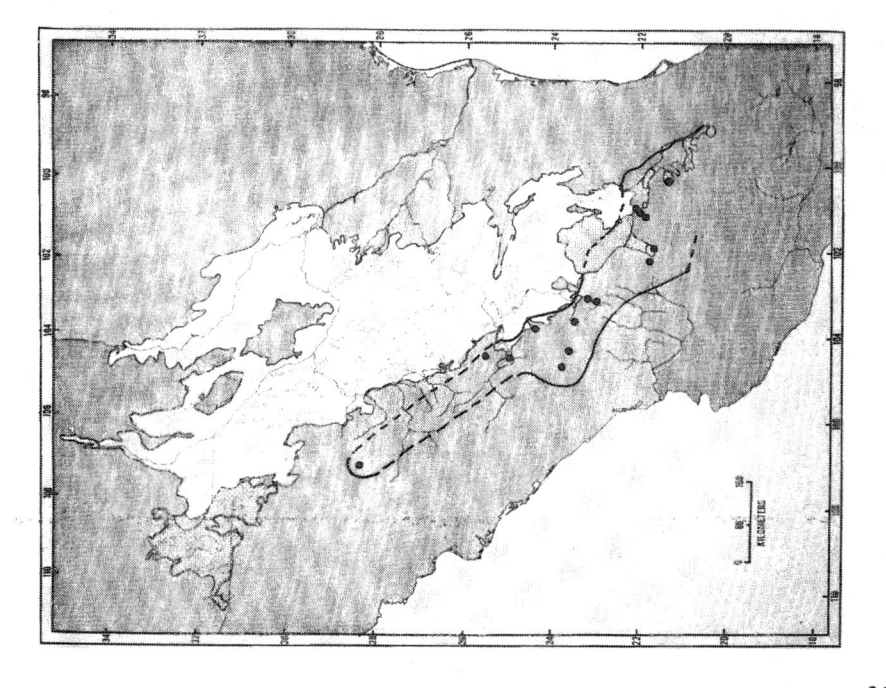

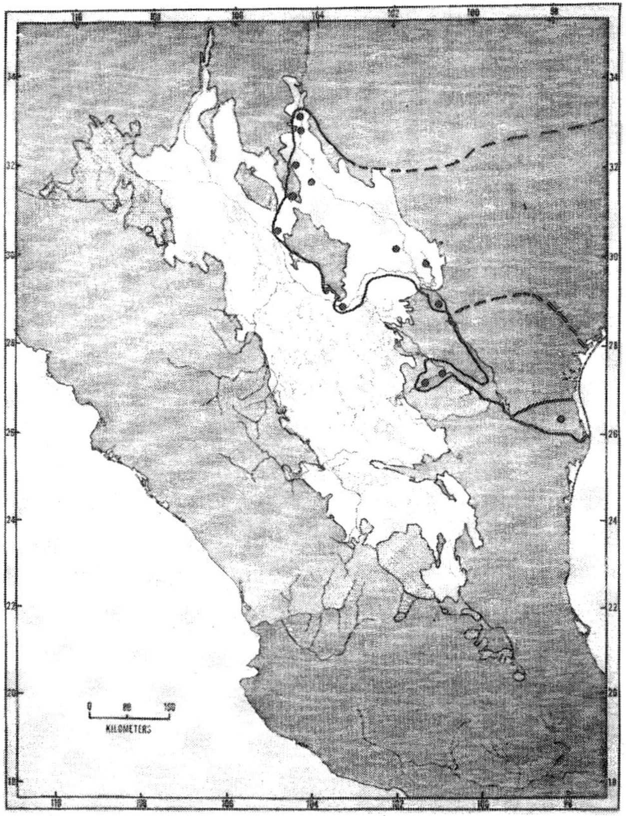

Map 28.   *Chrysemys concinna.*

Map 27.   *Kinosternon integrum.*

255

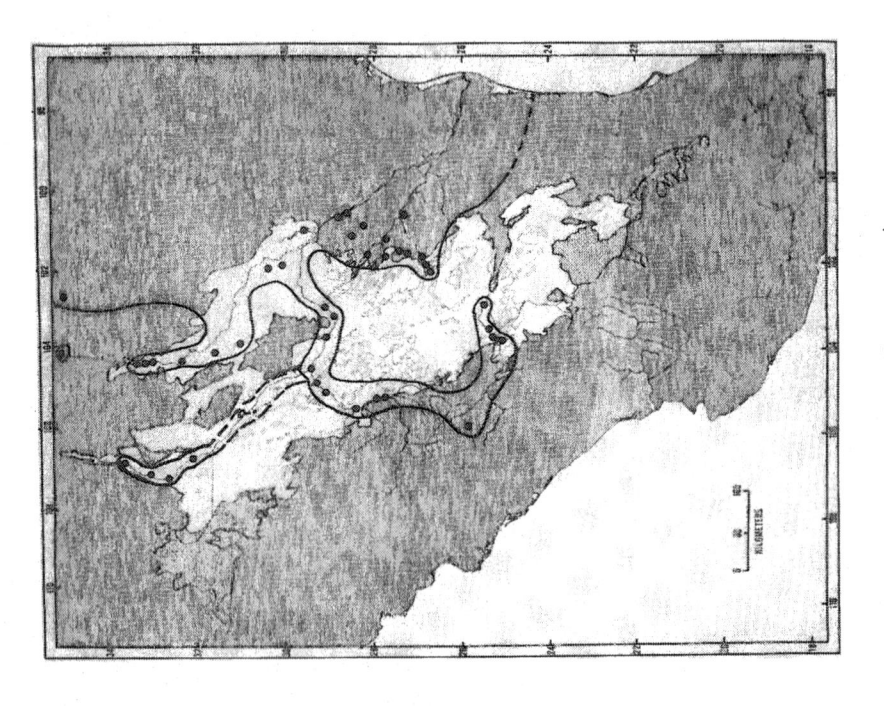

Map 30.   Chrysemys scripta.

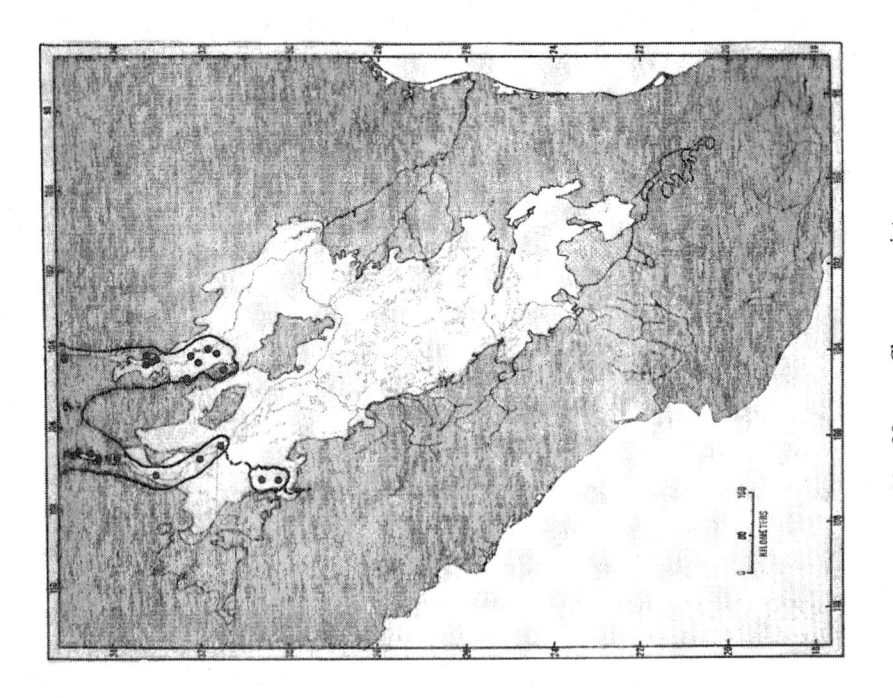

Map 29.   Chrysemys picta.

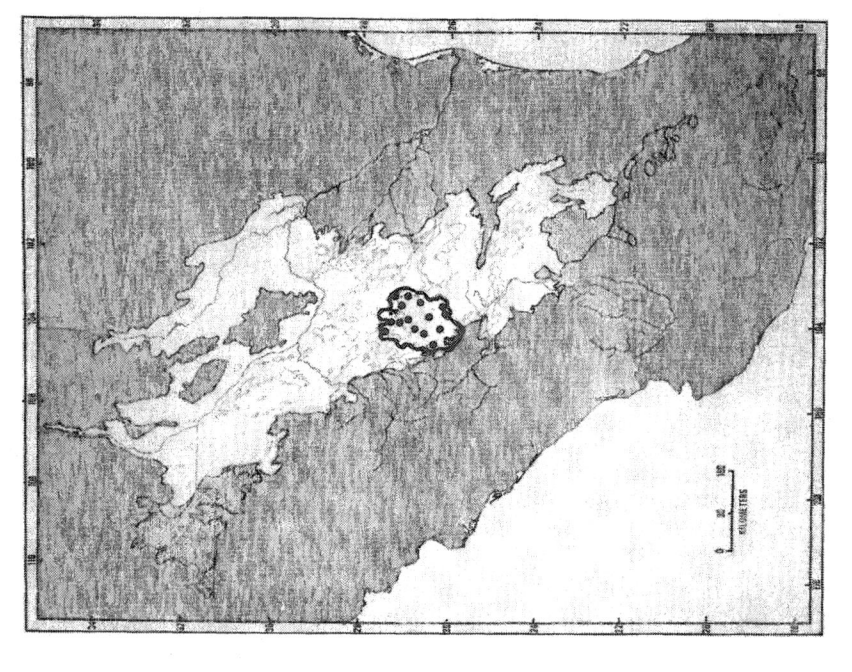

Map 32.   *Gopherus flavomarginatus.*

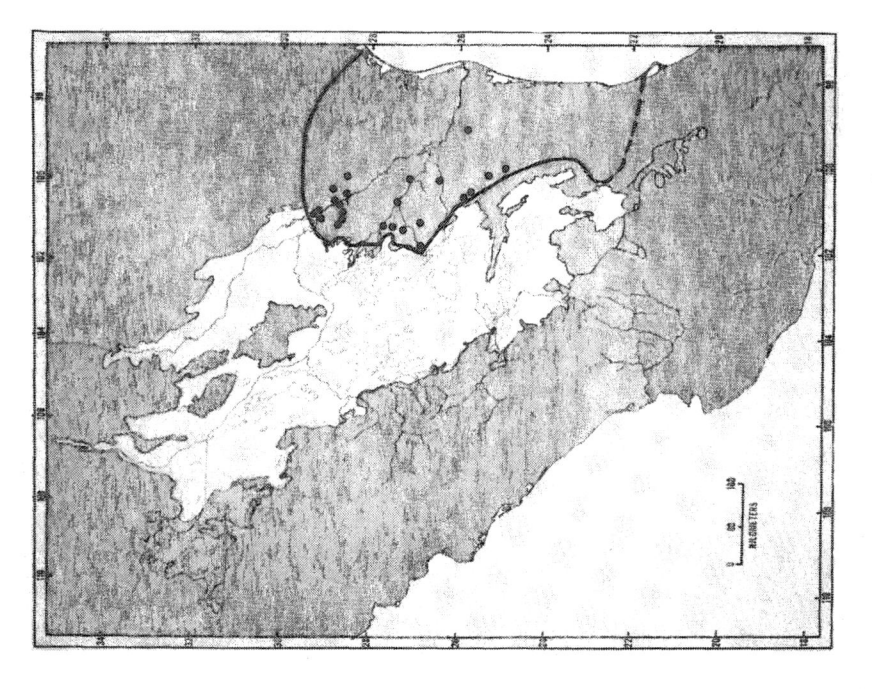

Map 31.   *Gopherus berlandieri.*

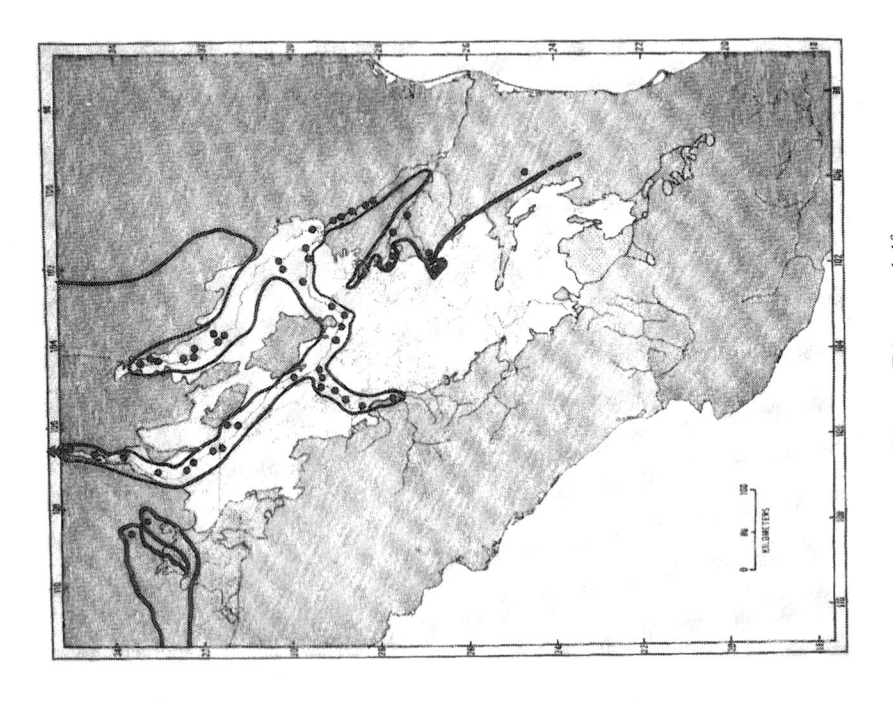

Map 34.   *Trionyx spiniferus.*

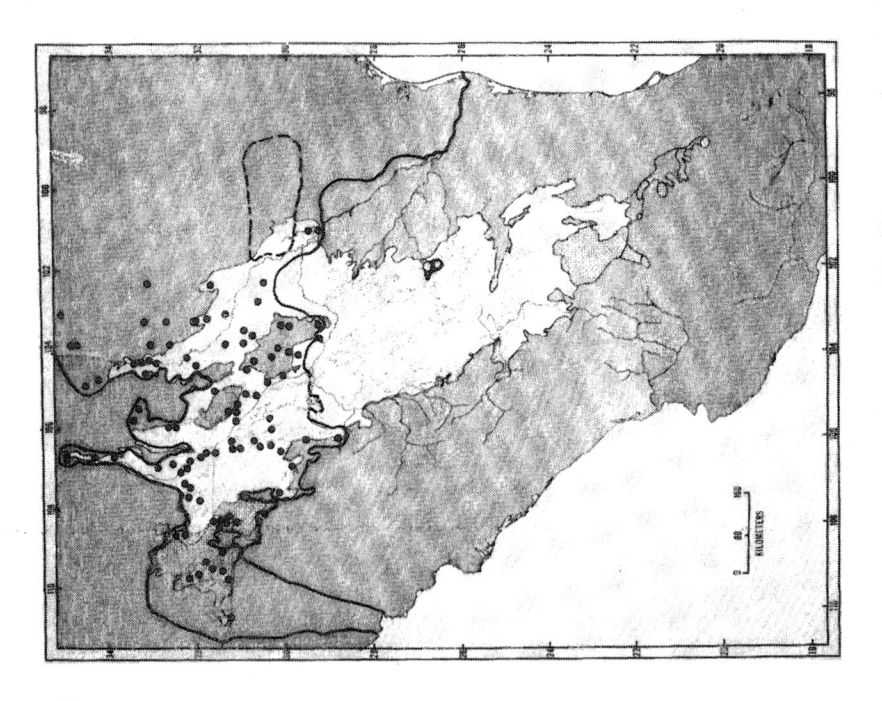

Map 33.   ○ *Terrepene coahuila.*   ● *Terrepene ornata.*

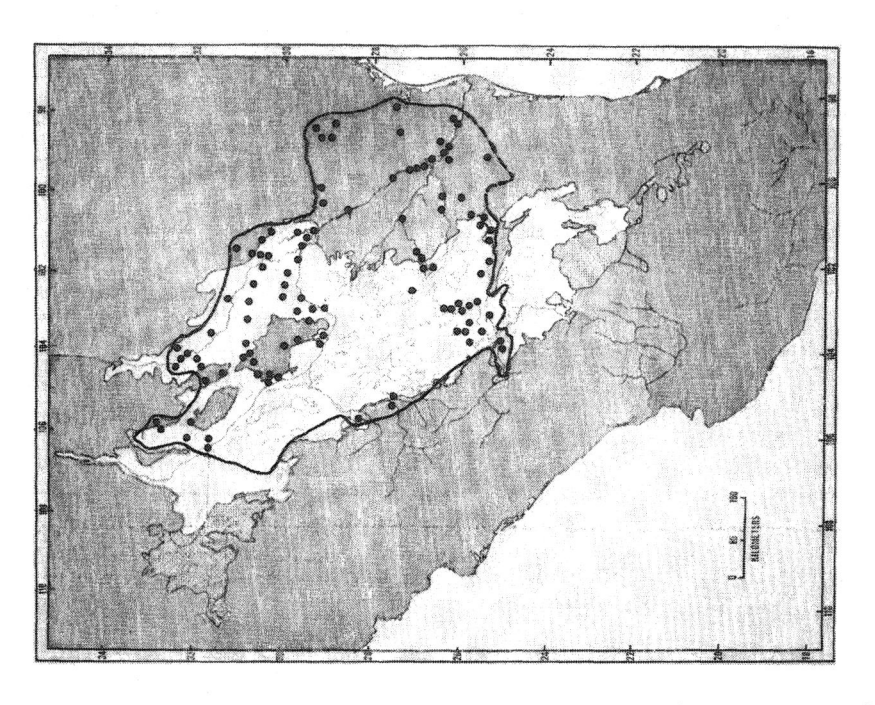

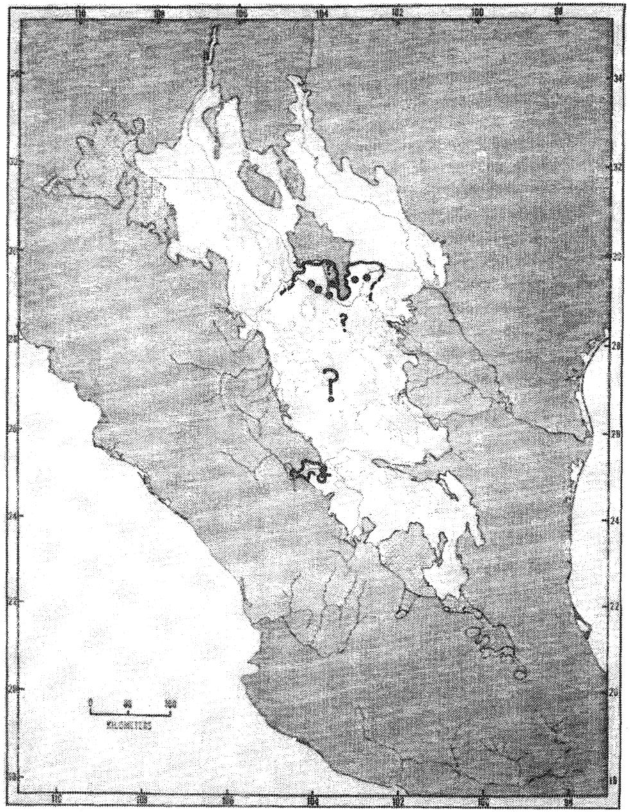

Map 36.   Coleonyx reticulatus.

Map 35.   Coleonyx brevis.

259

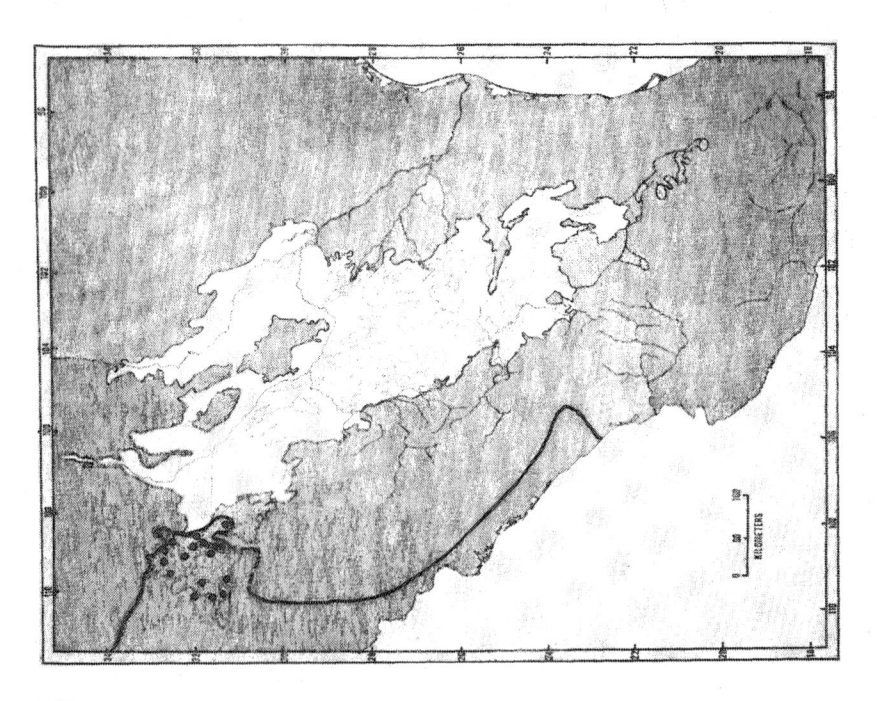

Map 37.   Coleonyx variegatus.

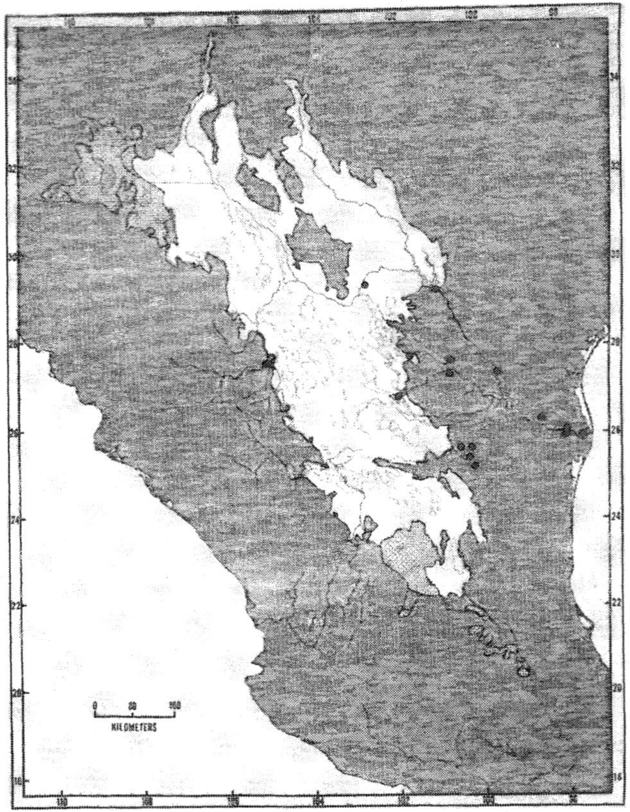

Map 38.   Hemidactylus turicus, (Introduced – local pockets only).

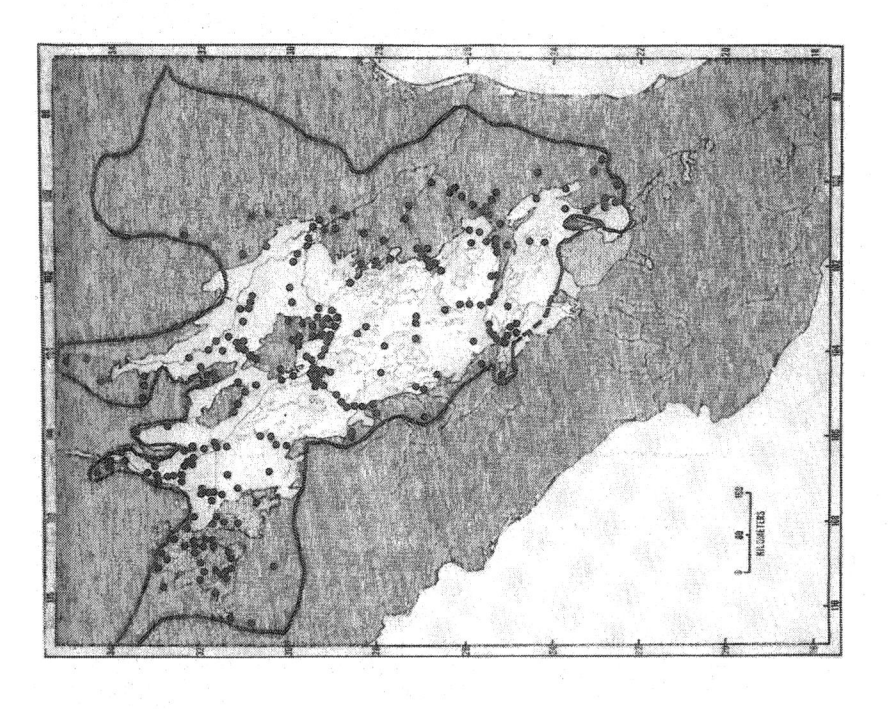

Map 40.   *Cophosaurus texanus.*

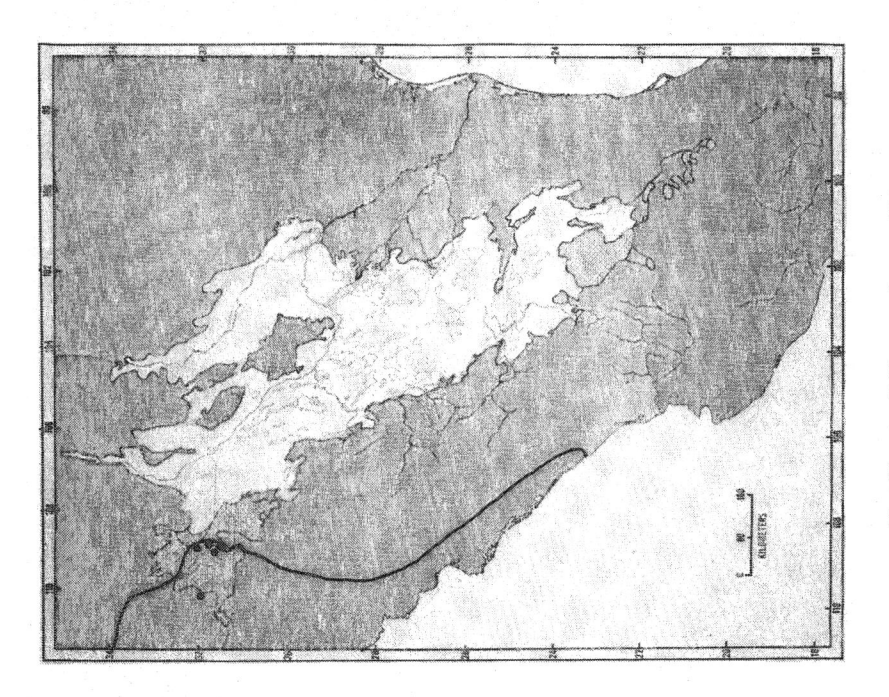

Map 39.   *Callisaurus draconoides.*

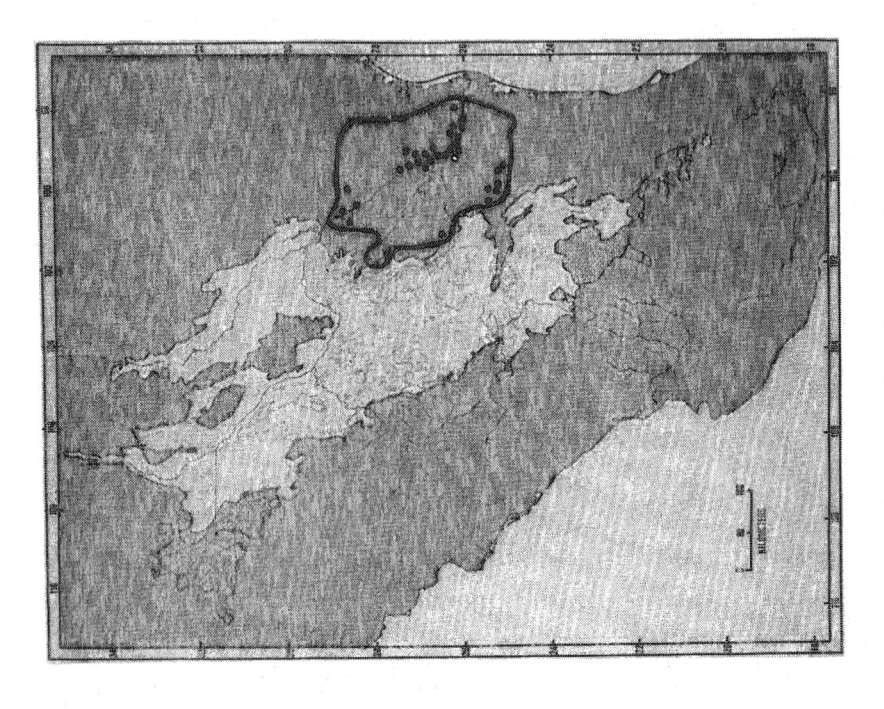

Map 42.   *Crotaphytus reticulatus.*

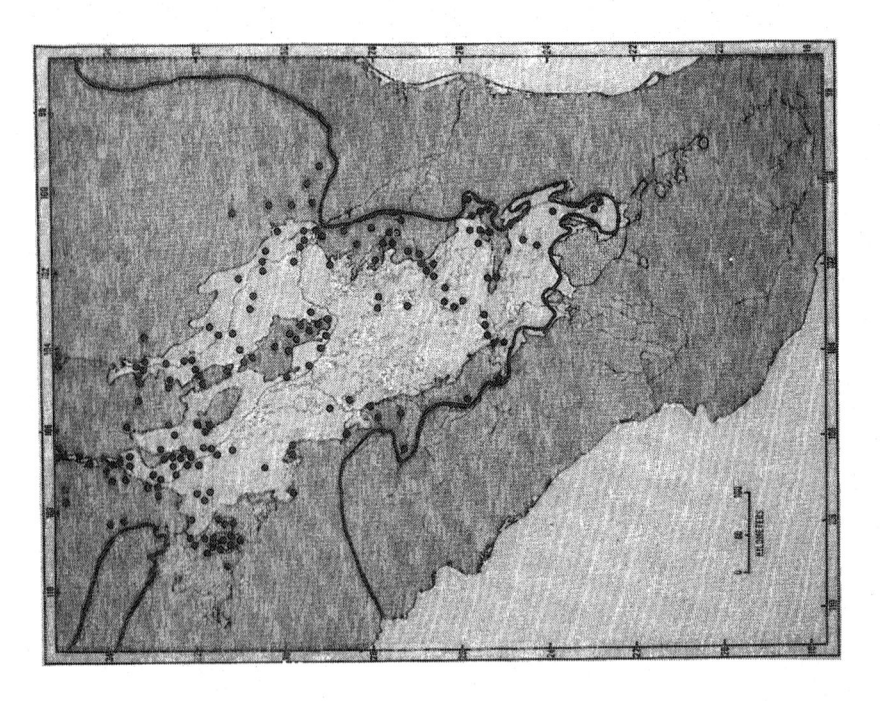

Map 41.   *Crotaphytus collaris.*

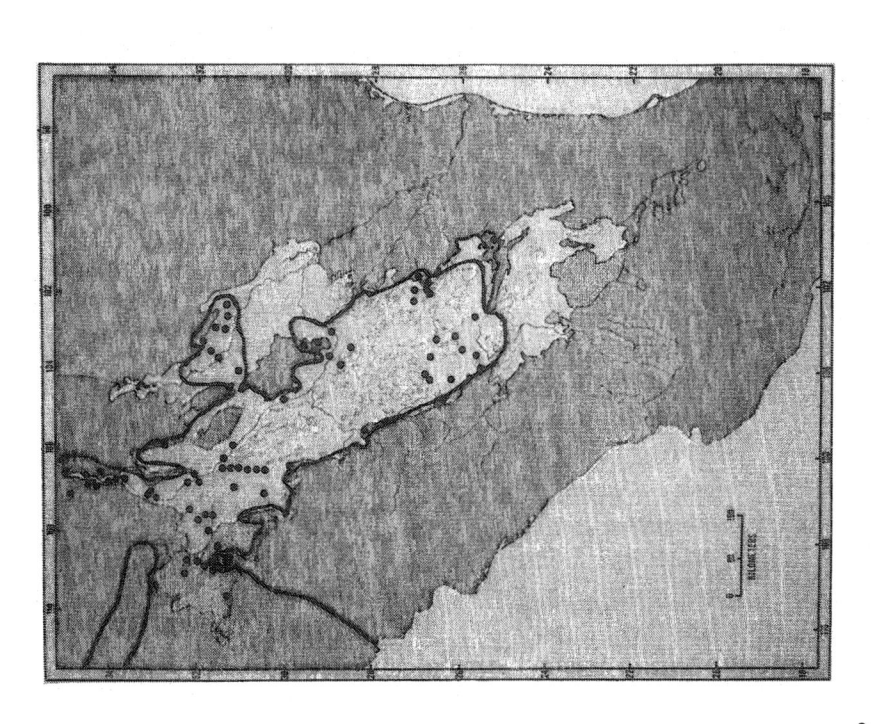

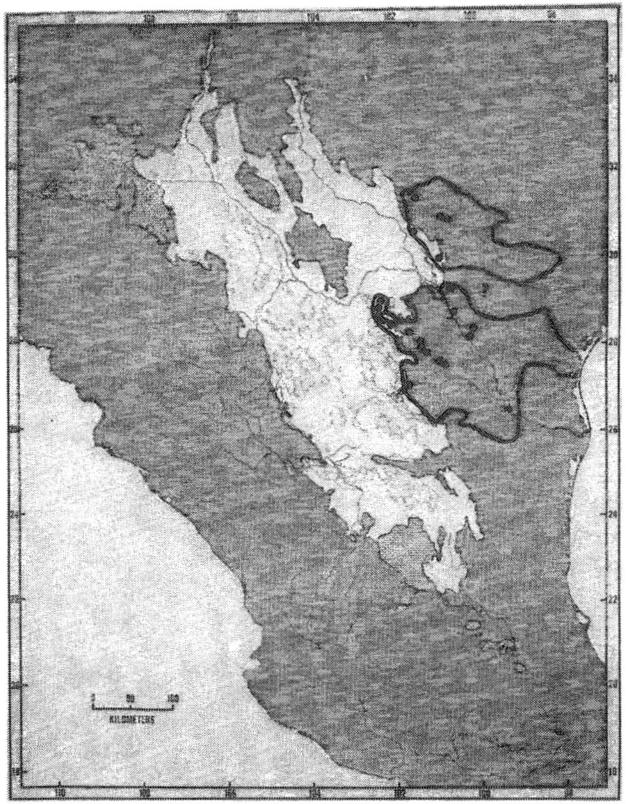

Map 44.   *Holbrookia lacerata.*

Map 43.   *Crotaphytus (Gambelia) wislizeni.*

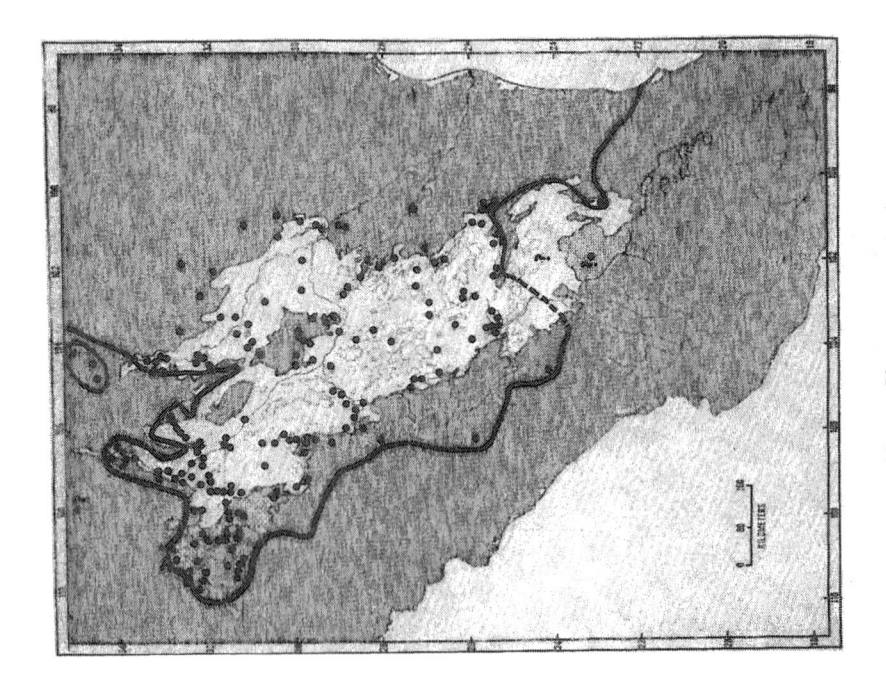

Map 46.   Phrynosoma cornutum.

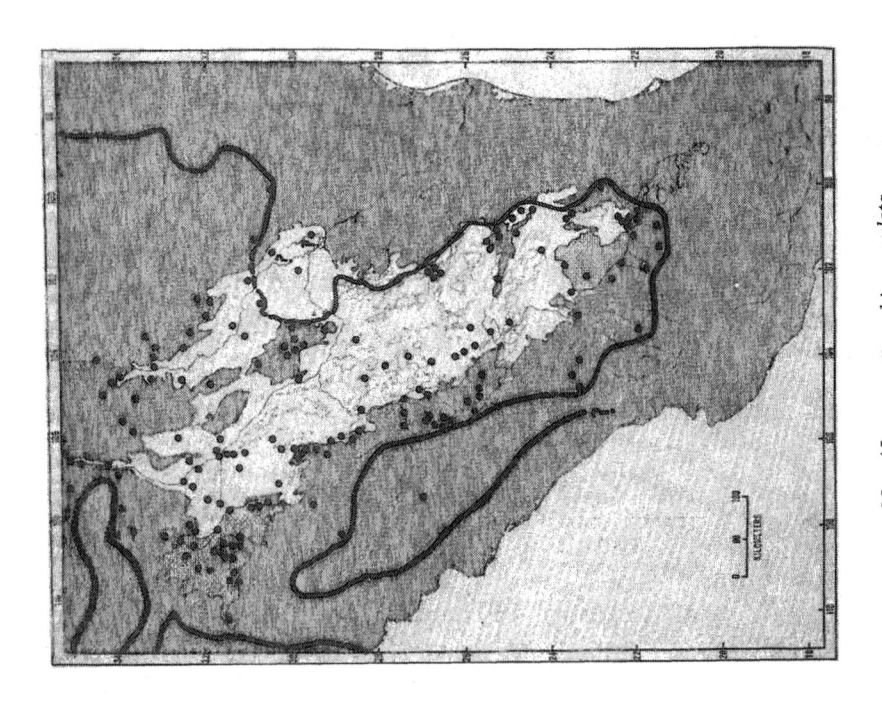

Map 45.   Holbrookia maculata.

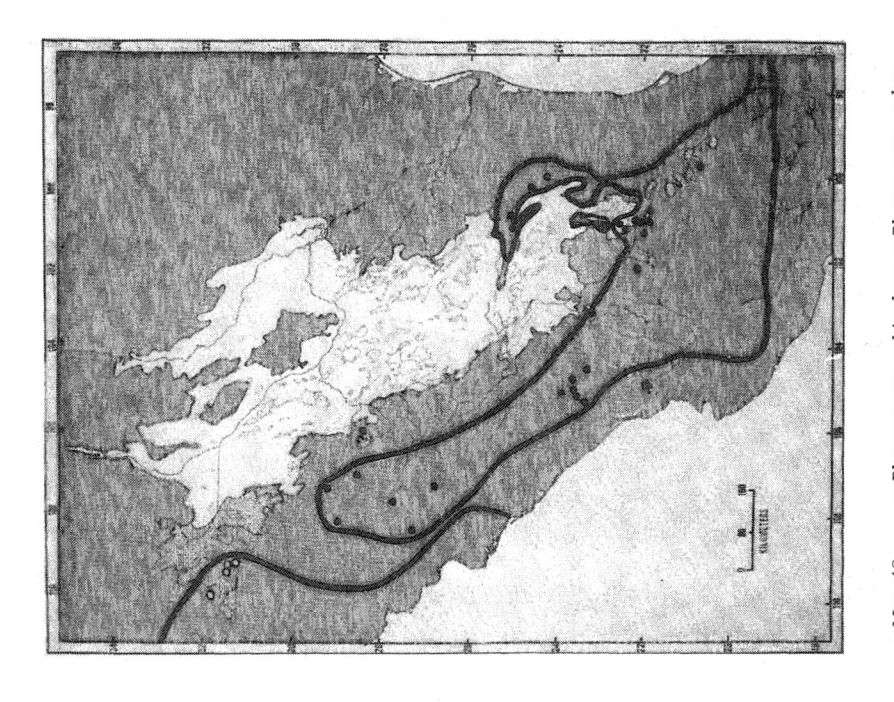

Map 48. ● *Phrynosoma orbiculare;* ○ *Phrynosoma solare.*

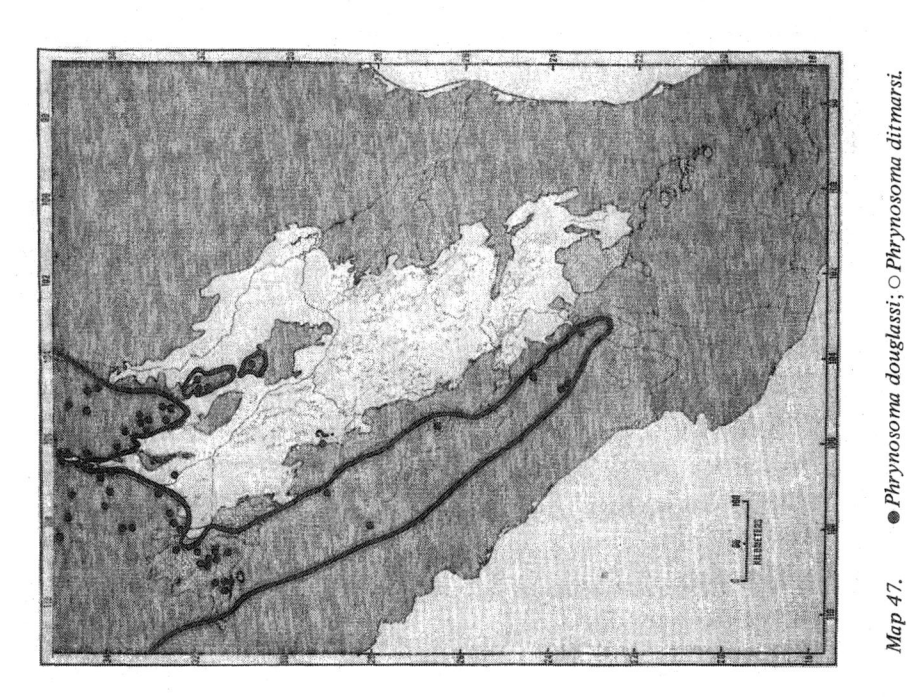

Map 47. ● *Phrynosoma douglassi;* ○ *Phrynosoma ditmarsi.*

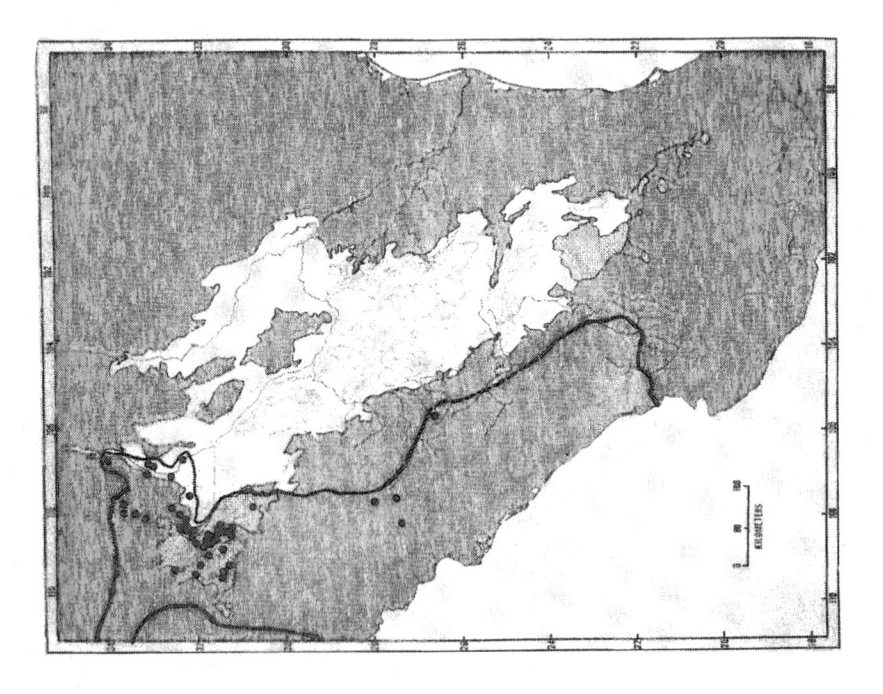

Map 50. *Sceloporus clarki.*

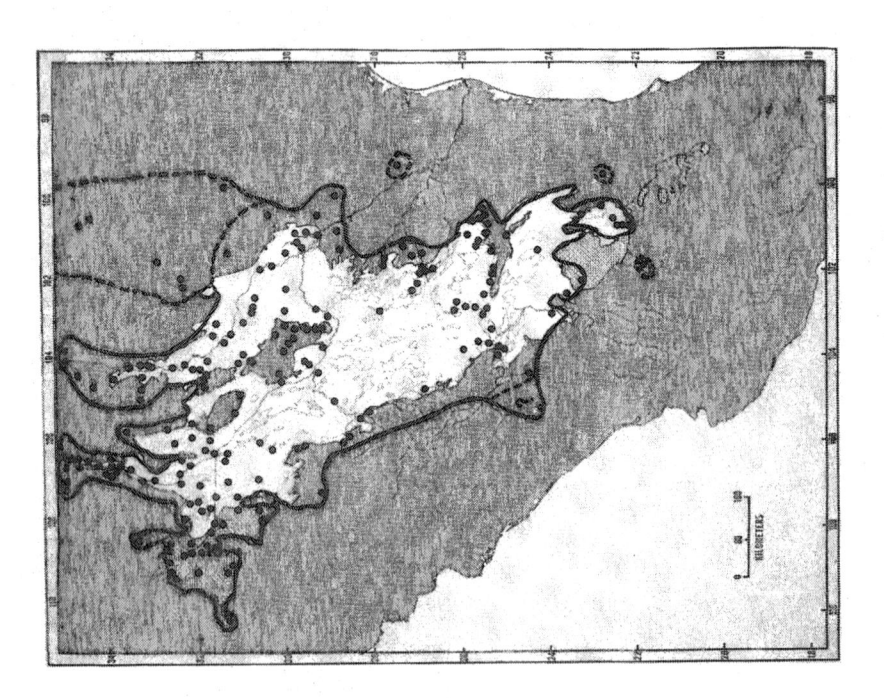

Map 49. *Phrynosoma modestum.*

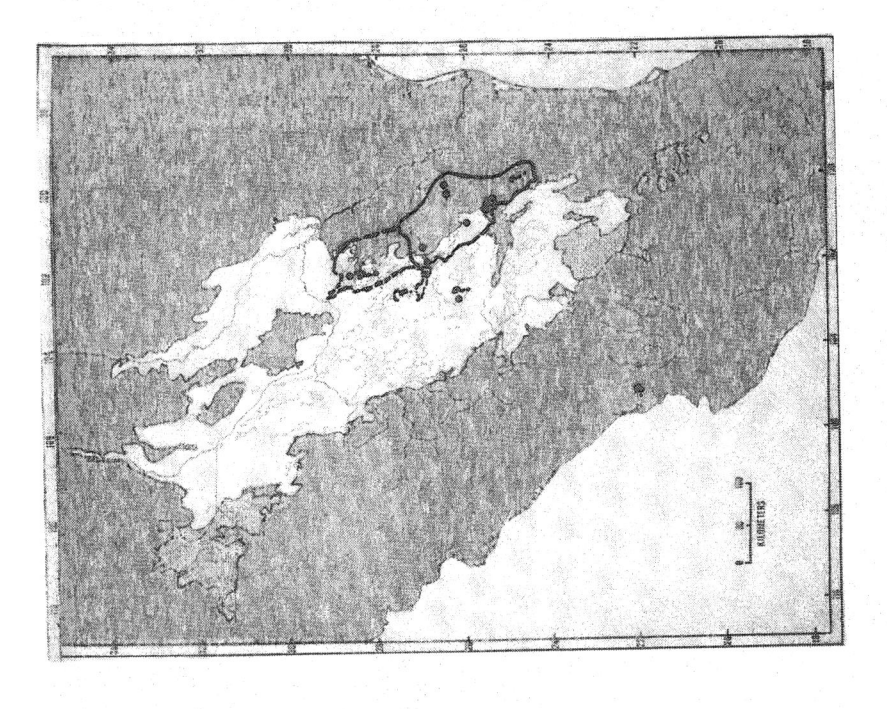

Map 52. *Sceloporus couchi.*

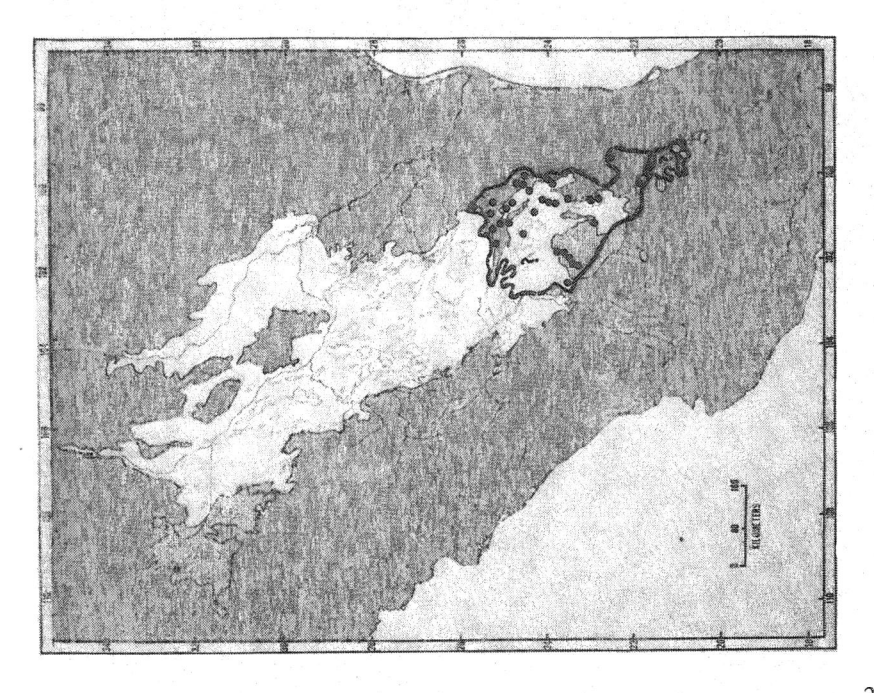

Map 51. ● *Sceloporus cautus;* ○ *Sceloporus exsul.*

267

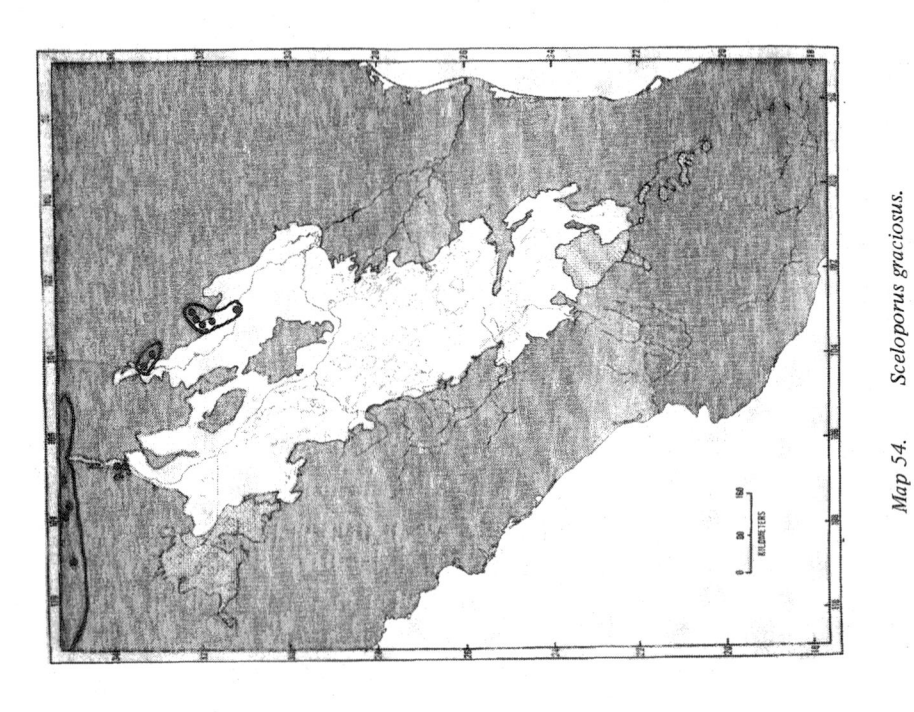

Map 54. *Sceloporus graciosus.*

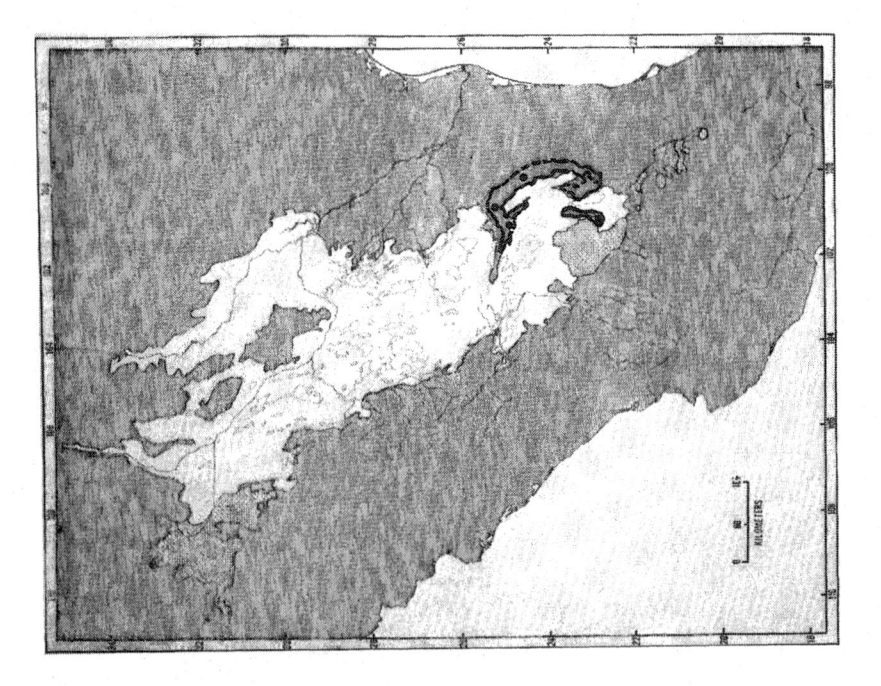

Map 53. *Sceloporus goldmani.*

268

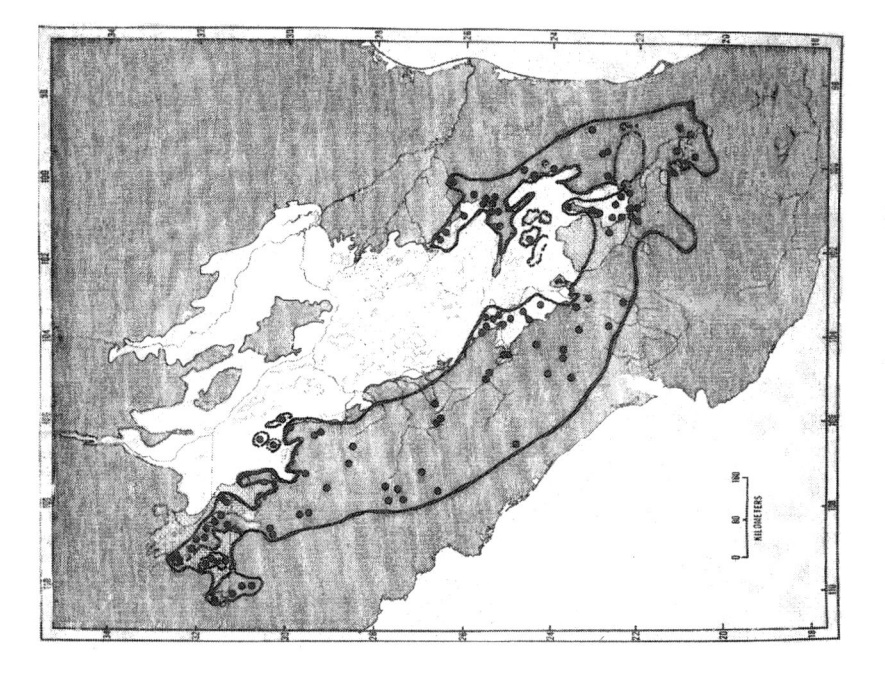

Map 56.   Sceloporus jarrovi,

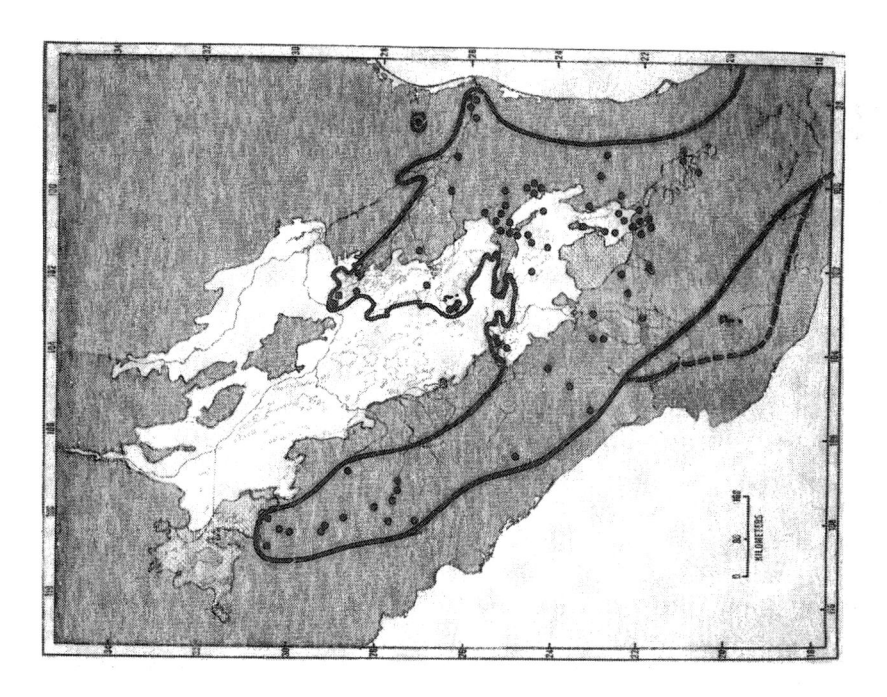

Map 55.   Sceloporus grammicus.

269

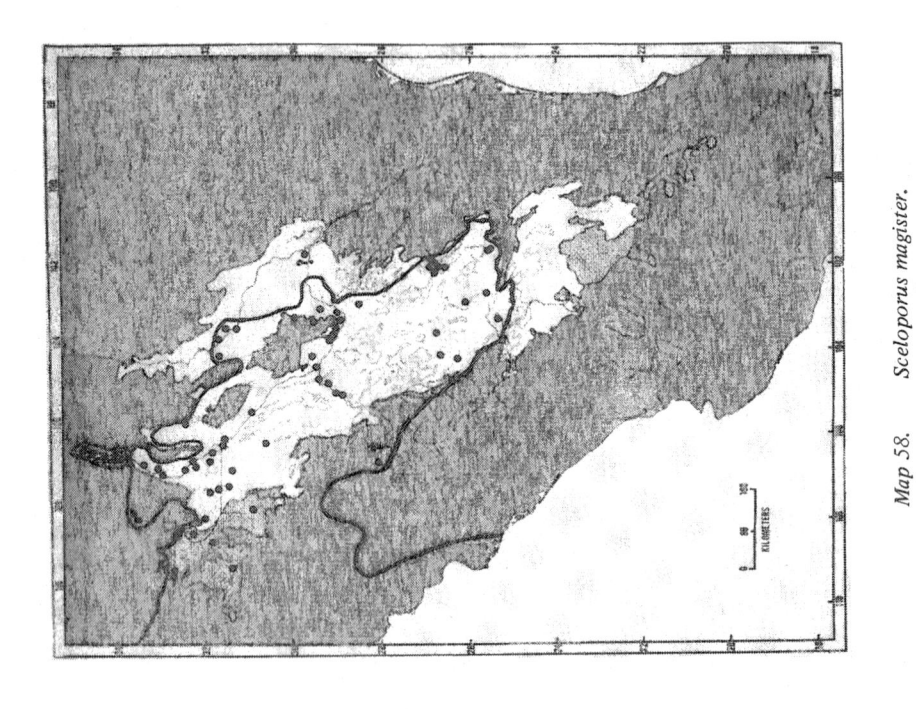

Map 58.   Sceloporus magister.

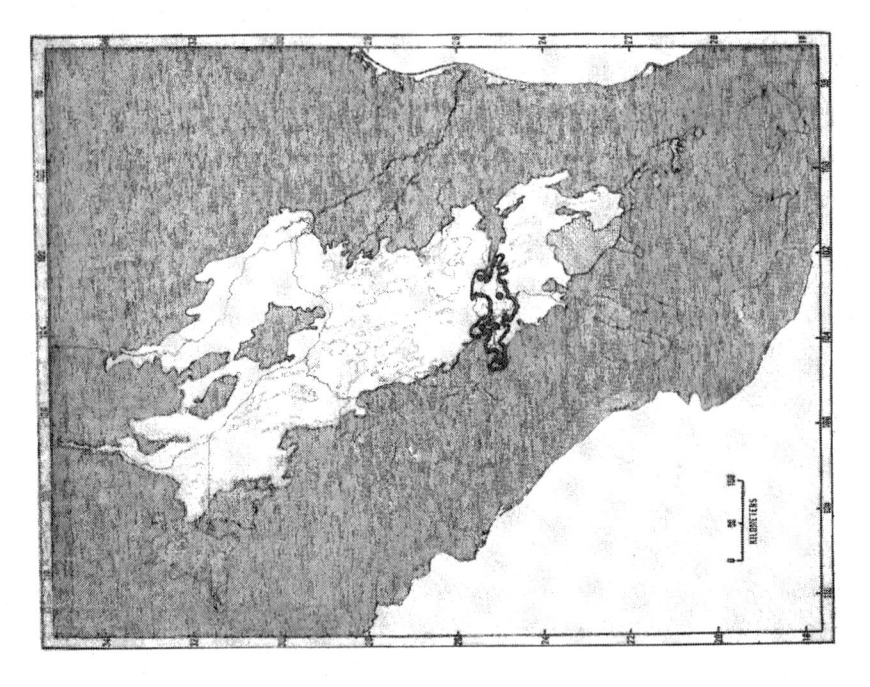

Map 57.   Sceloporus maculosus.

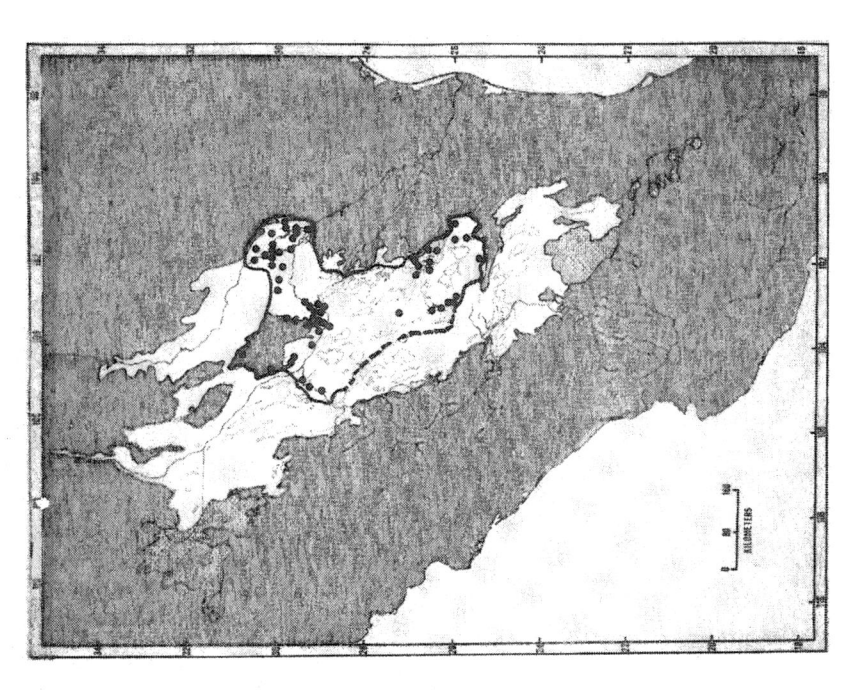

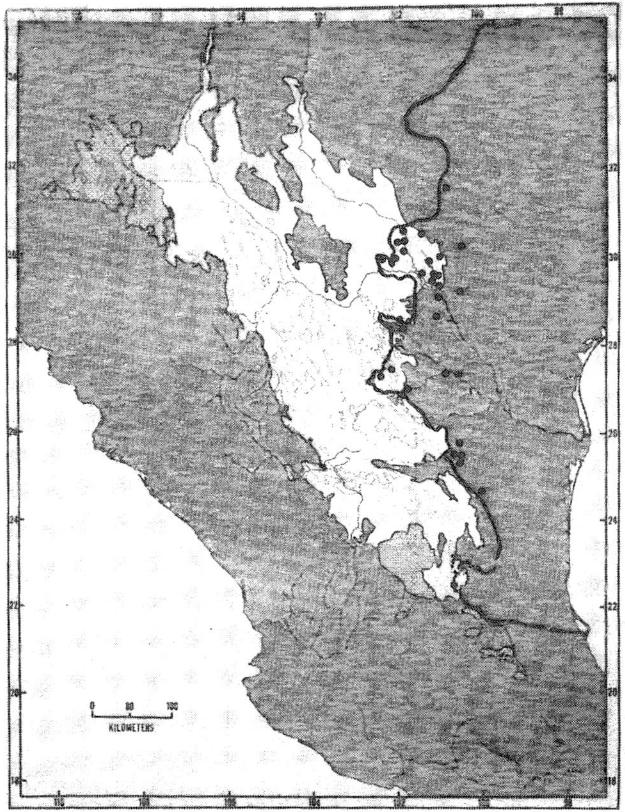

Map 60. *Sceloporus olivaceus.*

Map 59. *Sceloporus merriami.*

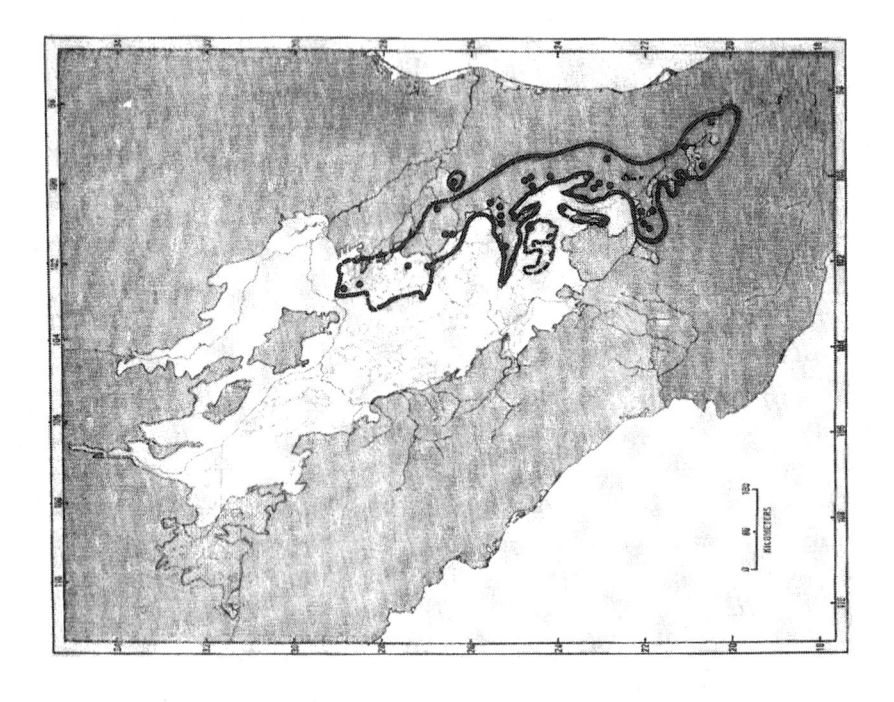

Map 62.   Sceloporus parvus.

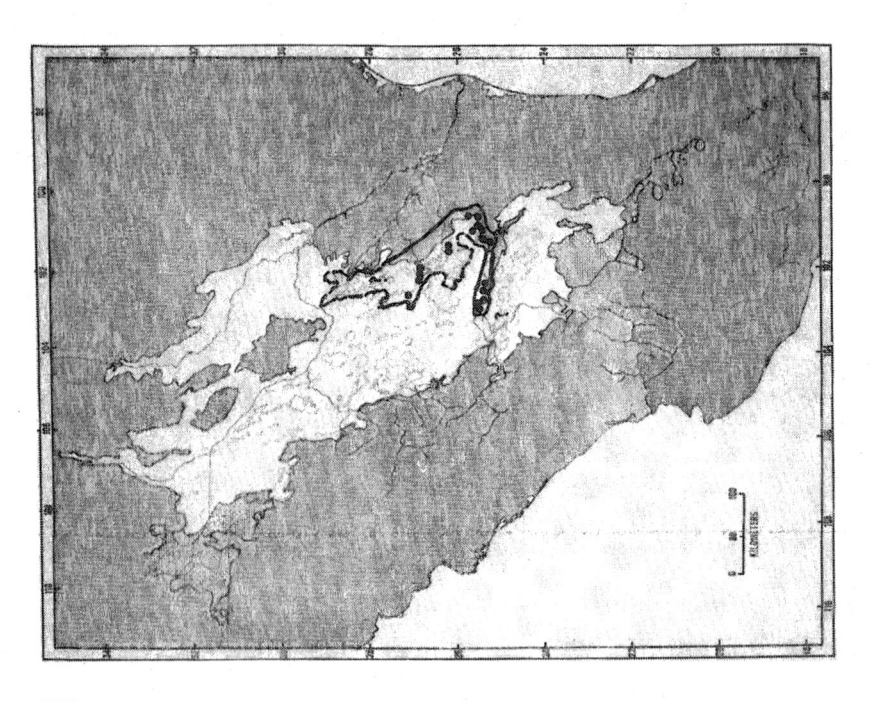

Map 61.   Sceloporus ornatus.

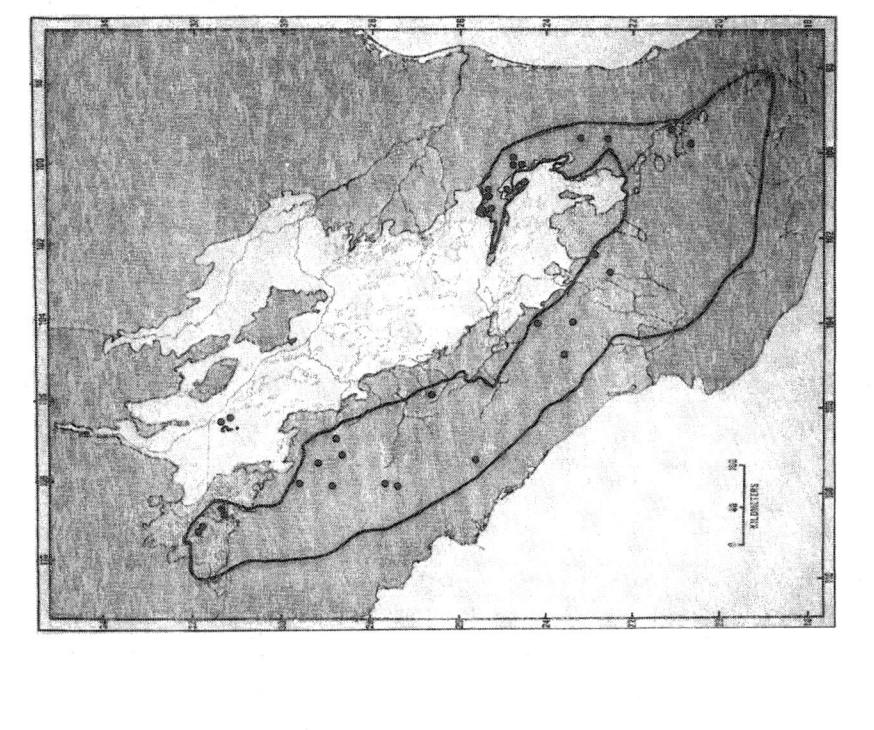

Map 63. *Sceloporus poinsetti.*

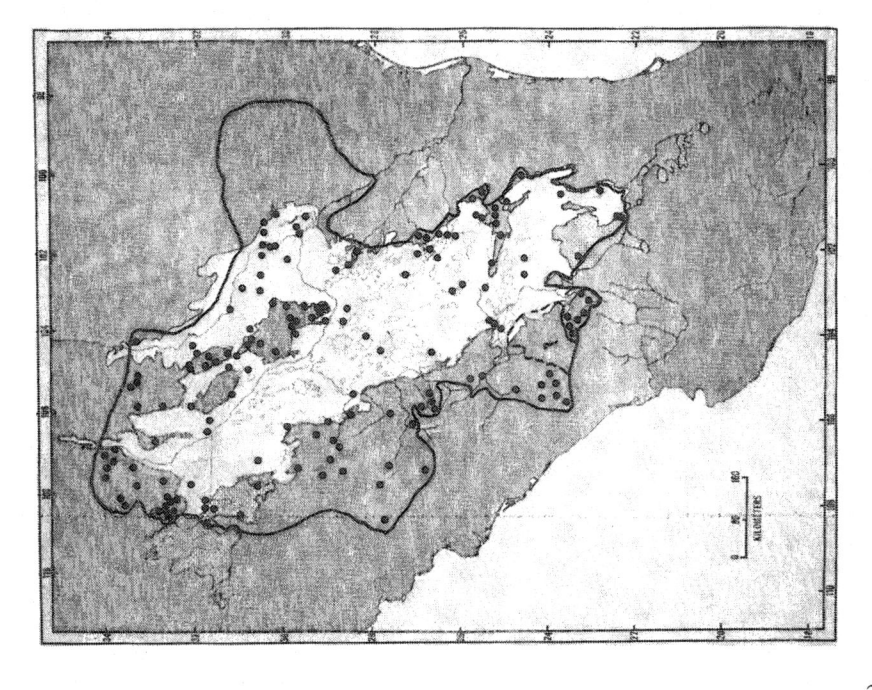

Map 64. *Sceloporus scalaris*

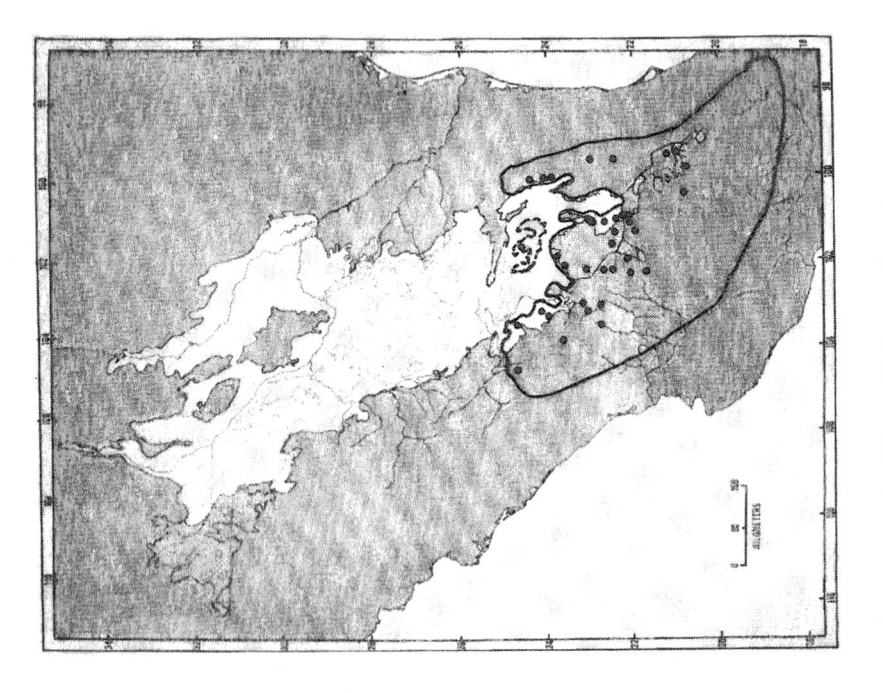

Map 66.   Sceloporus torquatus.

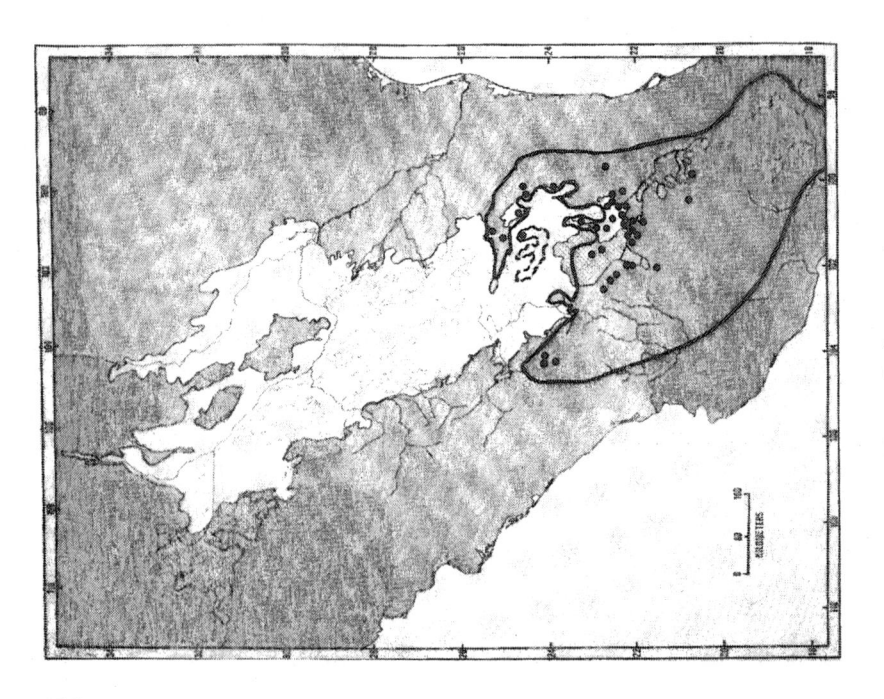

Map 65.   Sceloporus spinosus.

274

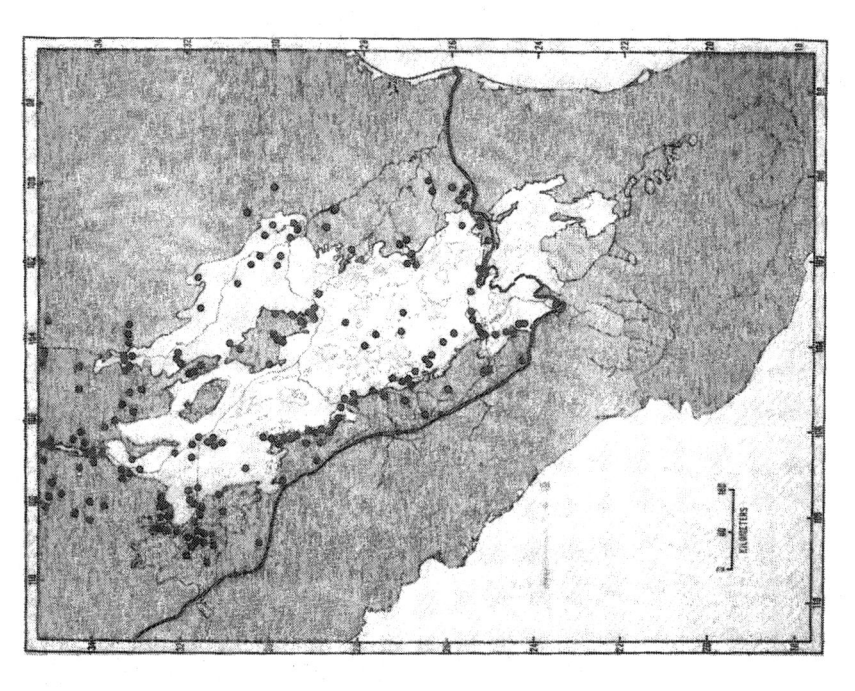

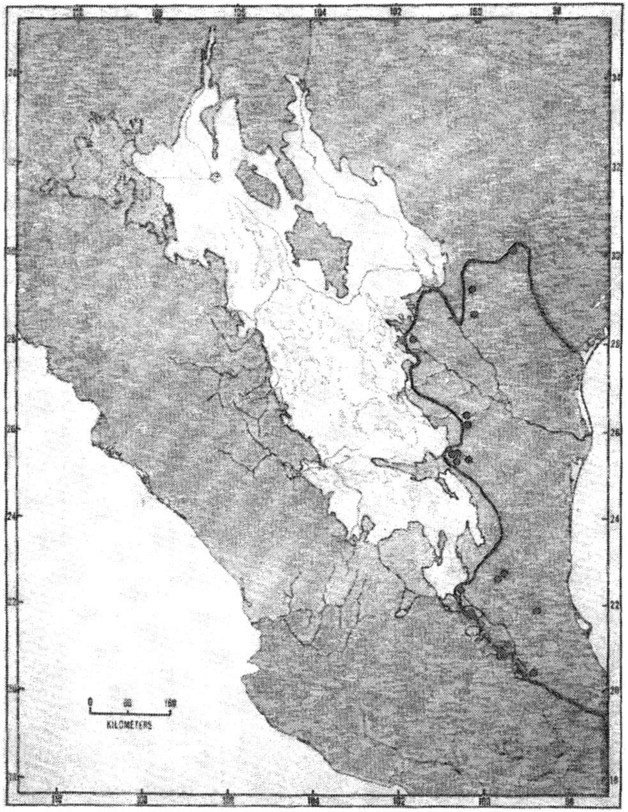

Map 68.   *Sceloporus variabilis.*

Map 67.   *Sceloporus undulatus.*

275

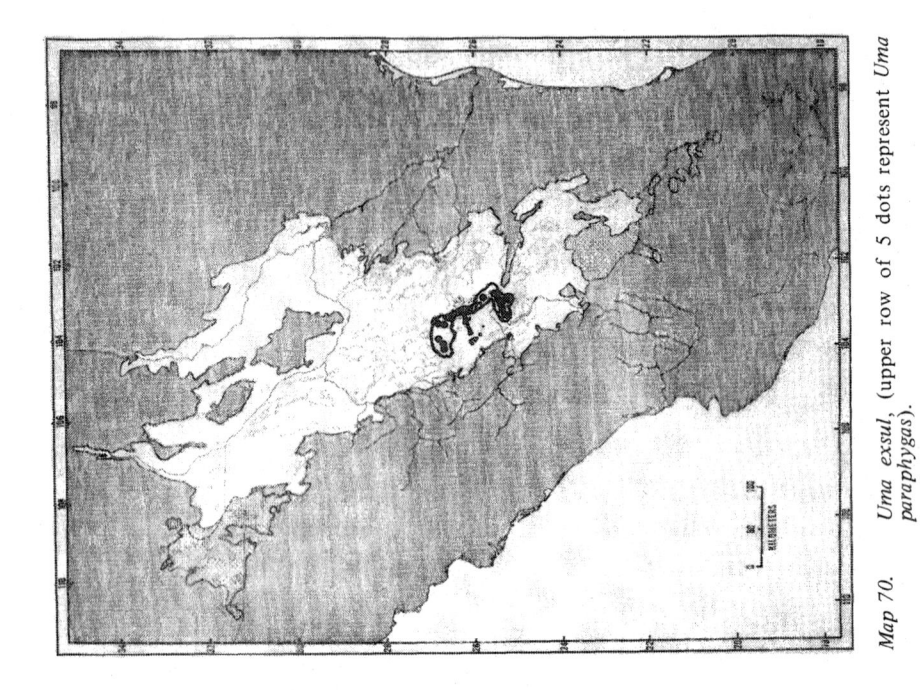

Map 70. *Uma exsul*, (upper row of 5 dots represent *Uma paraphygas*).

Map 69. *Sceloporus virgatus*.

276

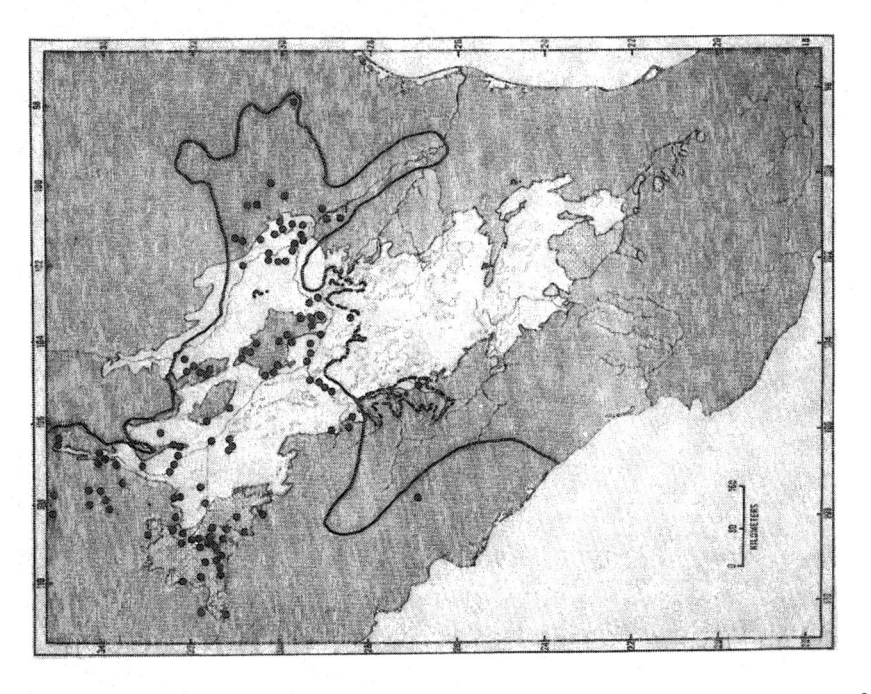

Map 71.   Urosaurus ornatus.

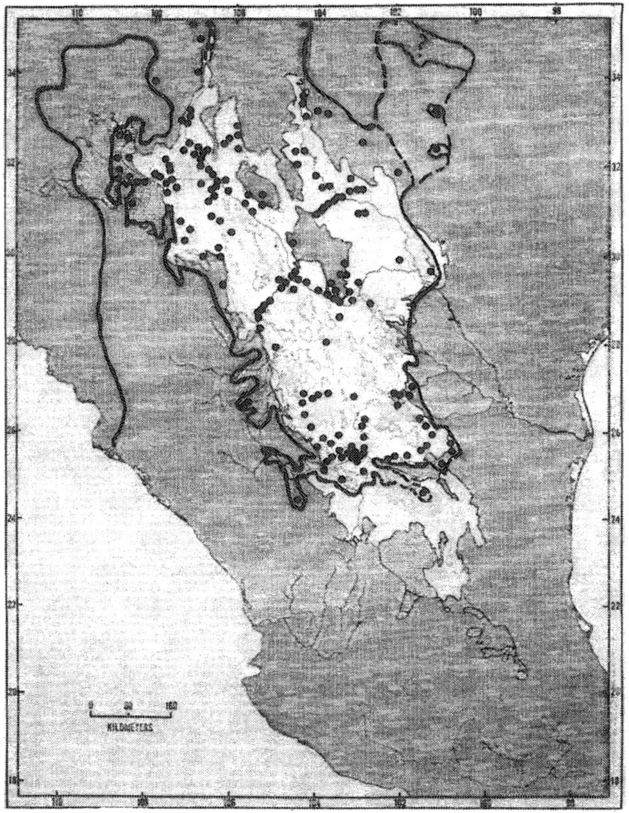

Map 72.   Uta stansburiana.

277

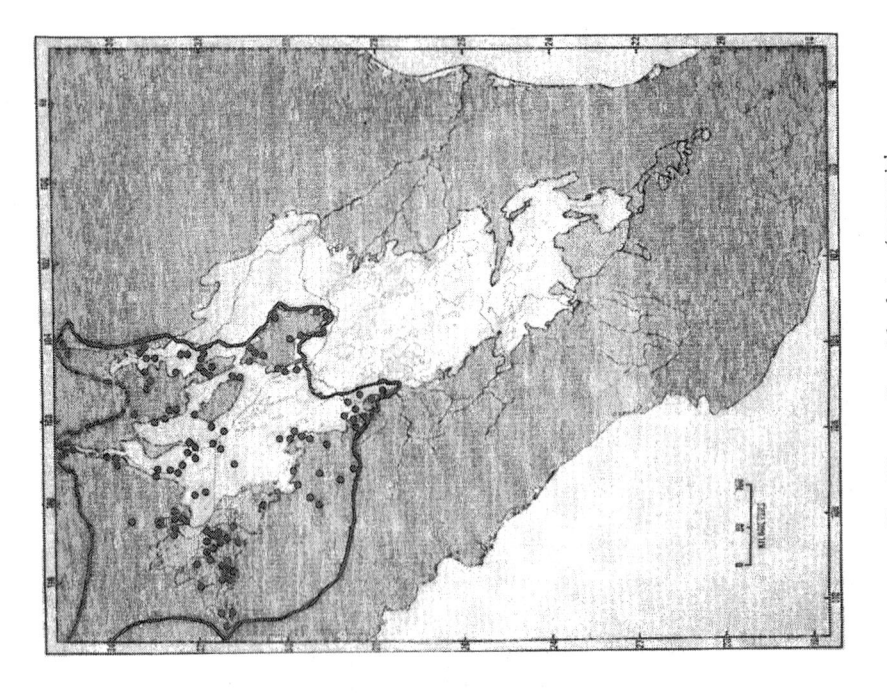

Map 74.   *Chemidophorus 'exanguis'.*

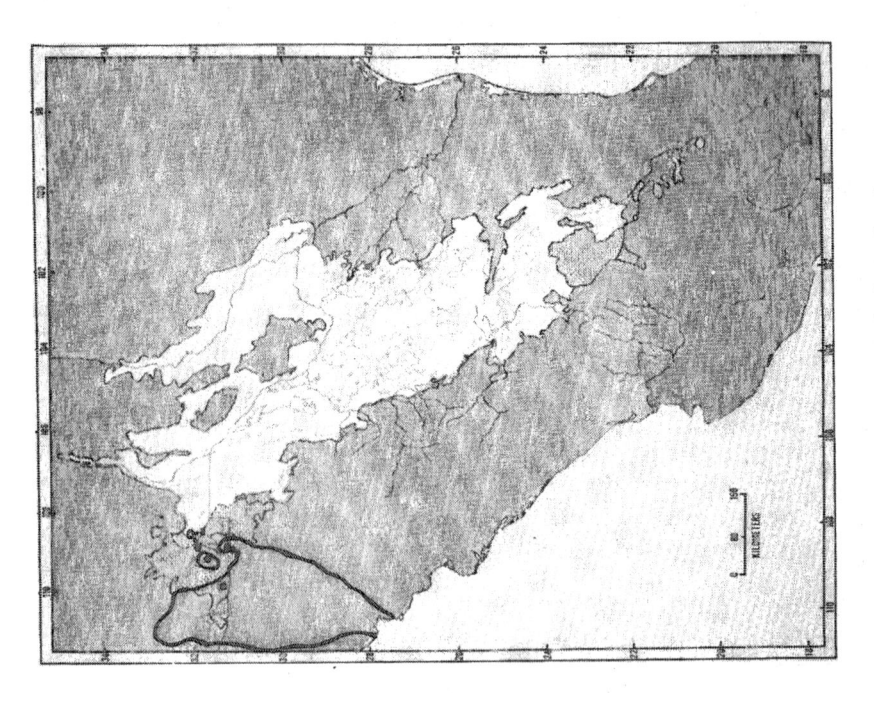

Map 73.   *Chemidophorus burti.*

278

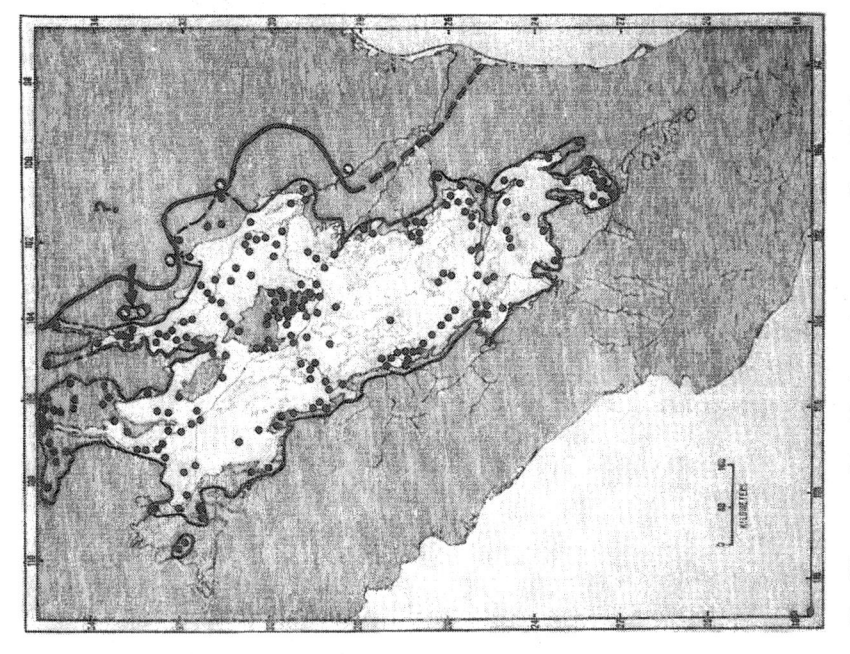

Map 76. ● *Chemidophorus inornatus.* ○ *Chemidophorus sex-lineatus.*

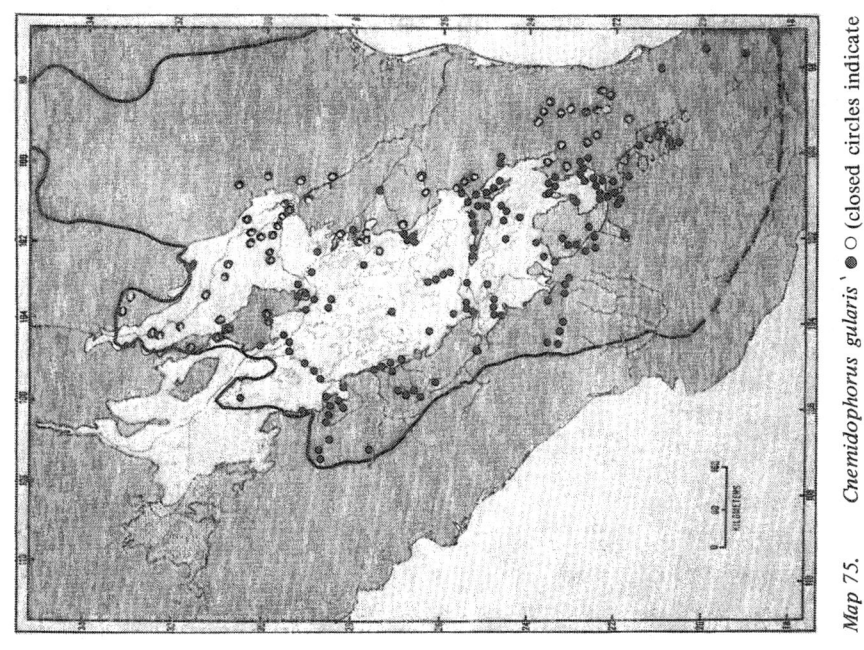

Map 75. *Chemidophorus gularis* ' ● ○ (closed circles indicate form – *scalaris*).

279

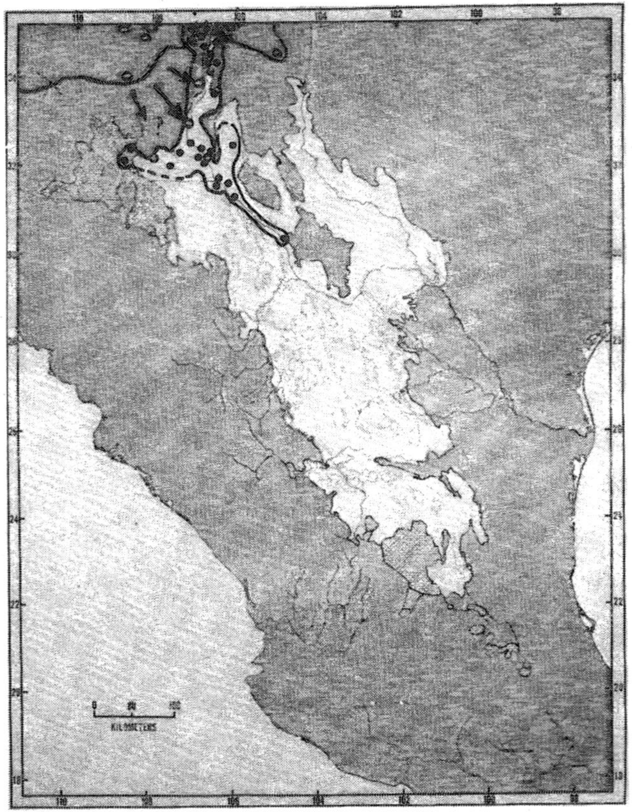

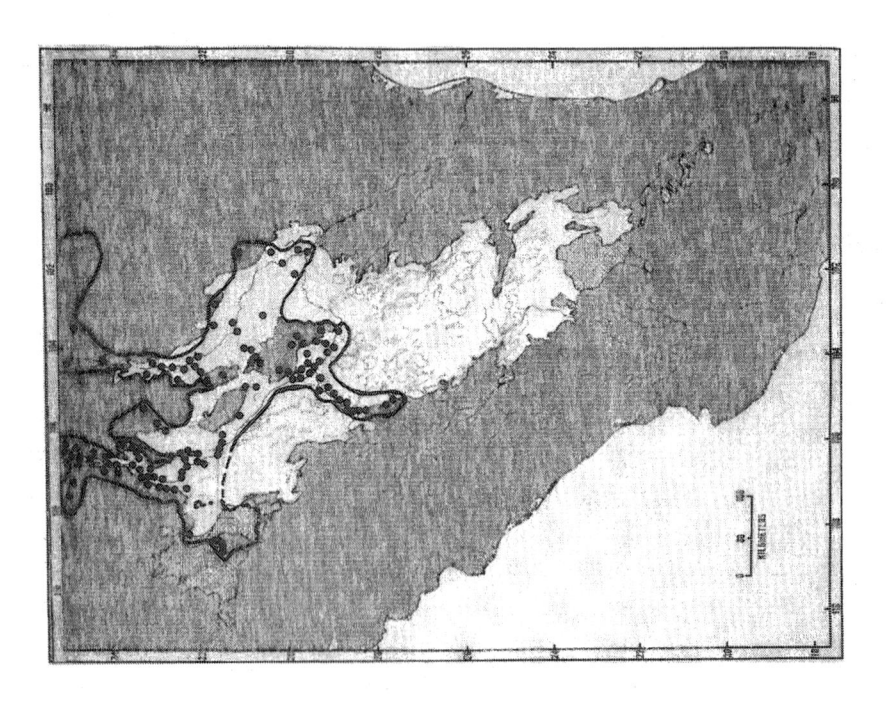

Map 78. *Cnemidophorus 'tesselatus'*

Map 77. ● *Cnemidophorus 'neomexicanus'*; ○ *Cnemidophorus 'velox'*.

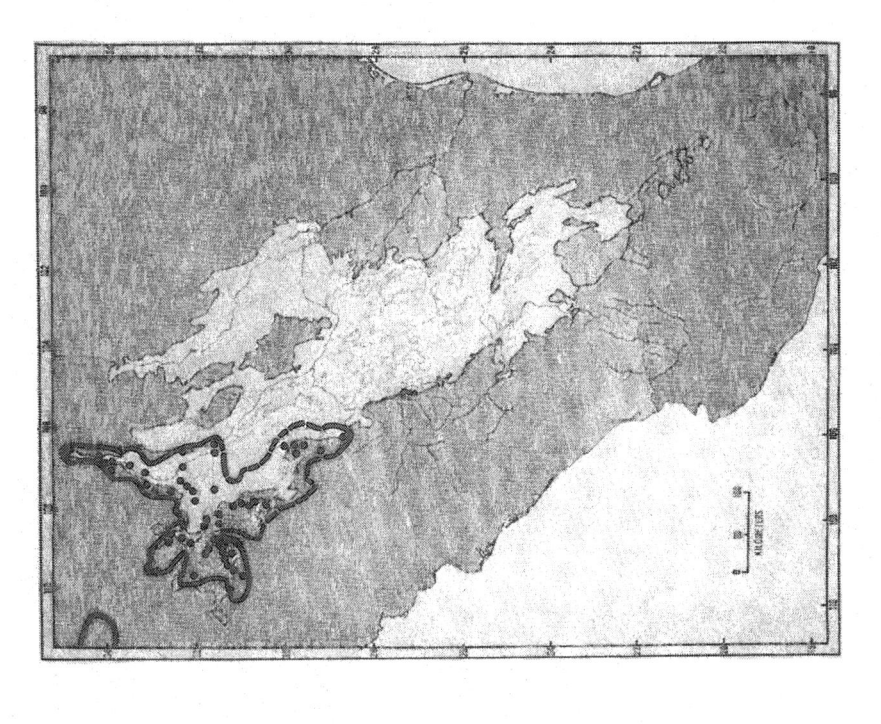

Map 80.   Cnemidophorus 'uniparens'.

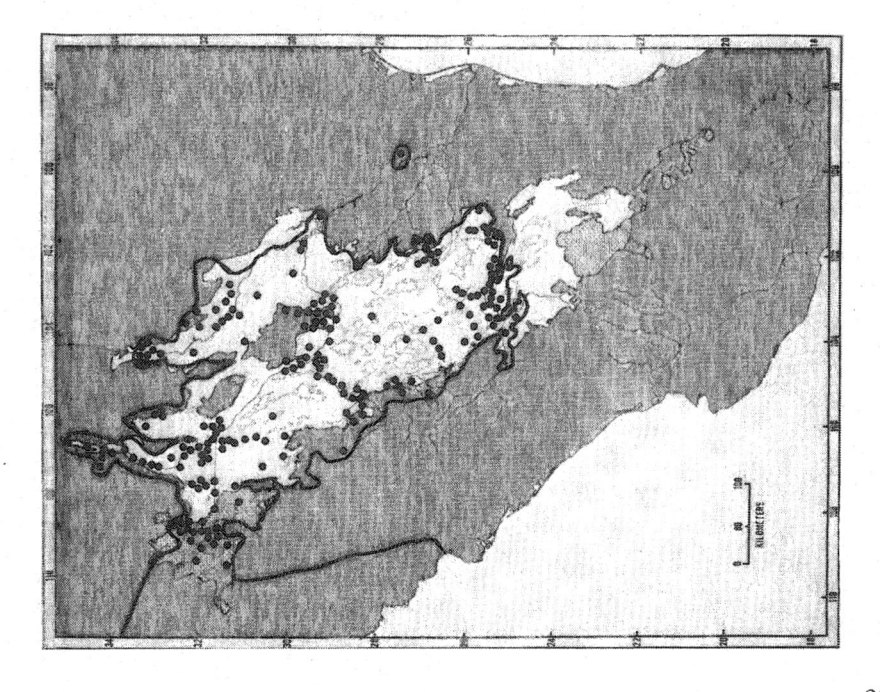

Map 79.   Cnemidophorus tigris.

281

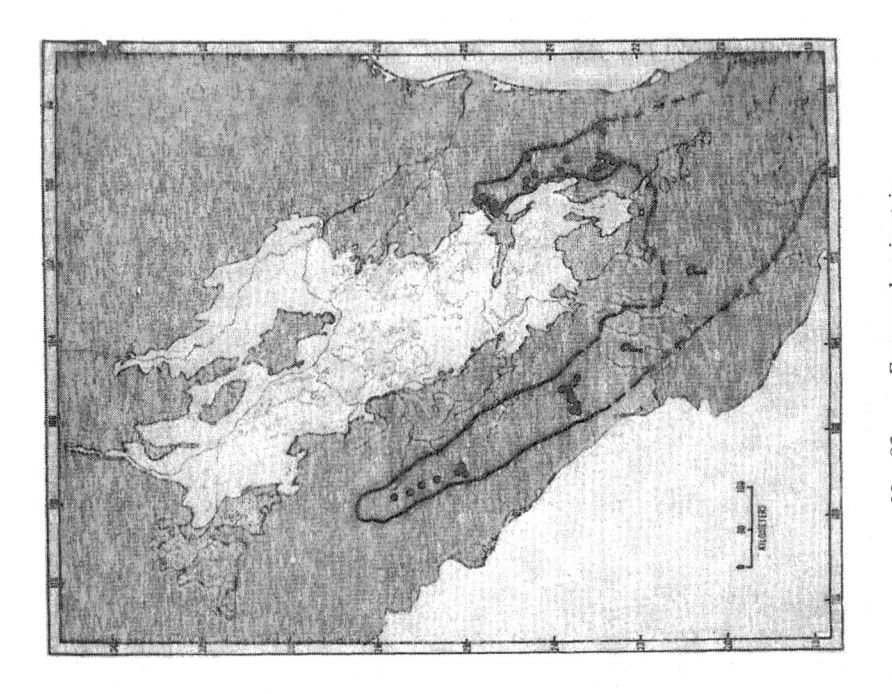

Map 82.   *Eumeces brevirostris.*

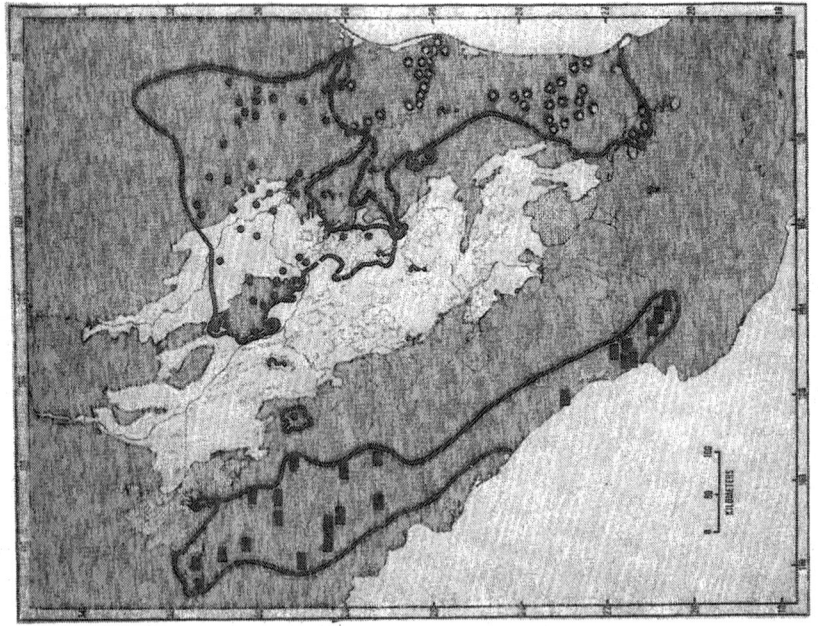

Map 81.   ● *Eumeces brevilineatus;* ○ *Eumeces tetragrammus;*
■ *Eumeces callicephalus.*

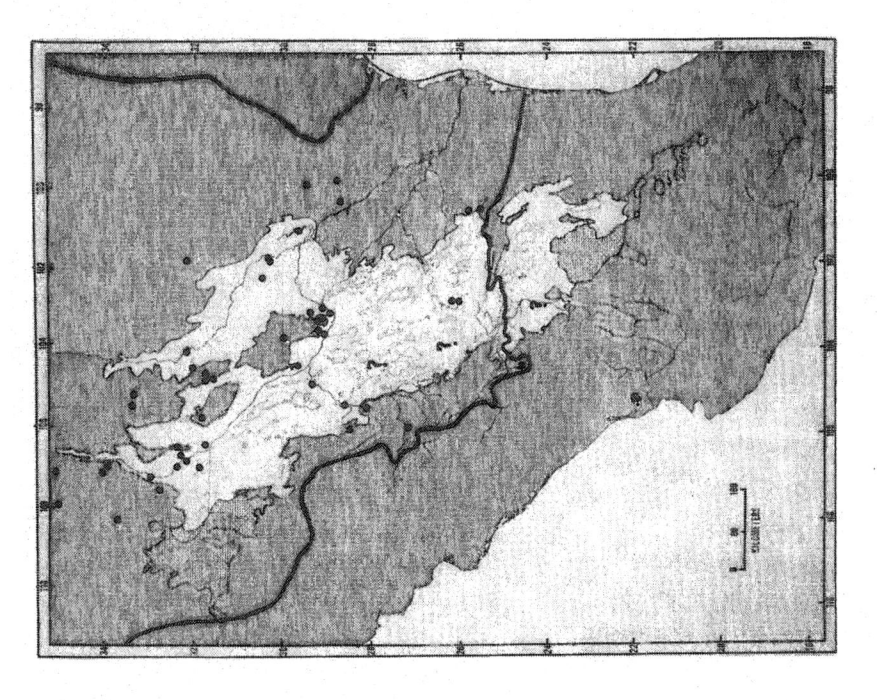

Map 84. *Eumeces obsoletus.*

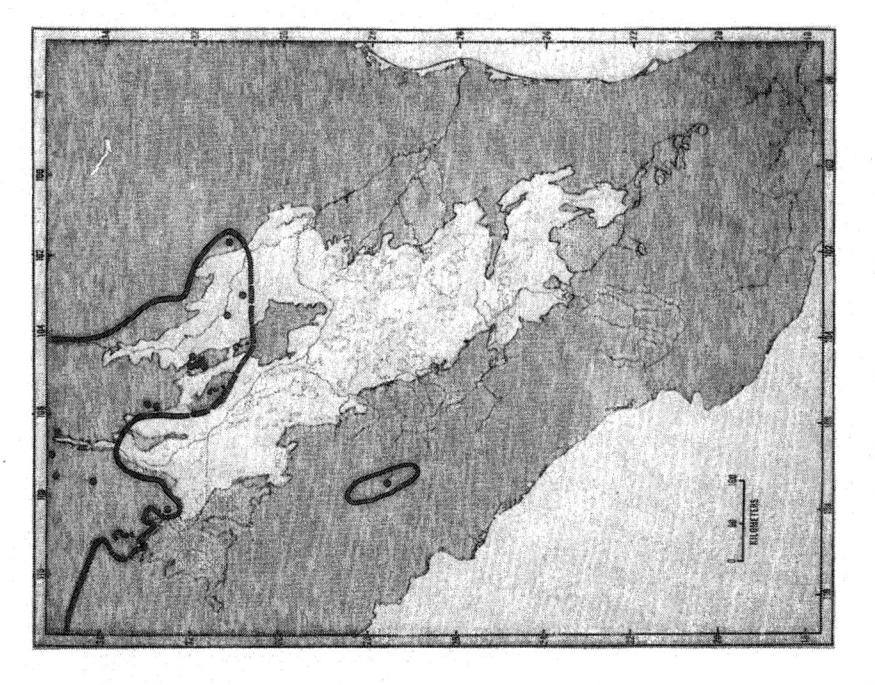

Map 83. *Eumeces multivirgatus.*

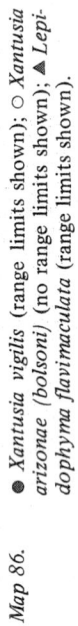

Map 86. ● *Xantusia vigilis* (range limits shown); ○ *Xantusia arizonae* (*bolsoni*) (no range limits shown); ▲ *Lepidophyma flavimaculata* (range limits shown).

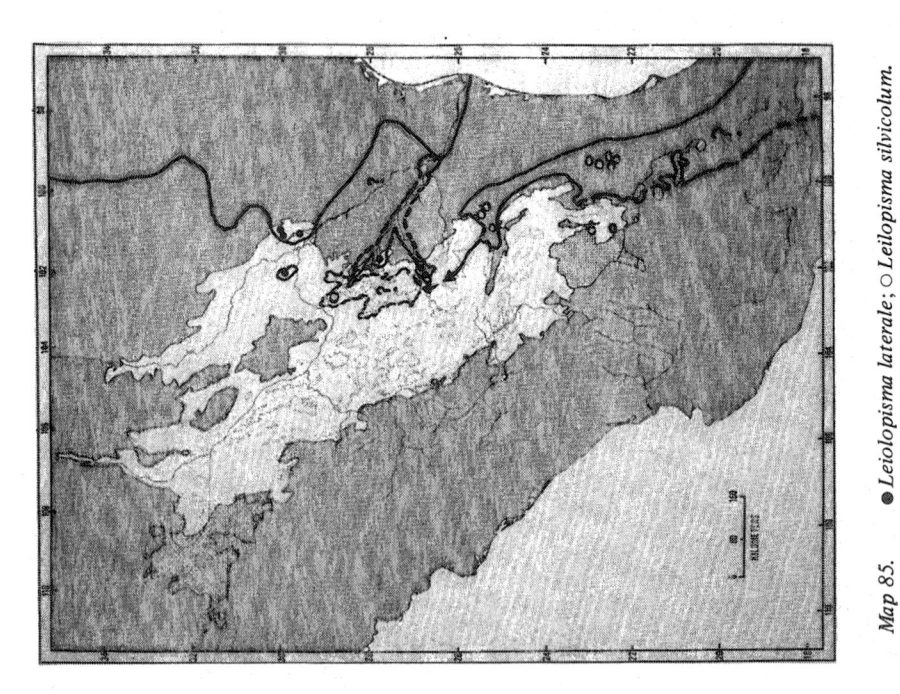

Map 85. ● *Leiolopisma laterale*; ○ *Leilopisma silvicolum*.

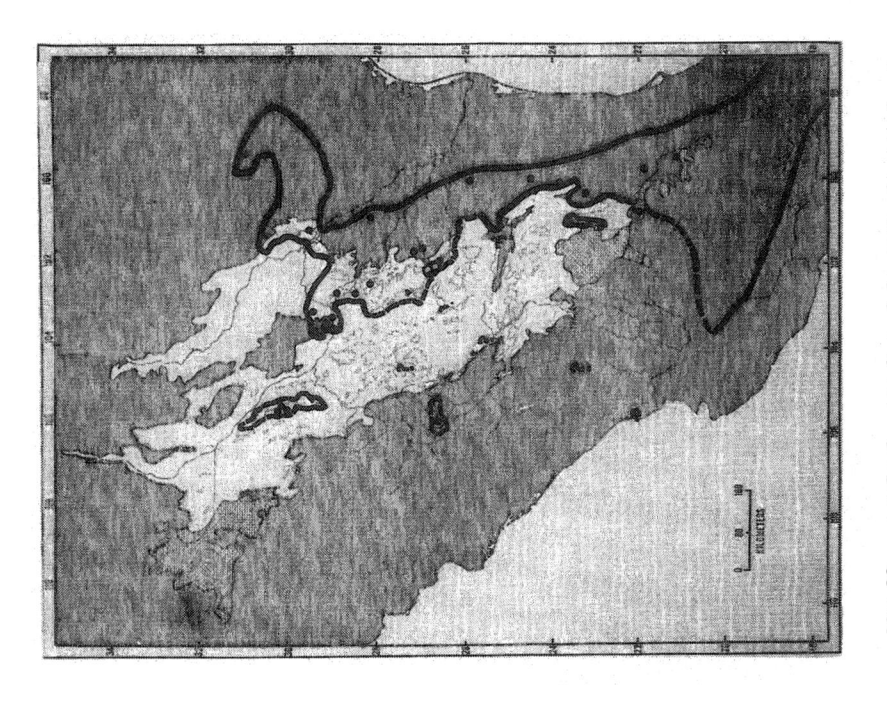

Map 88. ● *Gerrhonotus liocephalus.* ○ *Gerrhonotus lugoi.*

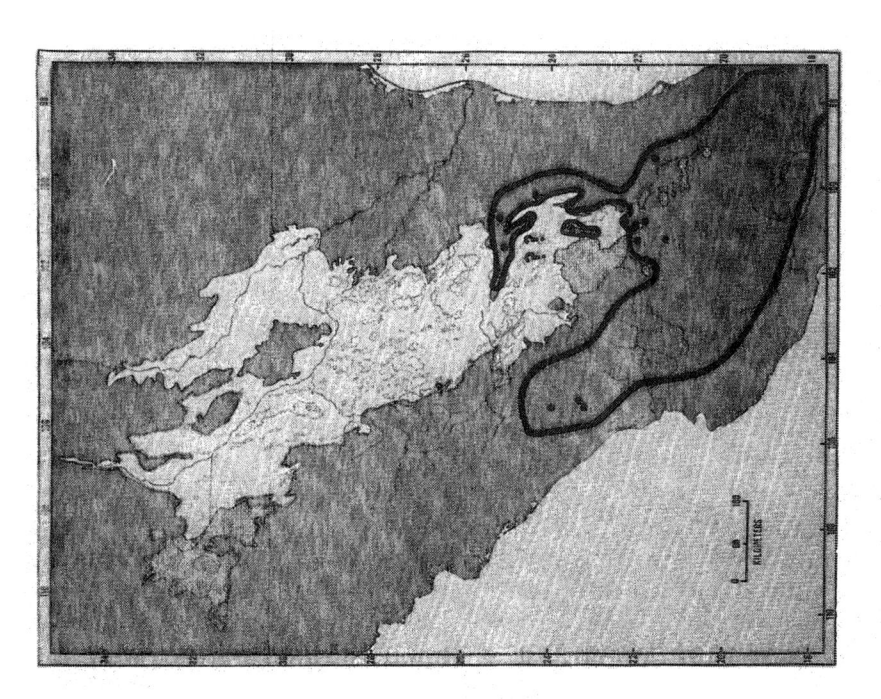

Map 87. *Gerrhonotus imbricatus.*

285

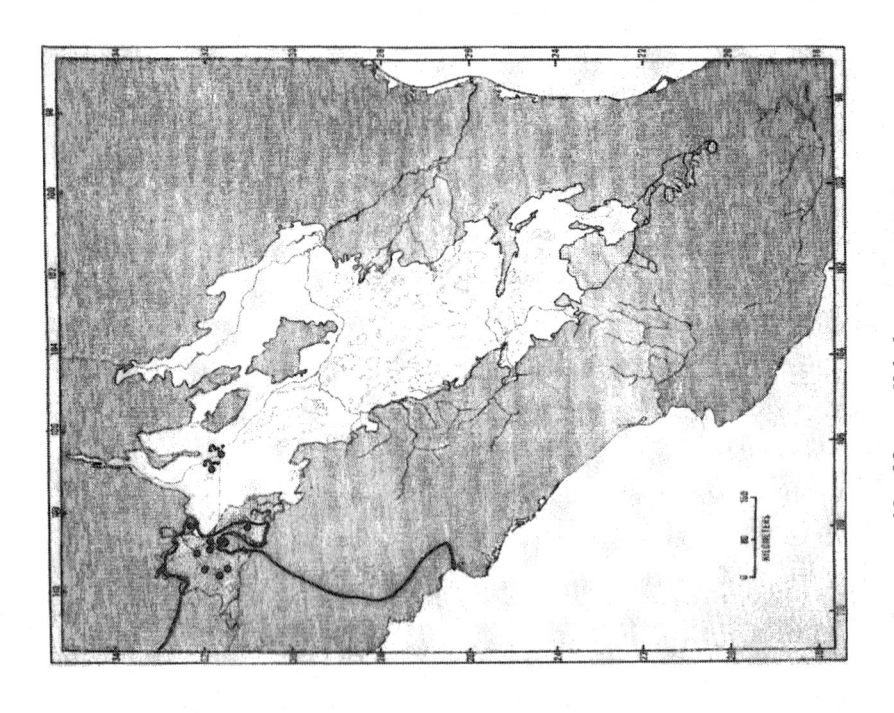

Map 90.   *Heloderma suspectum.*

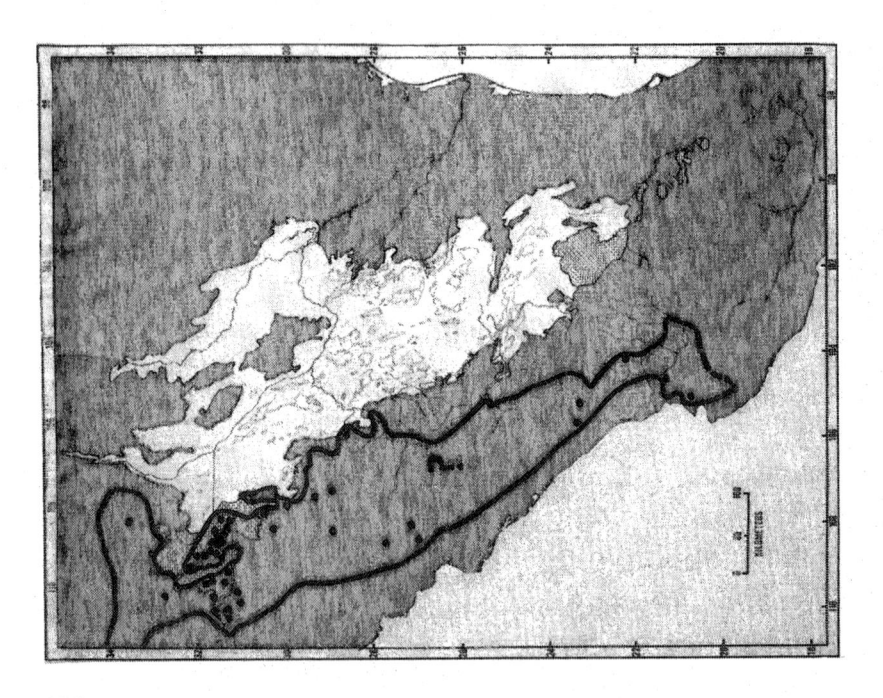

Map 89.   *Gerrhonotus kingi.*

286

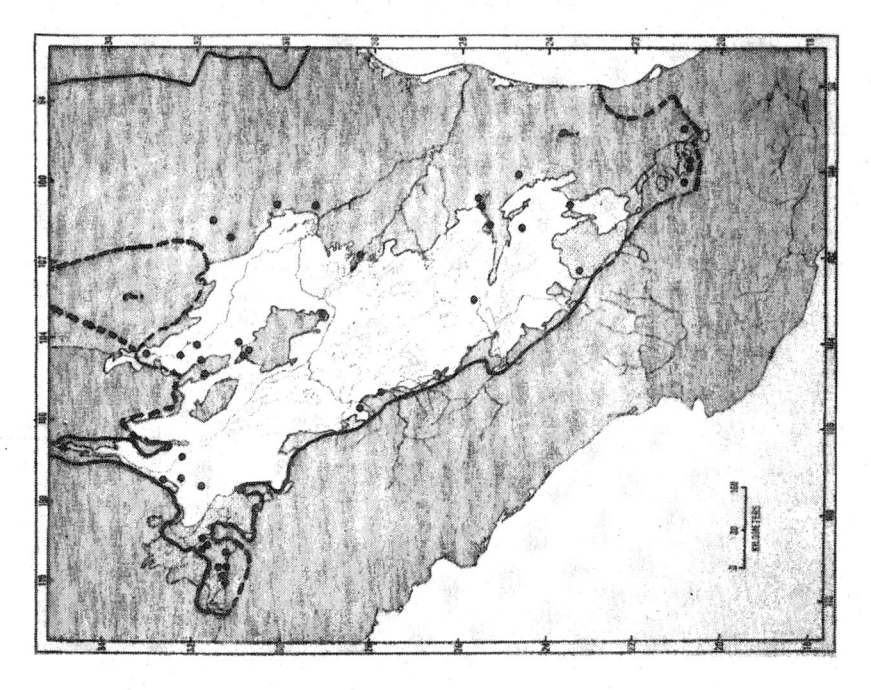

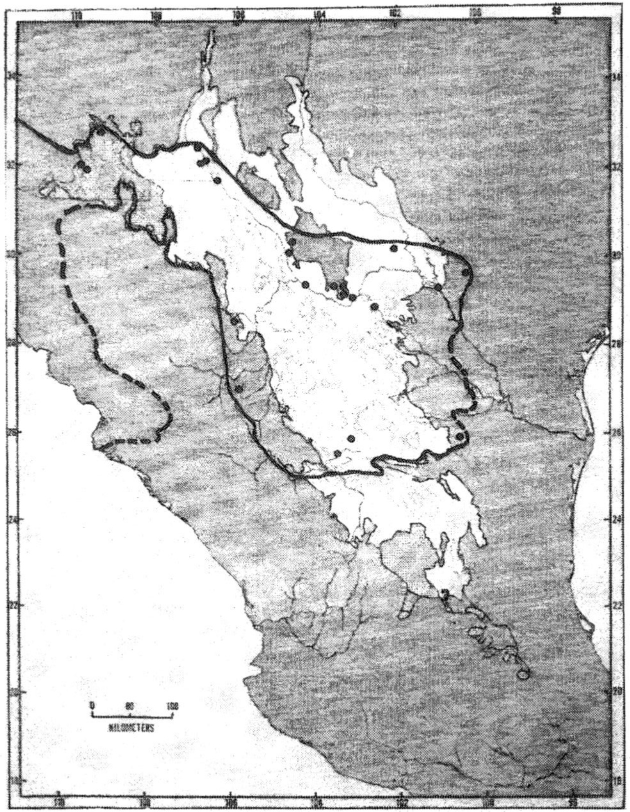

Map 92. *Leptotyphlops humilis.*

Map 91. *Leptotyphlops dulcis.*

287

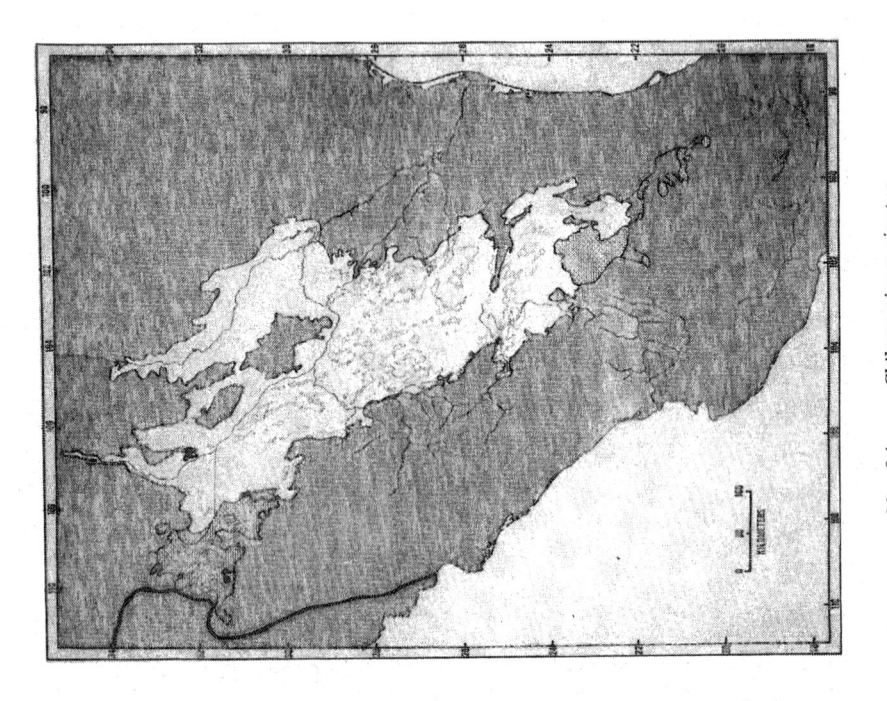

Map 94.   *Chilomeniscus cinctus.*

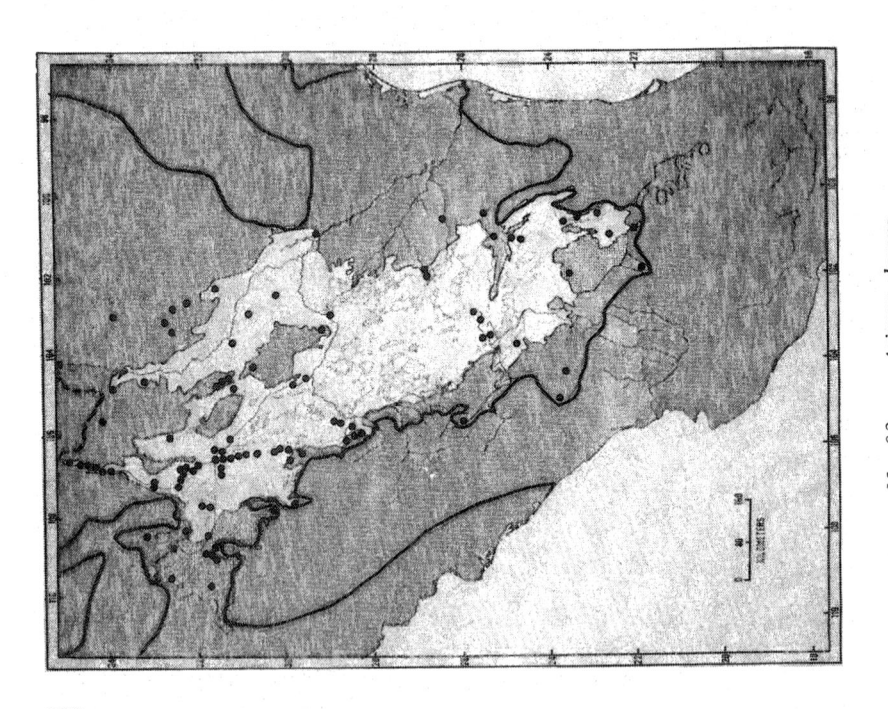

Map 93.   *Arizona elegans.*

288

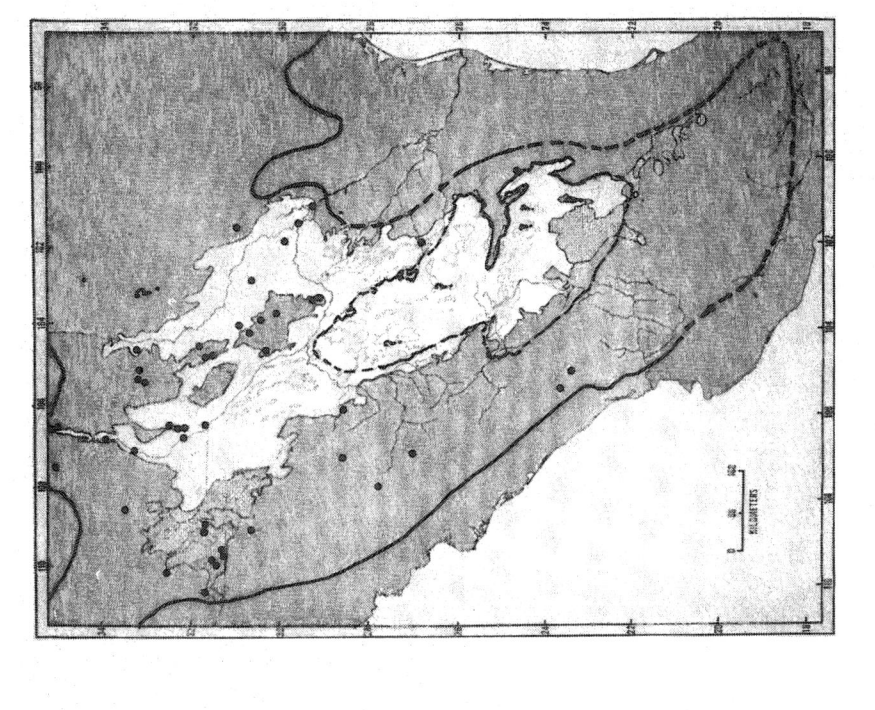

Map 96.   *Diadophis puctatus.*

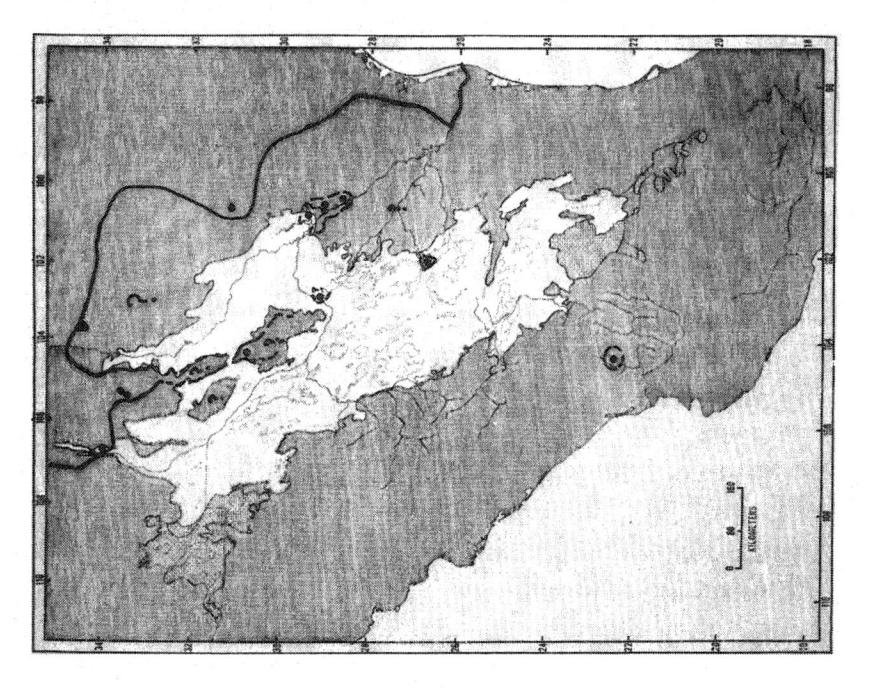

Map 95.   *Coluber constrictor.*

289

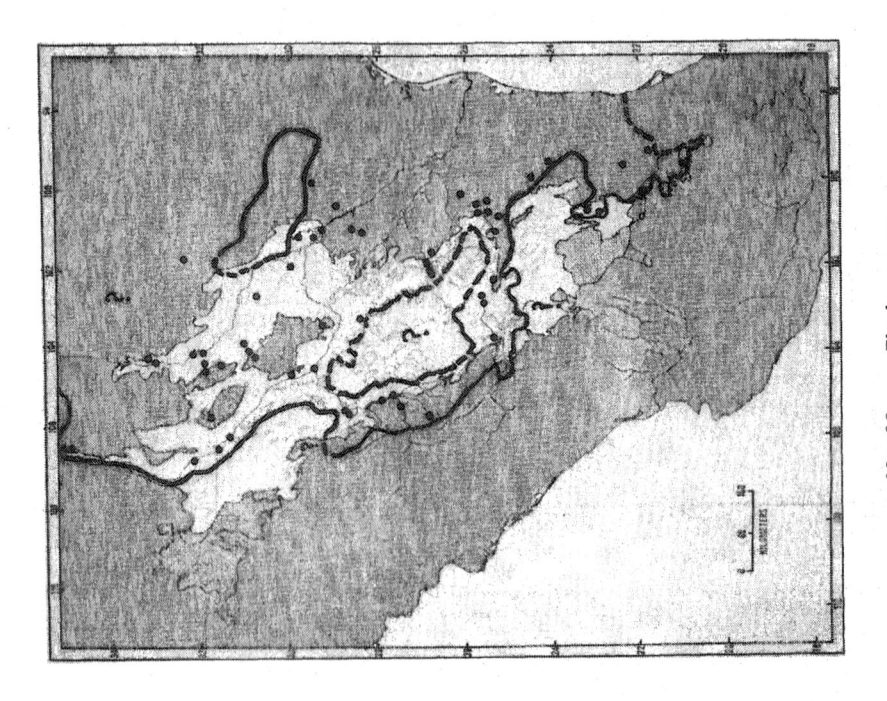

Map 98.   *Elaphe guttata.*

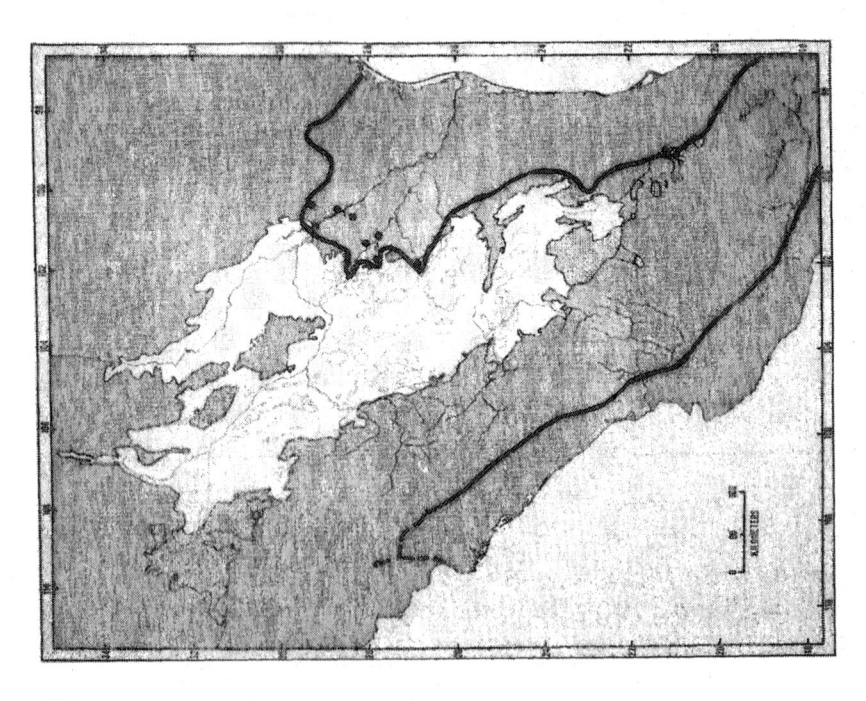

Map 97.   *Drymarchon corias.*

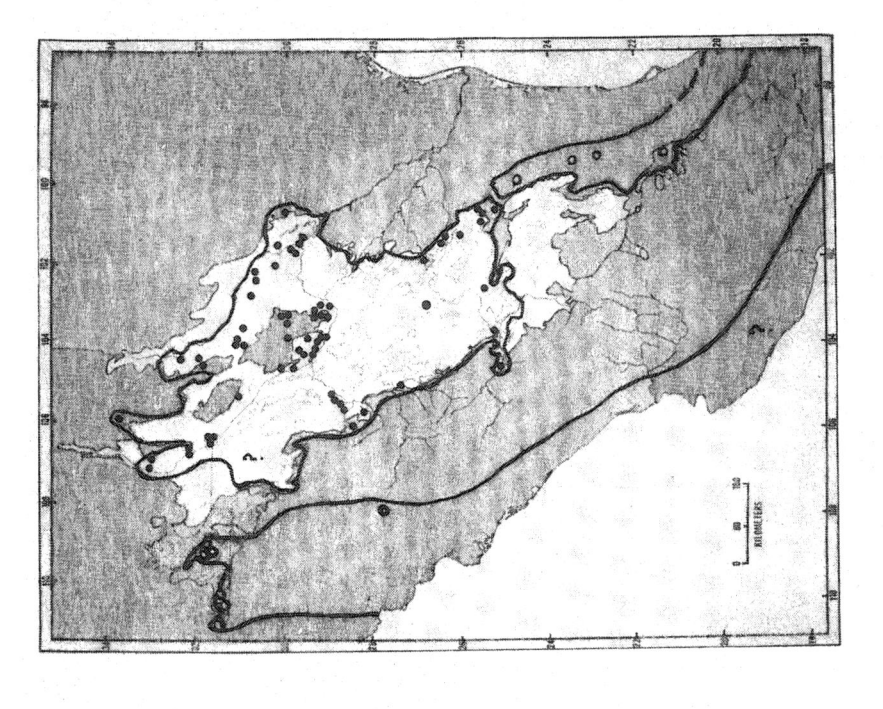

Map 100. ● *Elaphe subocularis;* ○ *Elaphe triaspis.*

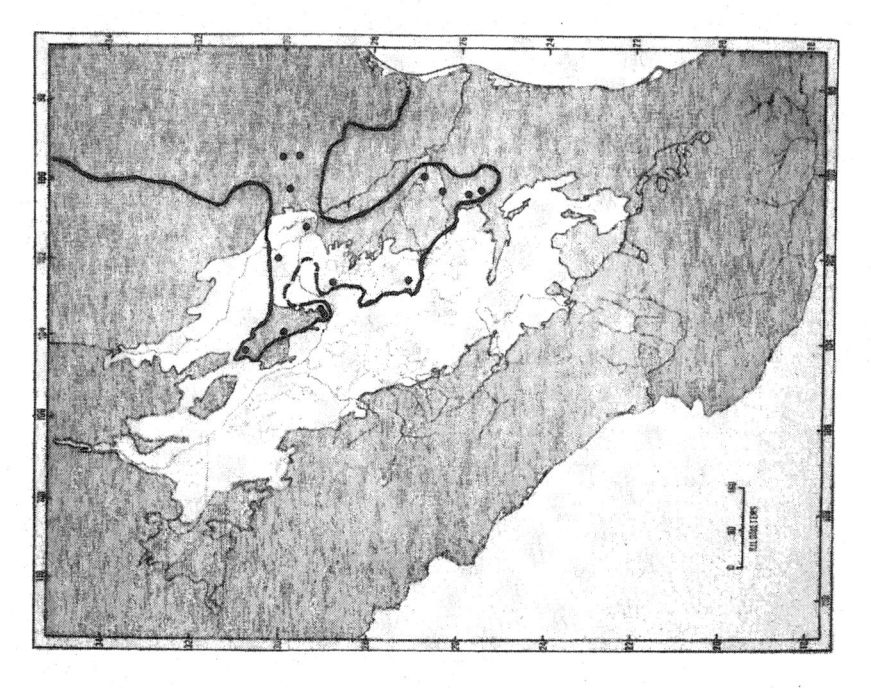

Map 99. *Elaphe obsoleta.*

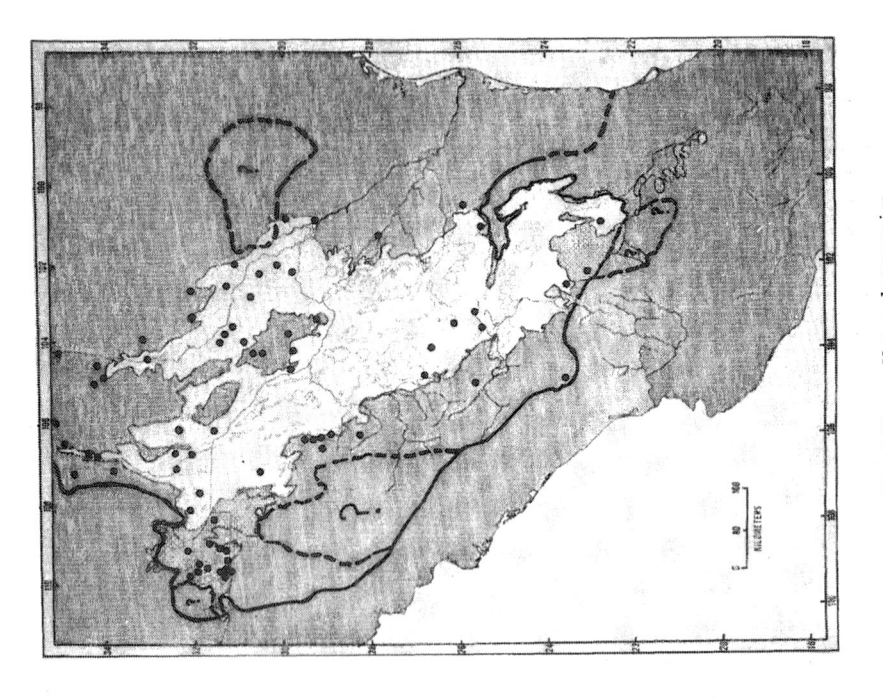

Map 102.   *Heterodon nascius.*

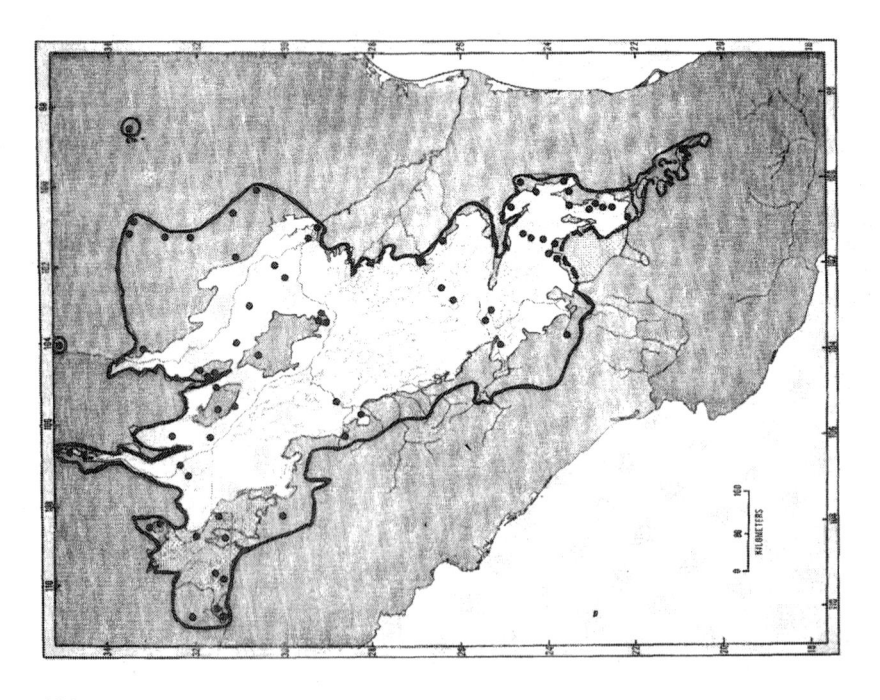

Map 101.   *Ficimia cana.*

292

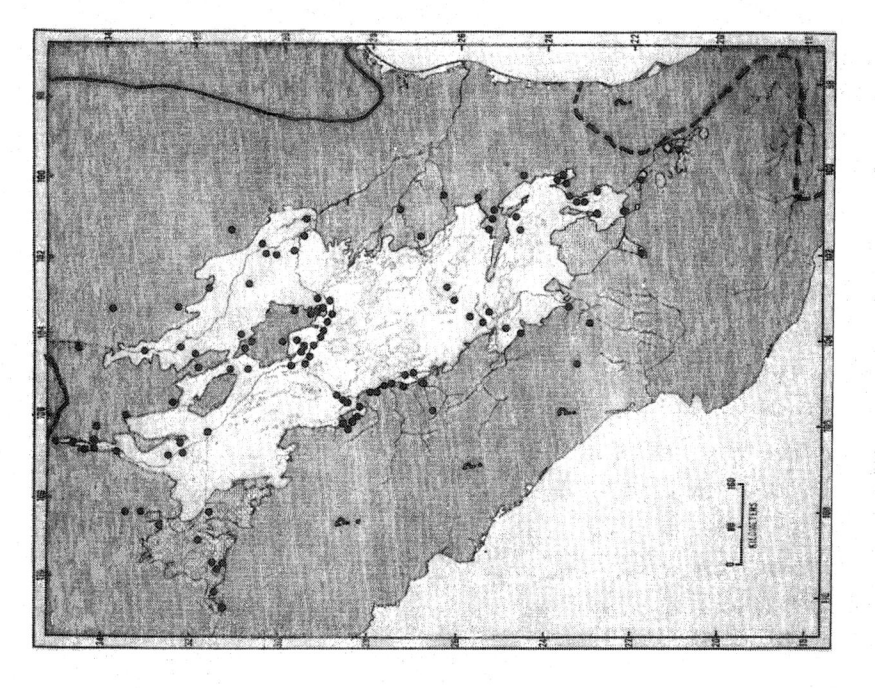

Map 103. ● *Hypsiglena torquata*; ○ *Hypsiglena tanzeri.*

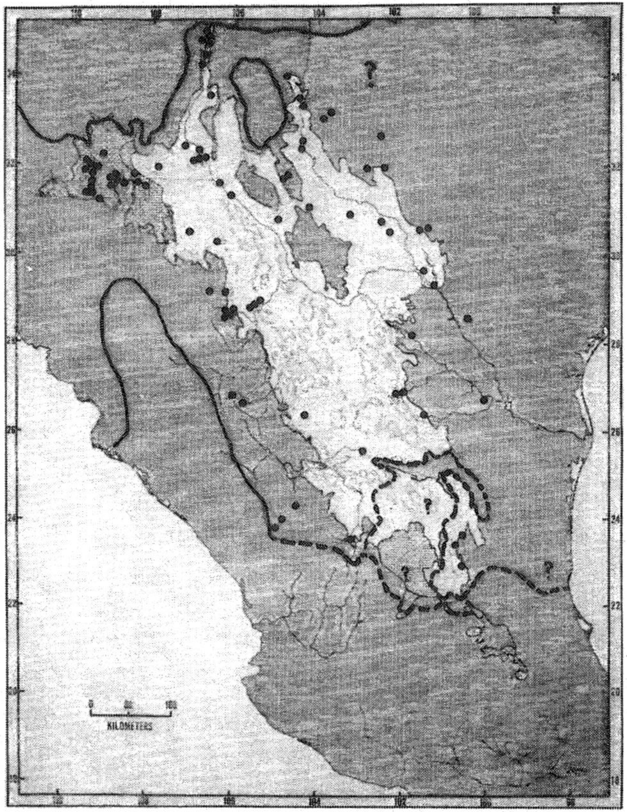

Map 104. *Lampropeltis getulus.*

293

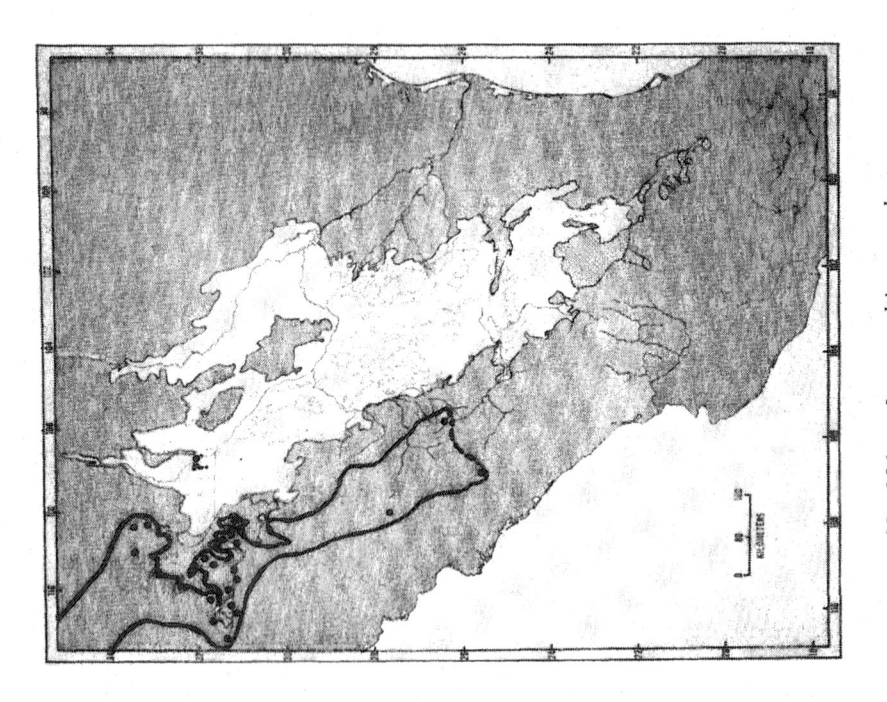

Map 106.   *Lampropeltis pyromelana.*

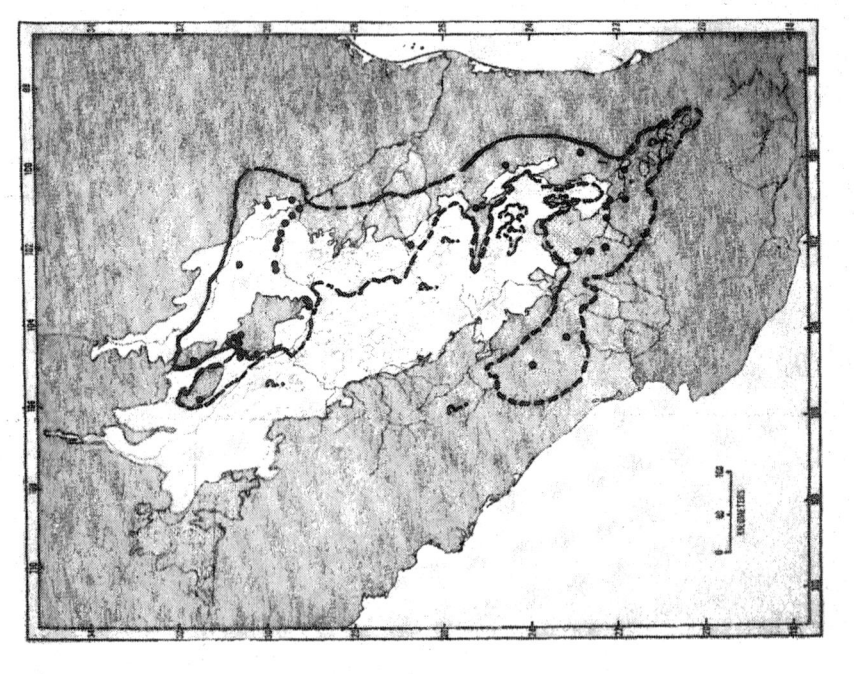

Map 105.   *Lampropeltis mexicana.*

294

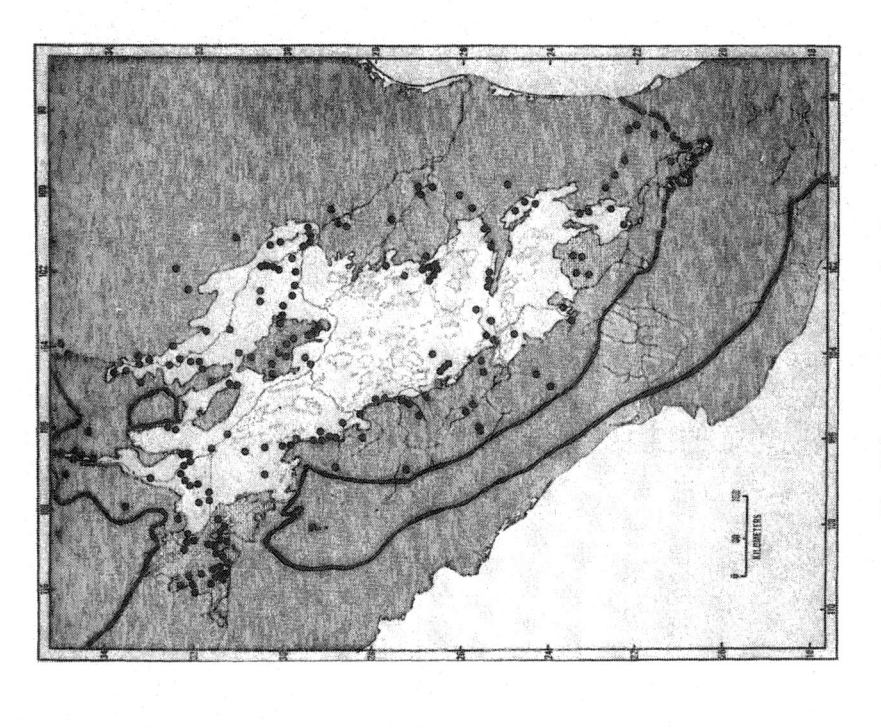

Map 108.   Masticophis flagellum.

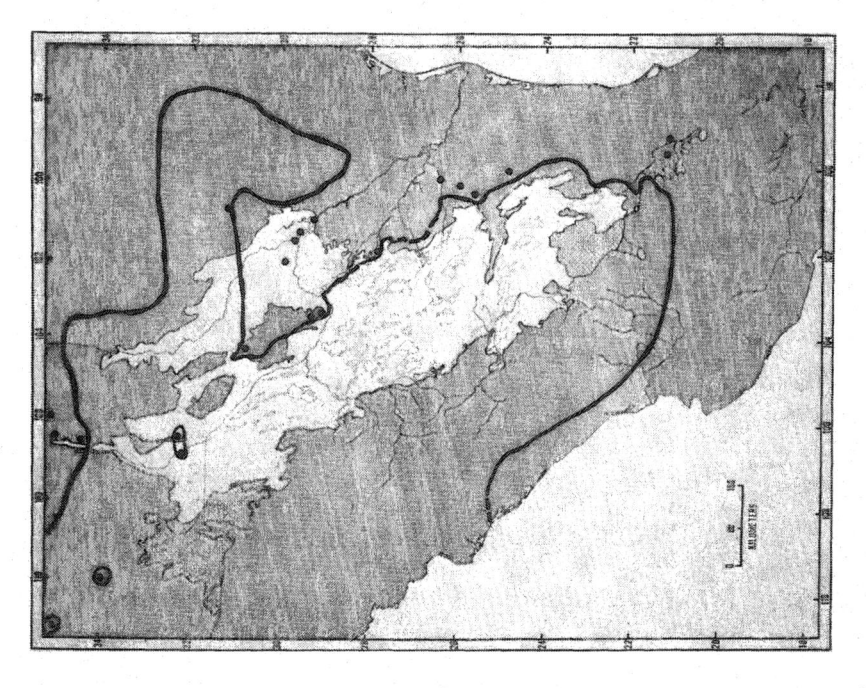

Map 107.   Lampropeltis triangulum.

295

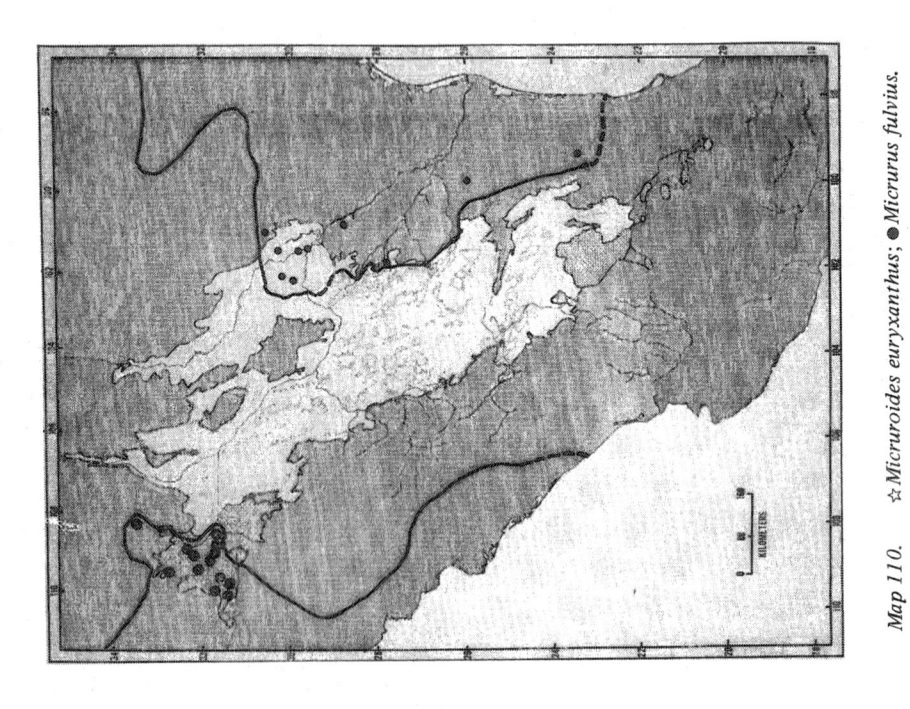

Map 110. ☆ *Micruroides euryxanthus;* ● *Micrurus fulvius.*

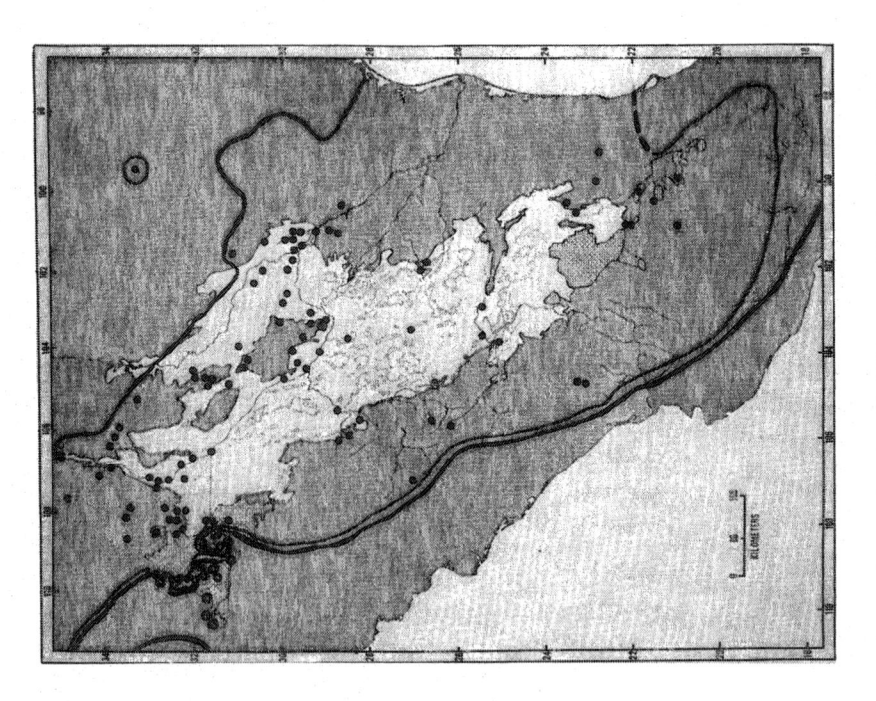

Map 109. ☆ *Masticophis bilineatus;* ● *Masticophis taeniatus.*

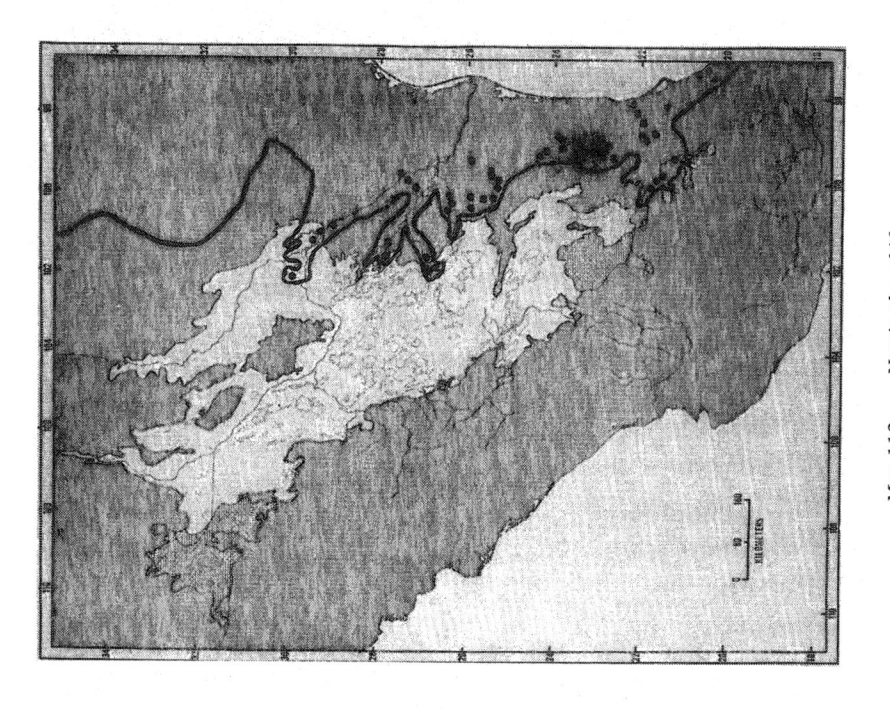

Map 112.   Natrix rhombifera.

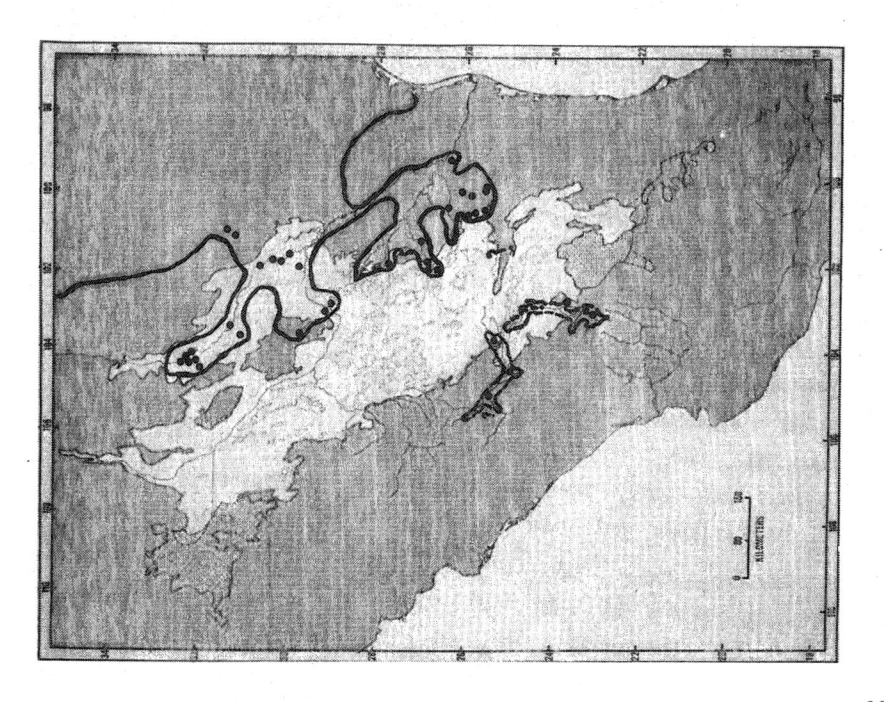

Map 111.   Natrix erythrogaster.

297

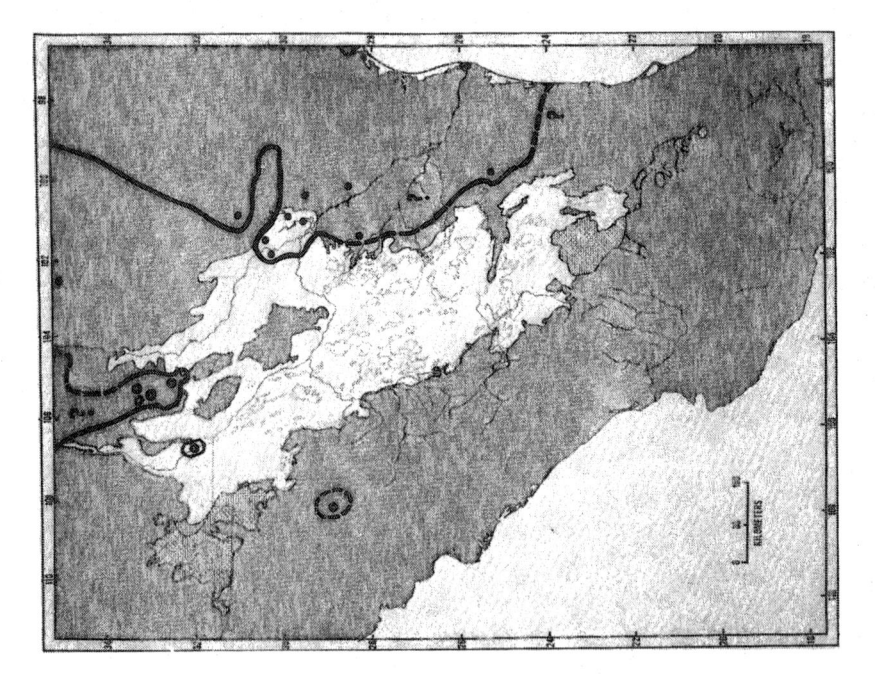

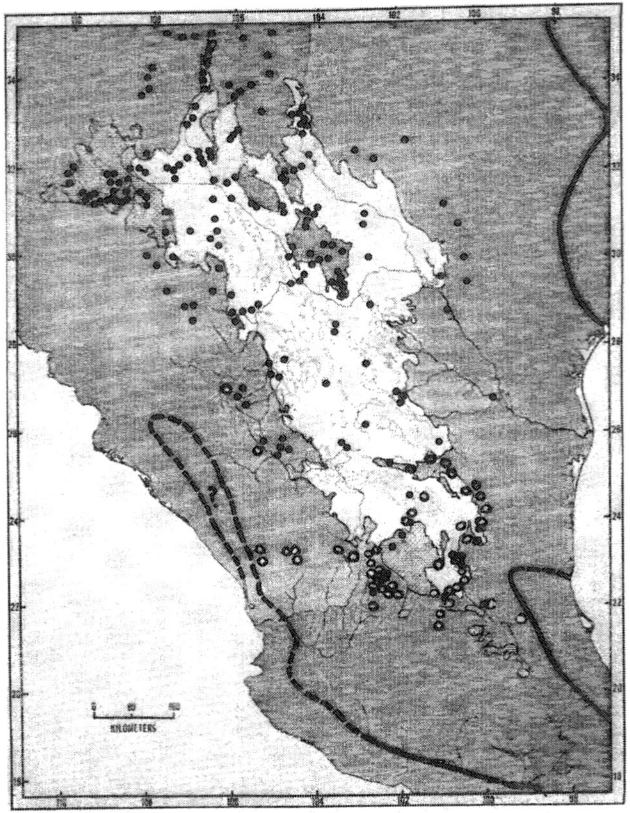

Map 114. ● *Pituophis melanoleucus;* ○ *(P. deppei have been considered as conspecific with P. melanoleucus)* ● *integrades.*

Map 113. ● *Opheodrys aestivus;* ○ *Opheodrys vernalis.*

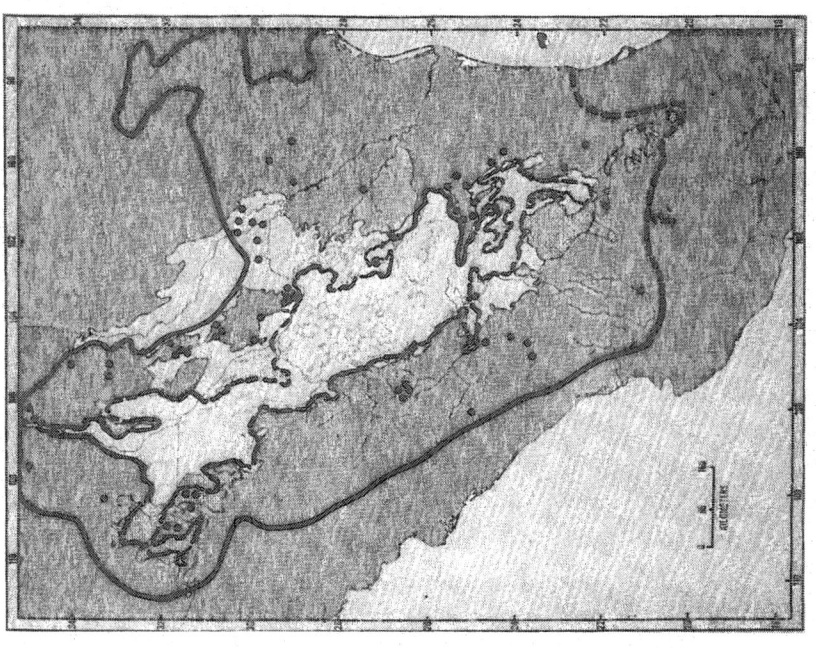

Map 116. *Salvadora grahamiae.*

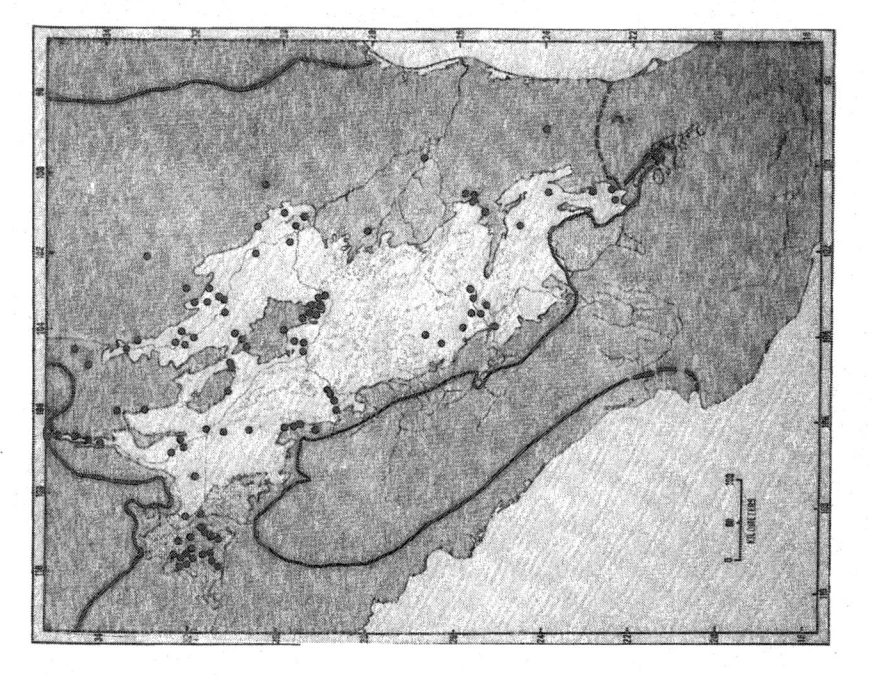

Map 115. *Rhinocheilus lecontei.*

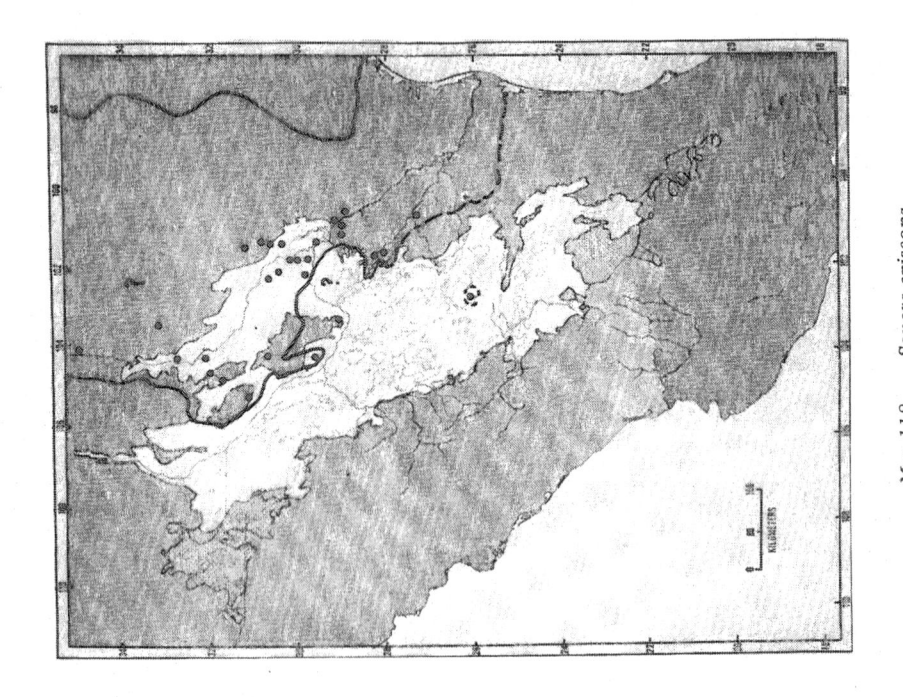

Map 118.   *Sonova episcopa.*

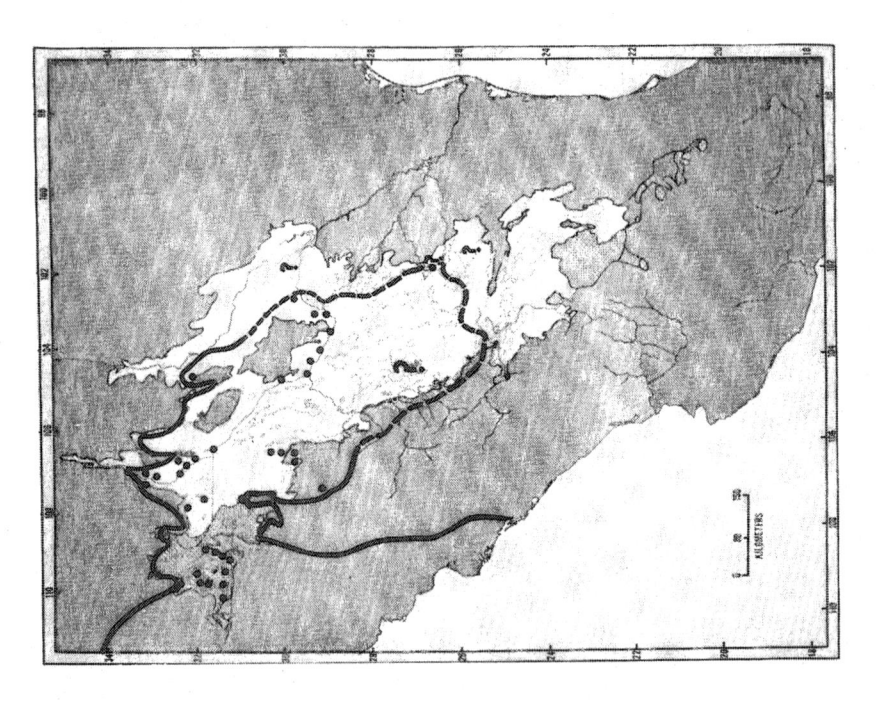

Map 117.   *Salvadora hexalepis.*

300

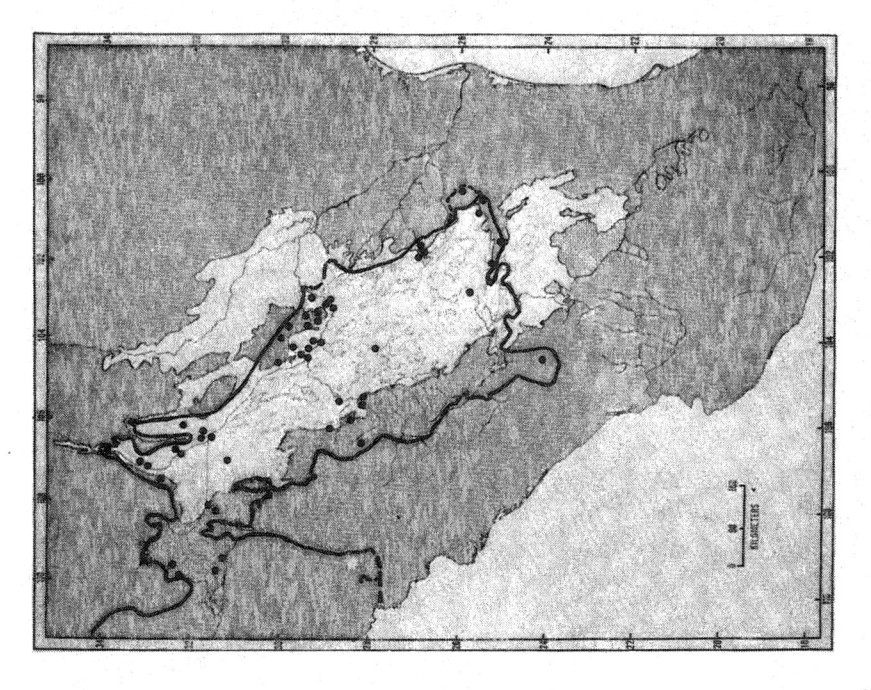

Map 119.   ● *Sonora semiannulata*.

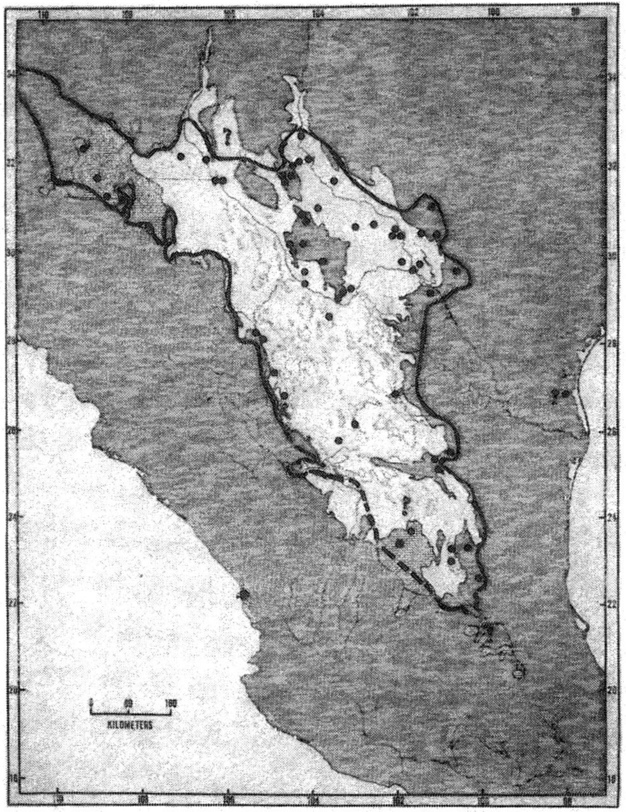

Map 120.   *Tantilla atriceps*.

301

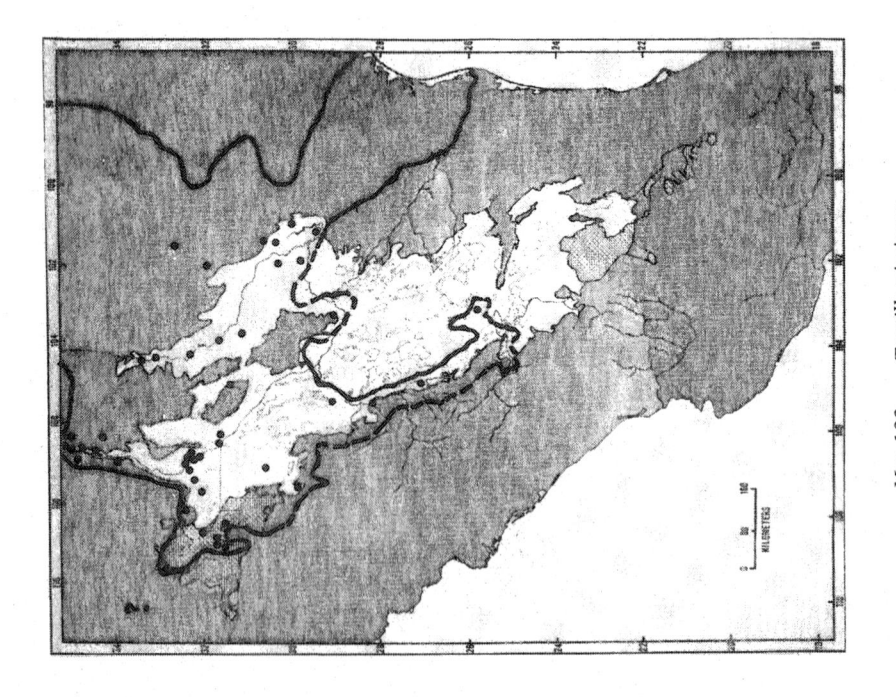

Map 122.   *Tantilla nigriceps.*

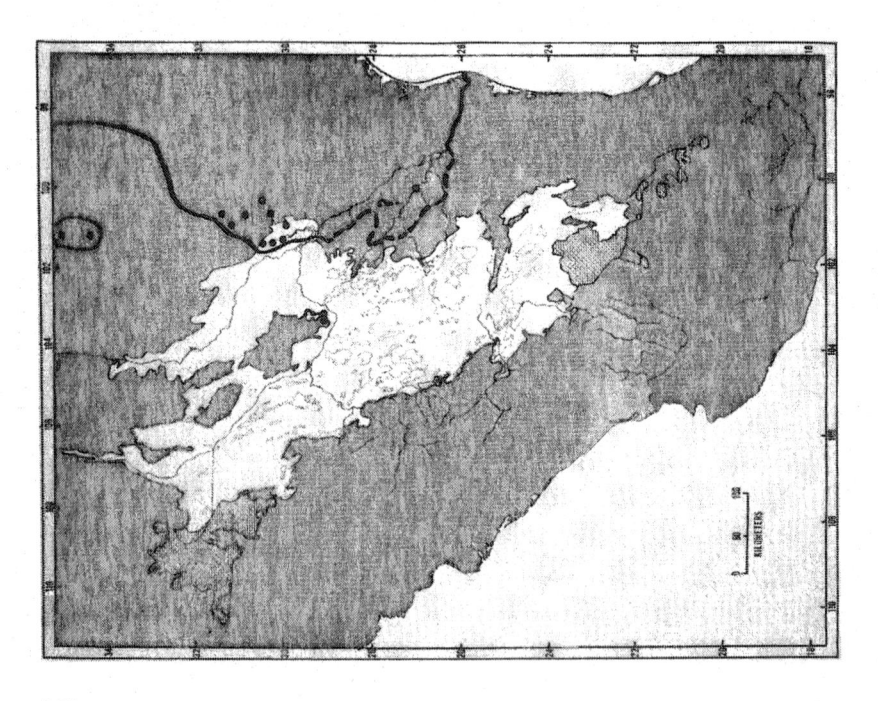

Map 121.   *Tantilla gracilis.*

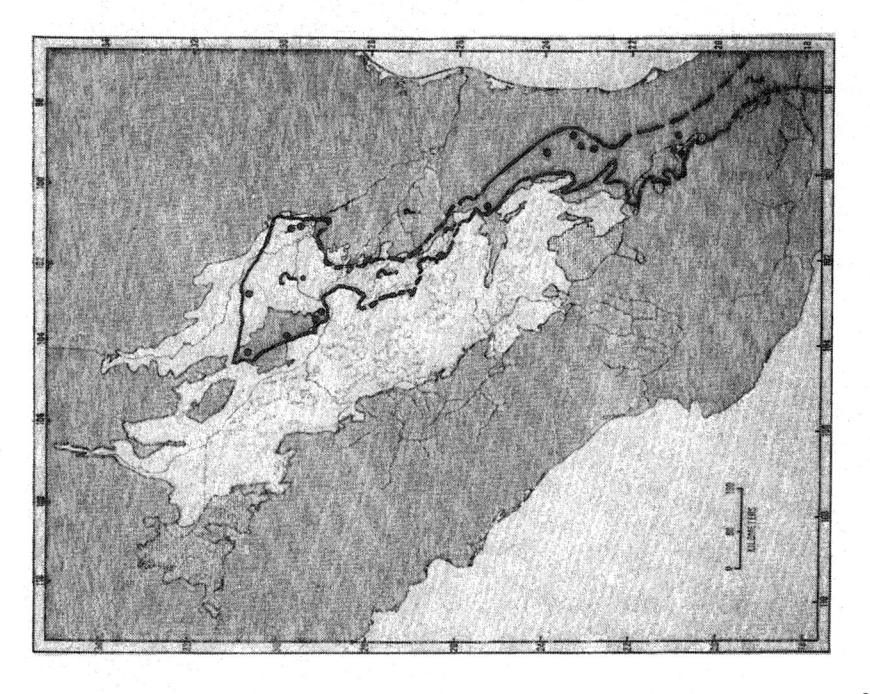

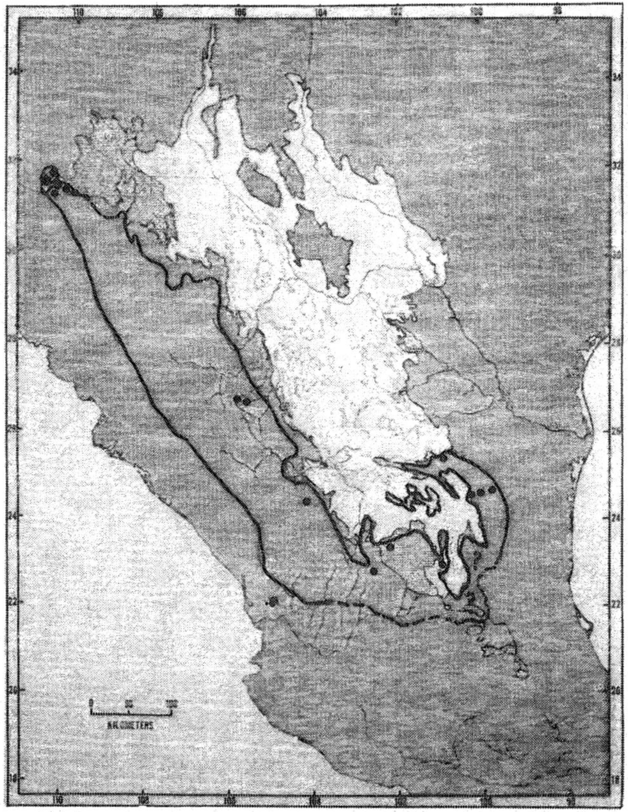

Map 124.   *Tantilla wilcoxi.*

Map 123.   *Tantilla rubra.*

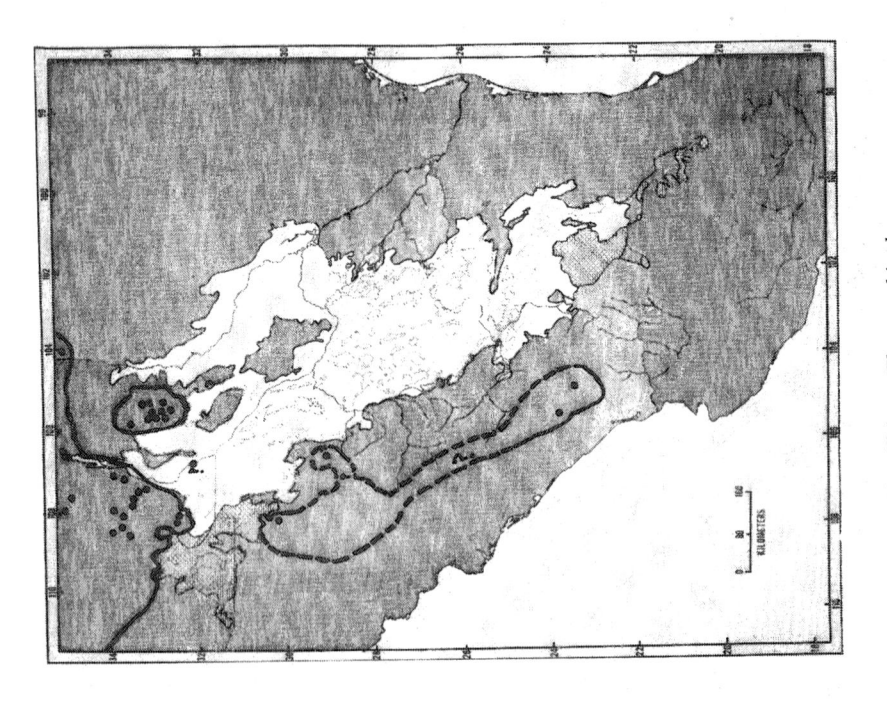

Map 126.   *Thamnophis elegans.*

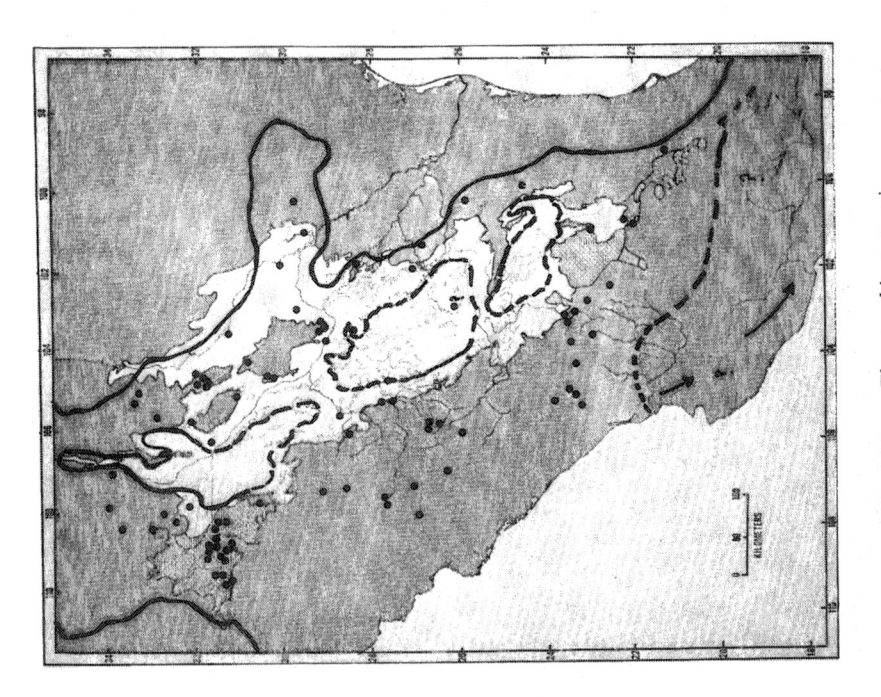

Map 125.   *Thamnophis cyrtopsis.*

304

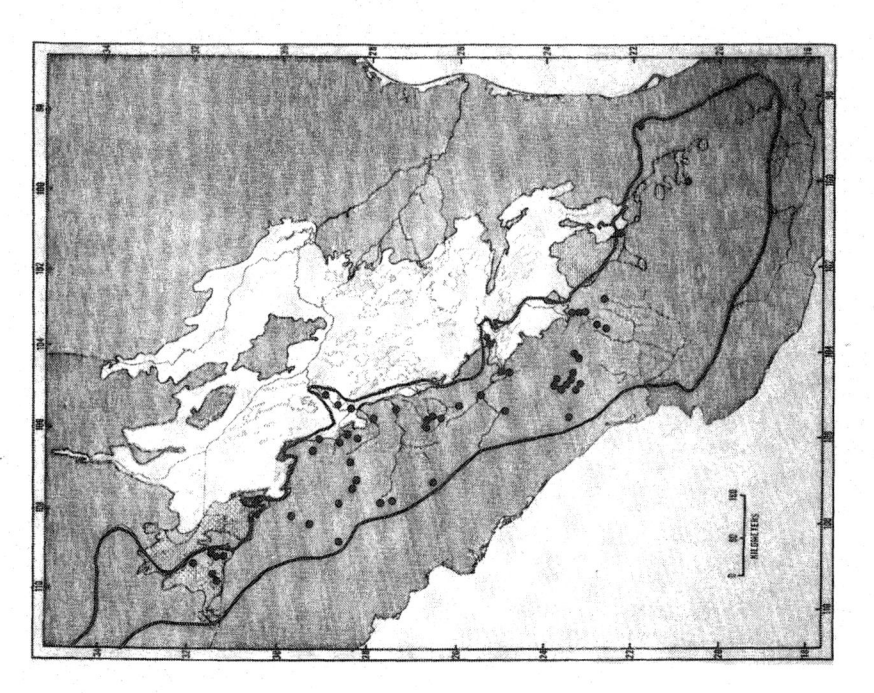

Map 127. *Thamnophis eques.*

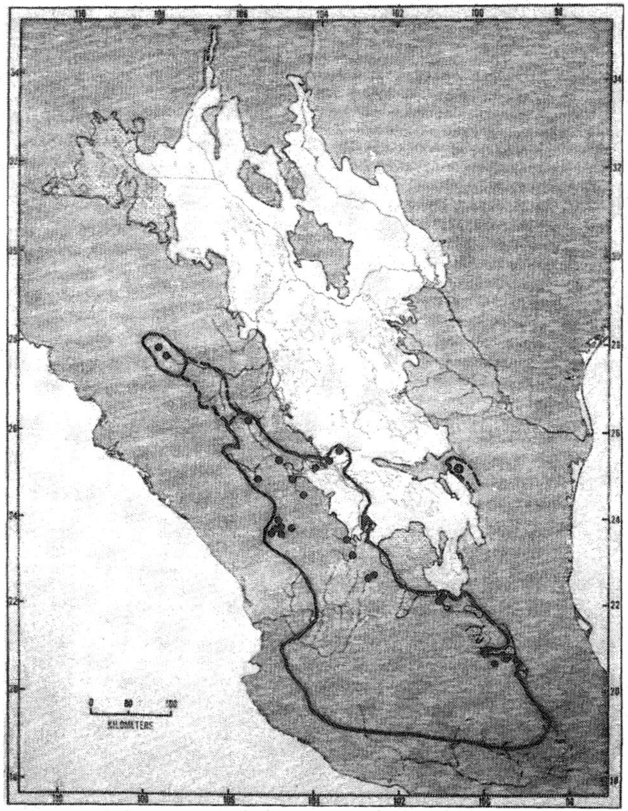

Map 128. ○ *Thamnophis exsul;* ● *Thamnophis melanogaster.*

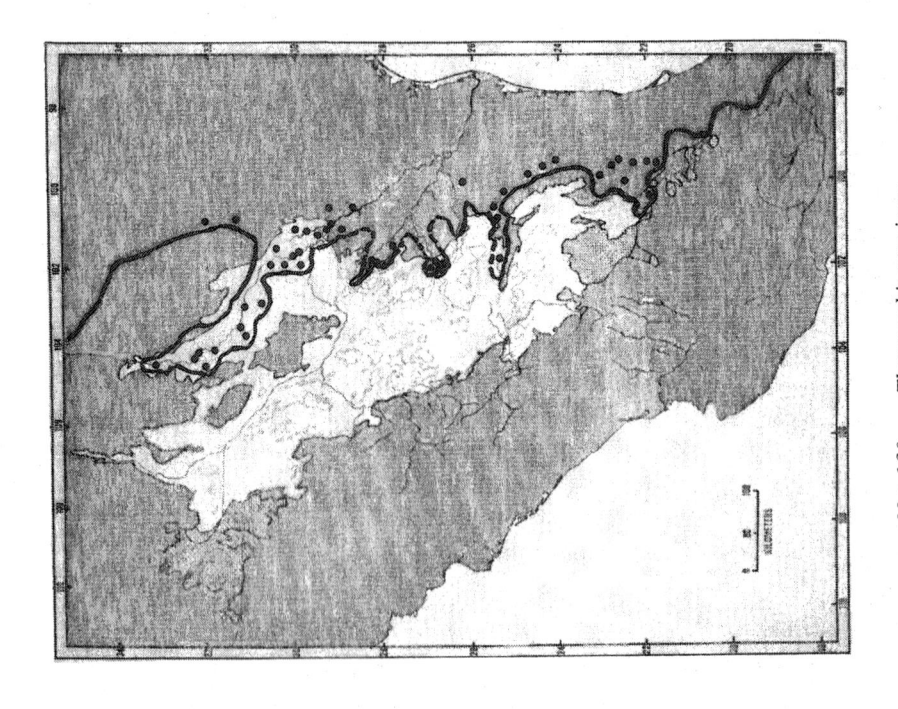

Map 130. *Thamnophis proximus.*

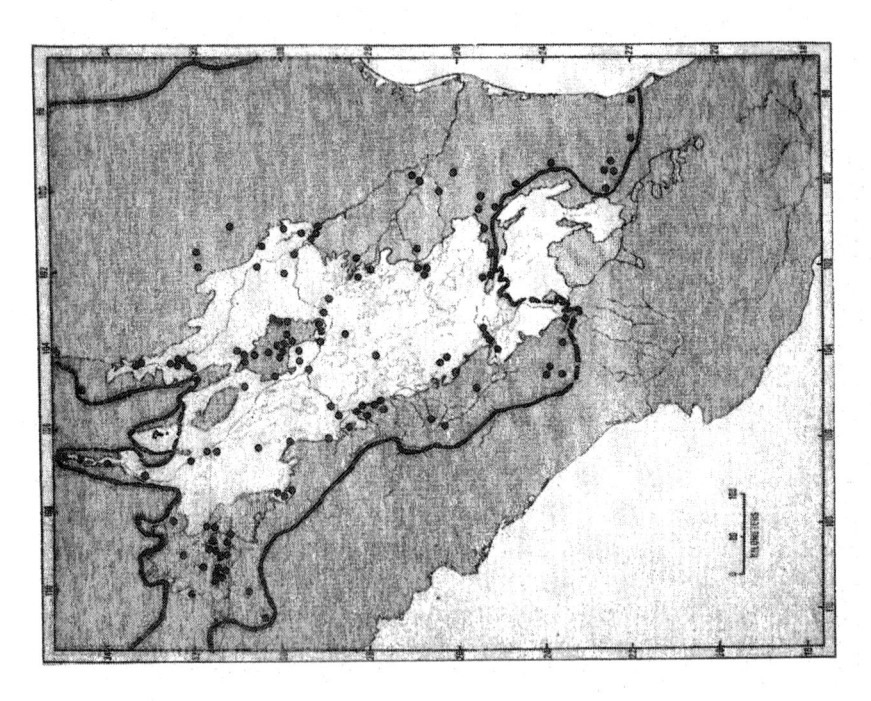

Map 129. *Thamnophis marcianus.*

306

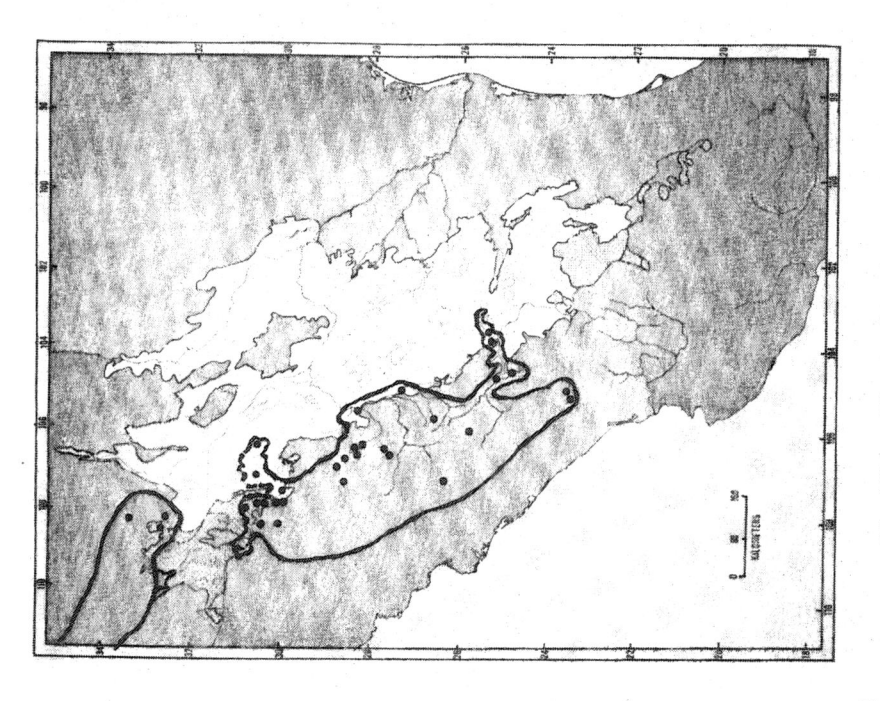

Map 131.   *Thamnophis rufipunctatus.*

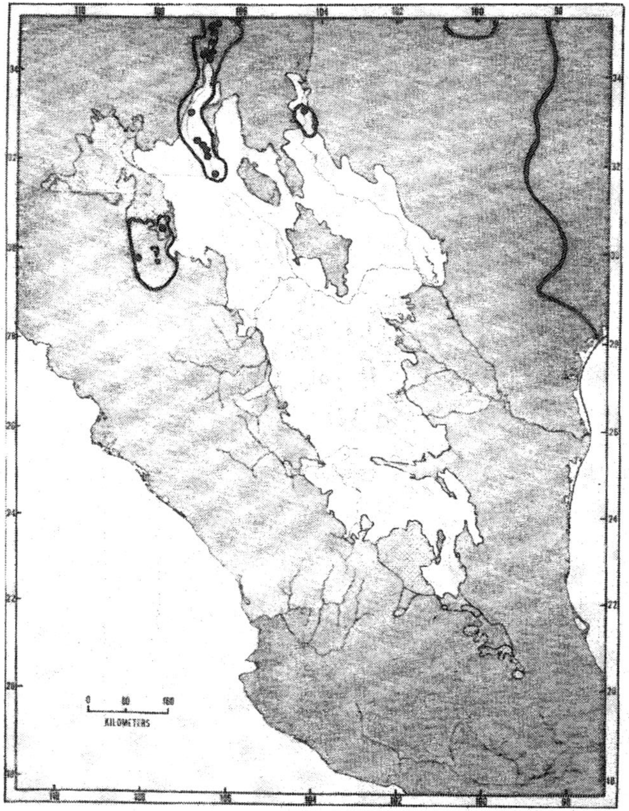

Map 132.   *Thamnophis sirtalis.*

307

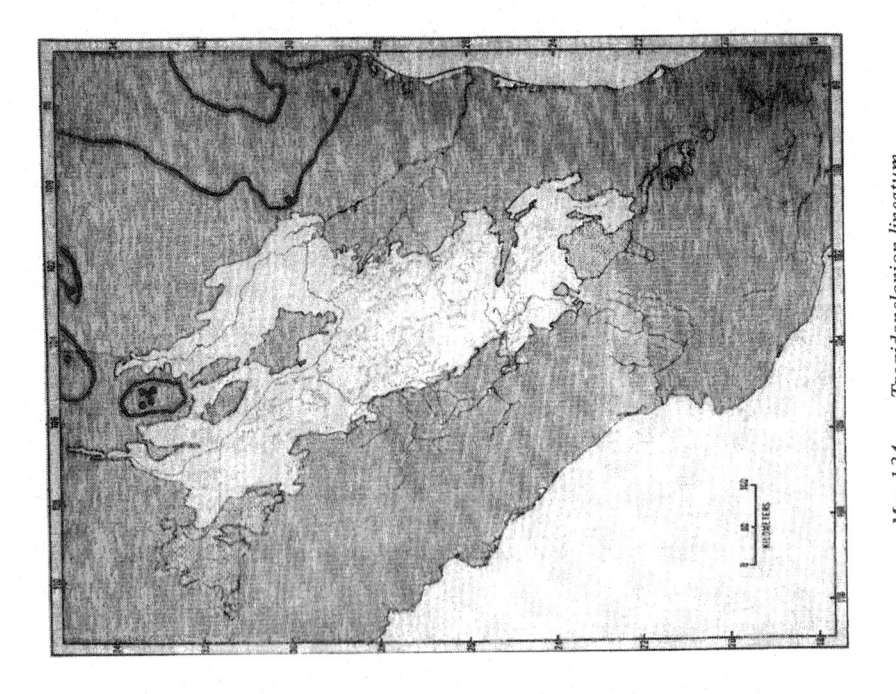

Map 134.   *Tropidonclonion lineatum.*

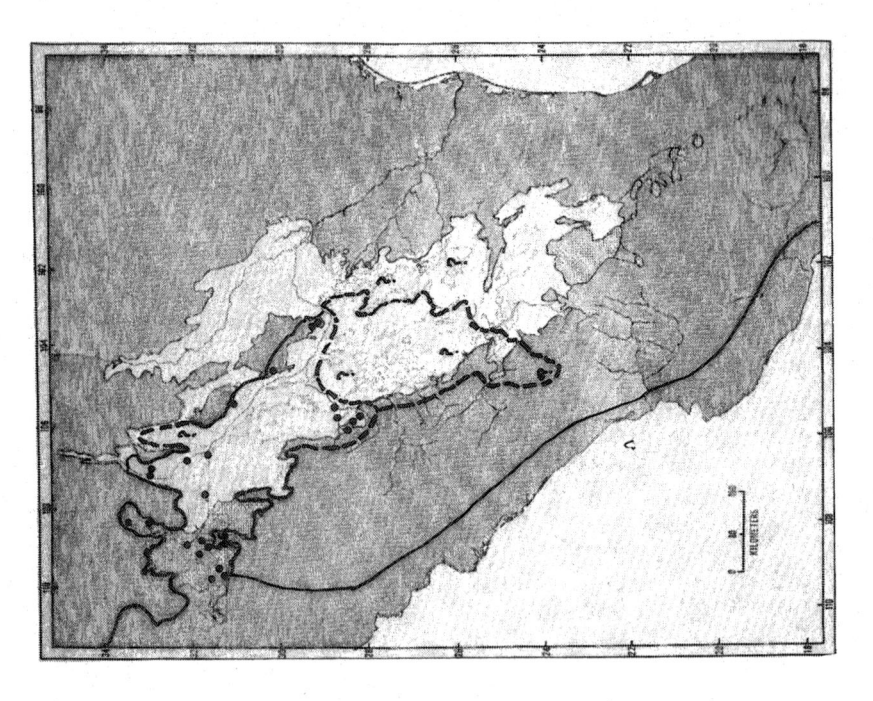

Map 133.   *Trimorphodon biscutatus.*

308

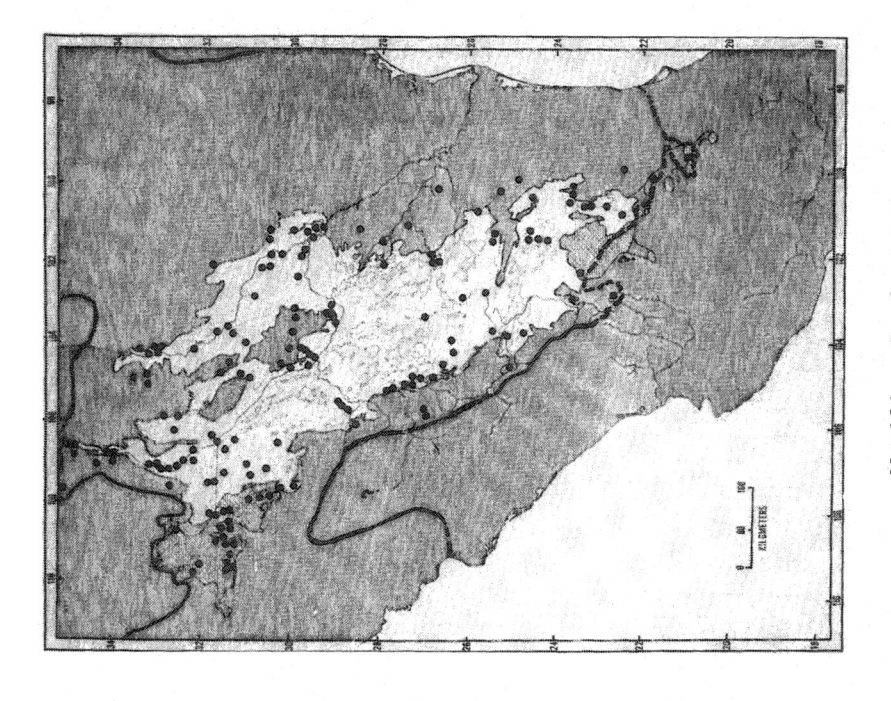

Map 136. Crotalus atrox.

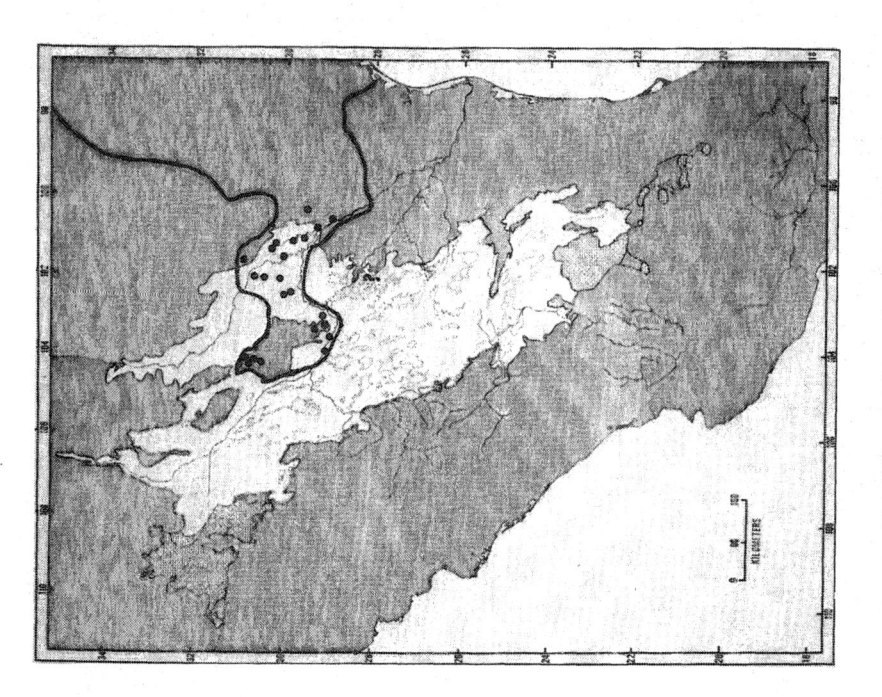

Map 135. Agkistrodon contortrix.

309

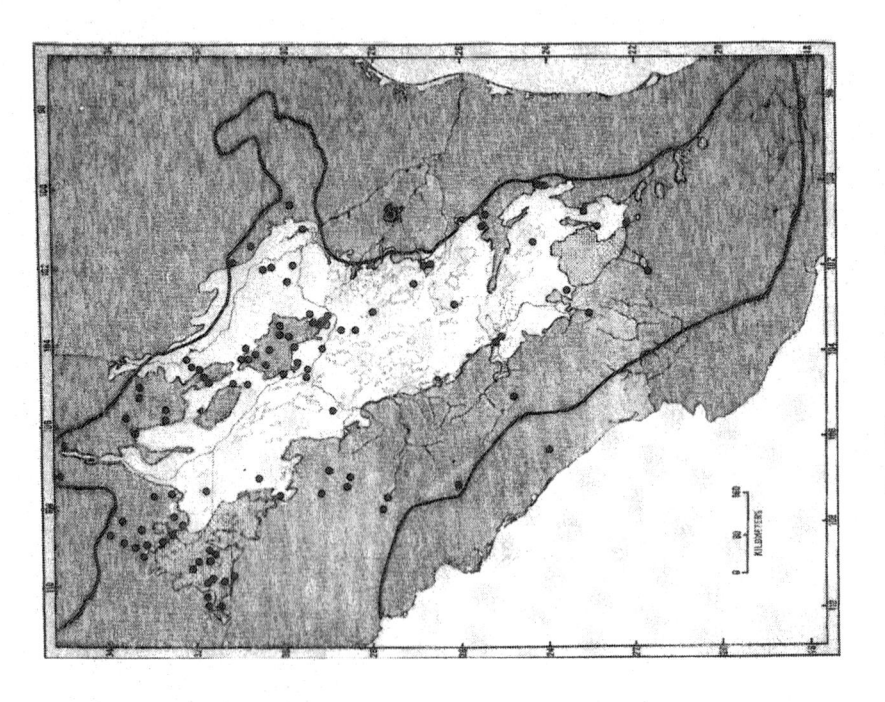

Map 138. *Crotalus molossus.*

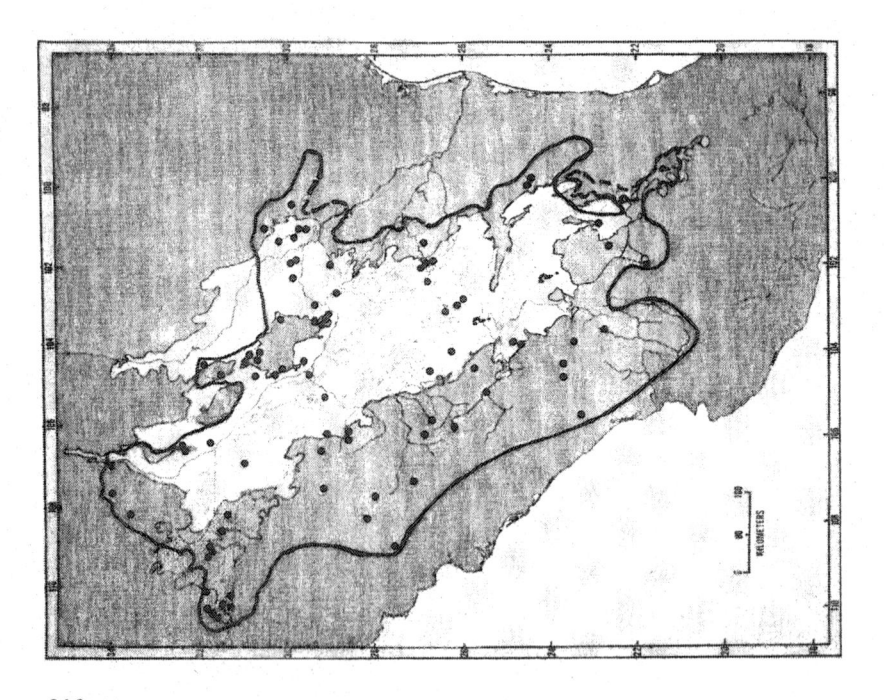

Map 137. *Crotalus lepidus.*

310

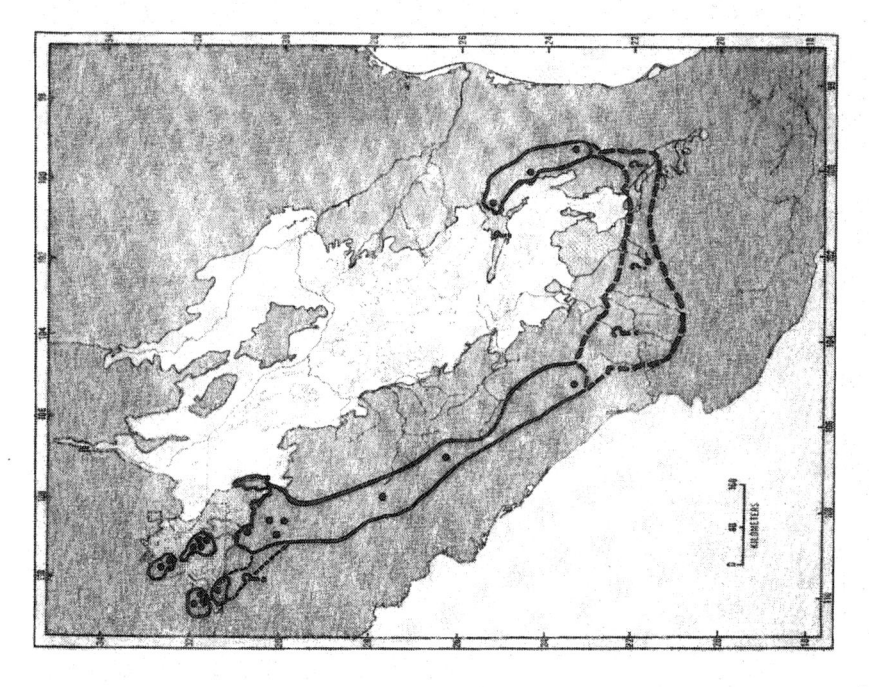

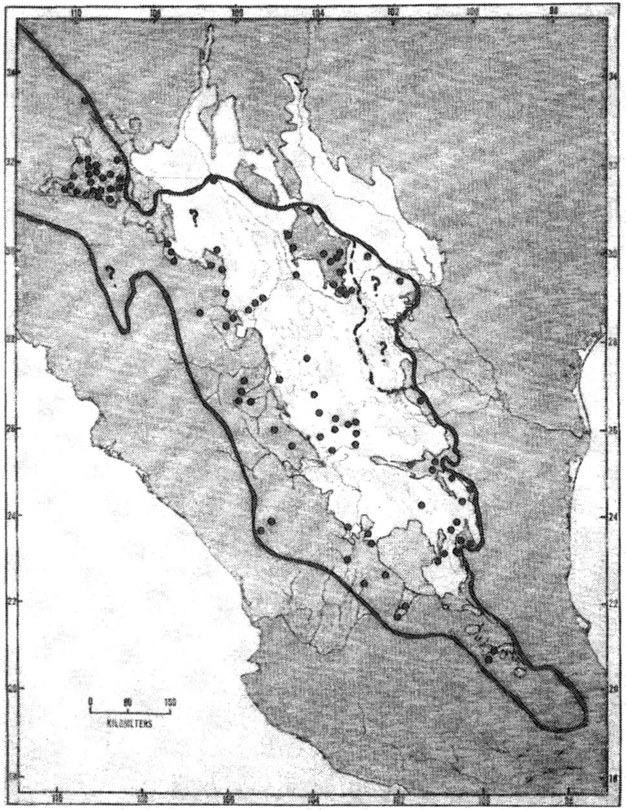

Map 140.   *Crotalus scutulatus.*

Map 139.   *Crotalus pricei.*

311

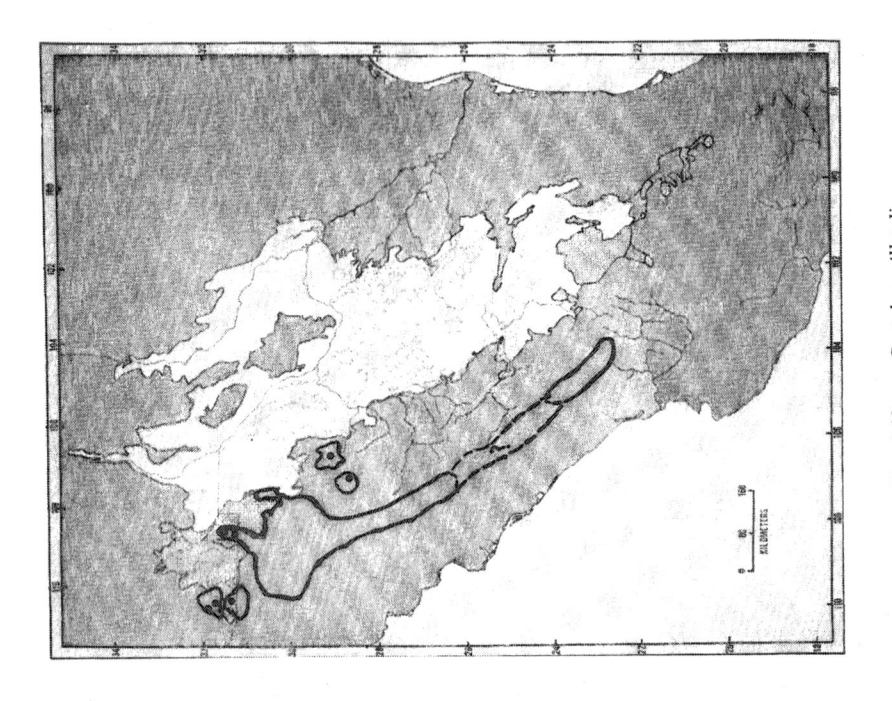

Map 142. Crotalus willardi.

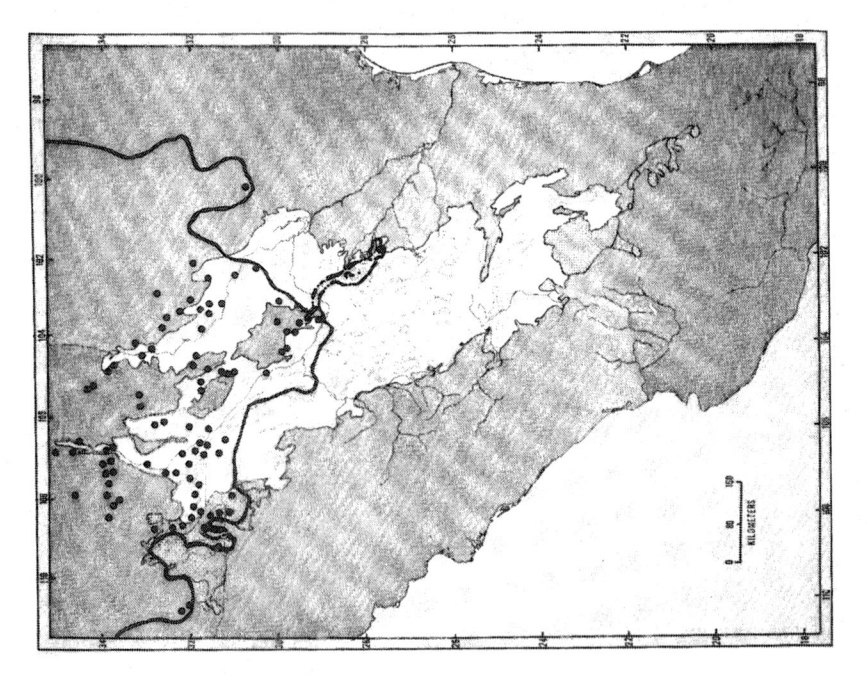

Map 141. Crotalus viridis.

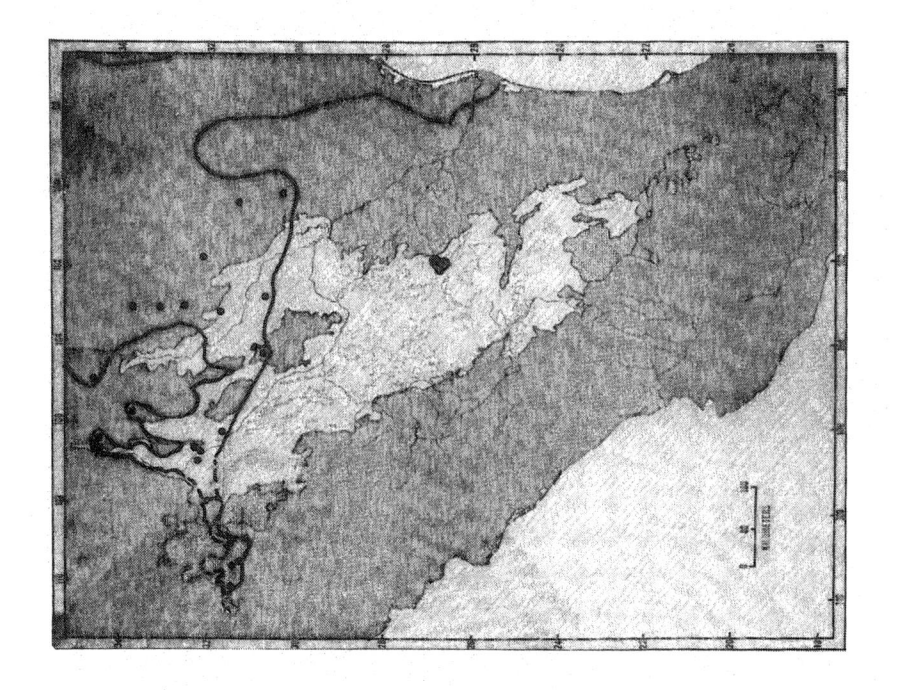

Map 143.   *Sistrurus catenatus.*